Nicotine and Related Alkaloids

Nicotine and Related Alkaloids

Absorption, distribution, metabolism and excretion

Edited by

J.W. Gorrod

Chelsea Department of Pharmacy
King's College, London, UK

and

J. Wahren

Department of Clinical Physiology
Karolinska Institute, Stockholm, Sweden

CHAPMAN & HALL
London · Glasgow · New York · Tokyo · Melbourne · Madras

Published by Chapman & Hall, 2–6 Boundary Row, London SE1 8HN

Chapman & Hall, 2–6 Boundary Row, London SE1 8HN, UK

Blackie Academic & Professional, Wester Cleddens Road, Bishopbriggs, Glasgow G64 2NZ, UK

Chapman & Hall Inc., 29 West 35th Street, New York NY10001, USA

Chapman & Hall Japan, Thomson Publishing Japan, Hirakawacho Nemoto Building, 6F, 1-7-11 Hirakawa-cho, Chiyoda-ku, Tokyo 102, Japan

Chapman & Hall Australia, Thomas Nelson Australia, 102 Dodds Street, South Melbourne, Victoria 3205, Australia

Chapman & Hall India, R. Seshadri, 32 Second Main Road, CIT East, Madras 600 305, India

First edition 1993

Typeset in 10/12 Palatino by Acorn Bookwork, Salisbury, Wiltshire
Printed by in Great Britain by St Edmundsbury Press, Bury St. Edmunds, Suffolk

ISBN 0 412 55740 1

A catalogue record for this book is available from the British Library

Library of Congress Cataloging-in-Publication data available

∞ Printed on permanent acid-free text paper, manufactured in accordance with the proposed ANSI/NISO Z 39.48-199X and ANSI/NISO Z 39.48-1984

Dedication

This volume is dedicated to the fond memory of Professor Herbert (Herb) McKennis and Professor Edward (Ed) Leete with much affection. Herb and Ed greatly influenced our thinking on the degradation and biosynthesis of nicotine and related alkaloids.

Herb McKennis, together with his colleagues in the Department of Pharmacology at the Medical College of Virginia, undertook pioneering studies on the metabolism of nicotine in mammals. The state of the art in analytical techniques in the 1960s gave rise to many problems of detection associated with elucidating the metabolic pathways of a highly toxic substance. Not for Herb the transient response of a recorder indicating the presence of a compound; he would insist on isolating the metabolite in crystalline form and fully characterizing the urinary metabolite or a derivative and comparing its properties with synthetic material. From these data he proposed metabolic pathways and confirmed them by administering the putative metabolites. In this way, a complex picture of nicotine metabolism has emerged which only now, with the advent of radiolabelling and other techniques, have we begun to probe. Herb laid down the foundation on which the contributors to this volume have been proud to build.

Ed Leete was English by birth and university training, but soon moved to Canada and from there to the USA. After a brief stay in California, in 1958 he moved to the University of Minnesota where he rose to become full Professor. Ed was a classical chemist who applied his eminent talent to problems on biosynthesis in plants. Ed would administer potential precursors to the plant, allowing it to grow and then isolate the alkaloid from plant material. If the radiolabel had been incorporated clearly, the administered compound was a precursor. This then required the position of the label in the isolated material to be established by painstaking experimental degradation procedures. From these data Ed was able to propose biosynthetic pathways and test his ideas by further experiments using the proposed intermediates until a full biosynthetic map was constructed. Ed then used these pathways to obtain unnatural analogues of plant alkaloids which had some potential as medicinal agents.

Both Herb and Ed were very quiet, modest scientists who willingly gave of their knowledge and expertise; both had the knack of encouraging, and never putting down. It was a pleasure to be in their company and their contributions to 'nicotine' science and life will long be remembered.

Professor H. McKennis
1916–1983

Professor E. Leete
1918–1992

Contents

Contributors

N.J. Balter
Department of Pharmacology
Georgetown University School of Medicine
3900 Reservoir Road NW
Washington DC 20007
USA

N. Benowitz
Departments of Medicine and Psychiatry
Division of Clinical Pharmacology
San Francisco General Hospital Medical Center
University of California
San Francisco, CA 94110
USA

H.R. Burton
Department of Agronomy
University of Kentucky
Lexington, KY 40546-0091
USA

L.P. Bush
Department of Agronomy
University of Kentucky
Lexington, KY 40546-0091
USA

N. Castagnioli Jr
Department of Chemistry
Virginia Polytechnic Institute and State University
Blacksburg, VA 24061-0212
USA

R.L. Chelvarajan
Department of Agronomy
University of Kentucky
Lexington, KY 40546-0091
USA

S. Cholerton
Pharmacogenetics Research Unit
Department of Pharmacological Sciences
The Medical School
The University
Newcstle upon Tyne NE2 4HH
UK

P.A. Crooks
College of Pharmacy
University of Kentucky
Lexington, KY 40536-0091
USA

M. Curvall
Reserca AB
S-118 84 Stockholm
Sweden

F.F. Fannin
Department of Agronomy
University of Kentucky
Lexington, KY 40546-0091
USA

Y. Funae
Laboratory of Chemistry
Osaka City University Medical School
1-4-54 Asahamachi
Abeno-ku, Osaka 545
Japan

J. Gabrielsson
Kabi Pharmacia Therapeutics AB
Department of Pharmacokinetics
S-751 82 Uppsala
Sweden

M. Gastonguay
Department of Pharmacology
Georgetown University School of Medicine
3900 Reservoir Road NW
Washington DC 20007
USA

J.W. Gorrod
Chelsea Department of Pharmacy
King's College London
Manresa Road
London SW3 6LX
UK

M. Gumbleton
Department of Pharmaceutical Sciences
Royal College
University of Strathclyde
204 George Street
Glasgow G1 1XW
UK

J.R. Idle
Pharmacogenetics Research Unit
Department of Pharmacological Sciences
The Medical School
The University
Newcastle upon Tyne NE2 4HH
UK

S. Imaoka
Laboratory of Chemistry
Osaka City University Medical School
1-4-54 Asahamachi
Abeno-ku, Osaka 545
Japan

P. Jacob
Departments of Medicine and Psychiatry
Division of Clinical Pharmacology
San Francisco General Hospital Medical Center
University of California
San Francisco, CA 94110
USA

E. Kazemi Vala
Reserca AB
S-118 84 Stockholm
Sweden

L. Kita
Department of Pharmacology
Nara Medical University
Kashihara, Nara 634
Japan

X. Liu
Department of Chemistry
Virginia Polytechnic Institute and State University
Blacksburg, VA 24061-0212
USA

N.W. McCraken
Pharmacogenetics Research Unit
Department of Pharmacological Sciences
The Medical School
The University
Newcastle upon Tyne NE2 4HH
UK

T. Nakashima
Department of Pharmacology
Nara Medical University
Kashihara, Nara 634
Japan

H. Nakayama
Department of Pharmacology
Nara Medical University
Kashihara, Nara 634
Japan

G.B. Neurath
Institut für Biopharmazeut.- und Mikroanalytik
Hexentwiete 32
D-2000 Hamburg 56
Germany

O. Pelkonen
University of Oulu
Department of Pharmacology and Toxicology
Kajaanintie 52 D
SF-90220 Oulu
Finland

D.E. Robinson
Department of Pharmacology
Georgetown University School of Medicine
3900 Reservoir Road NW
Washington DC 20007
USA

S.L. Schwartz
Department of Pharmacology
Georgetown University School of Medicine
3900 Reservoir Road NW
Washington DC 20007
USA

K. Vähäkangas
University of Oulu
Department of Pharmacology and Toxicology
Kajaanintie 52 D
SF-90220 Oulu
Finland

J. Wahren
Department of Clinical Physiology
Karolinska Hospital
PO Box 60500
S-104 01 Stockholm
Sweden

Preface

Nicotine is an alkaloid which is present, together with a number of minor alkaloids, in tobacco and a wide variety of other plants. The introduction of tobacco as a therapeutic agent against diverse pathological and physiological conditions resulted in the widespread exposure of people to nicotine, and the subsequent recognition of the pleasurable effects of tobacco consumption. Tobacco may be used for pleasure by smoking it in pipes, cigars or cigarettes or by taking it in unsmoked form as oral and nasal tobacco snuff. Nonsmokers are exposed to nicotine through plant material and side-stream tobacco smoke. This means that in humans nicotine is always utilized in the presence of a very large variety of natural compounds or their pyrolysis products, depending on the route of administration. These compounds may modify the absorption, distribution, metabolism and excretion of nicotine and hence alter the duration of its pharmacological action.

In recent years the use of nicotine in chewing gum and cutaneous patches has been developed as an aid to smoking cessation. The toxic properties of nicotine make it useful as an insecticide, which has led to its use in agriculture and horticulture. It has also recently been recognized that tobacco consumption may be beneficial in the prevention of Parkinson's disease or in alleviating inflammatory bowel syndrome.

The above observations have continued to stimulate research into the mode of action of this relatively simple molecule. This has given rise to numerous publications, symposia, books etc. on the pharmacology of nicotine. However, it seemed to me that insufficient attention had been paid to the ADME of nicotine and related compounds. There were numerous observations distributed throughout the scientific literature predominantly stemming from the *in vivo* work of McKennis and colleagues and a few reviews, but there was no single volume where specialists on individual areas were brought together to produce a contemporary picture of the ADME of nicotine and the pharmacological and physiological factors which control them. This volume sets out to remedy the situtation.

It is clear that considerable advances have been made in our understanding of the biotransformation of nicotine and related compounds. However, much remains to be done, in particular the enzymes involved in individual metabolic processes and the factors controlling them remain to be elucidated. Furthermore it is recognized that a considerable part of our present knowledge relates to data obtained from experimental animals, and this will need to be evaluated in terms of its relevance to humans. The recent discovery of the importance of hydroxylated metabolites of cotinine, and quaternization reactions involving nicotine and certain metabolites point the way to obtaining a total balance to account for nicotine intake. This in turn may well aid our understanding of nicotine pharmacology.

It is hoped that the data presented in this volume will act as a stimulus for further work in the field.

John W. Gorrod
London, 1993

Acknowledgements

The editors gratefully acknowledge the support and encouragement they have received from Fabriques de Tabac Réunies SA, Neuchâtel, Switzerland.

The conference that led to the production of this volume was supported by grants from:

British–American Tobacco Limited
Ciba-Geigy Limited
Fabriques de Tabac Réunies SA
Fiat Auto
Glaxo Research UK
Imperial Tobacco Limited
Rothmans International Tobacco Limited
Smith–Kline and Beecham Pharmaceuticals
Swedish Tobacco Company
The Royal Bank of Scotland

to whom we express our sincere thanks. The success of the meeting was in no small part due to the efforts of Dott. Italico Rota, Hotel Valentini, Salsomaggiore Terme, Italy, and his staff who embraced the idea of this conference with great enthusiasm. They translated our nebulous ideas and wishes into reality and provided an ideal tranquil setting for the maximum social interaction, discussions and exchange of ideas. The choice of the Hotel Valentini as the venue was an inspired suggestion of Professor L. Manzo, University of Pavia, Italy.

We acknowledge with thanks the help we received from The Royal Bank of Scotland (High Wycombe Branch) in managing the financial affairs of the conference.

We wish to thank Mrs D.M. Gorrod, who co-ordinated all aspects of the conference. Finally, we thank our contributors who responded so positively to the idea of the present volume and our editorial suggestions.

Biosynthesis and metabolism of nicotine and related alkaloids

1

L.P. Bush, F.F. Fannin, R.L. Chelvarajan, H.R. Burton

1.1 INTRODUCTION

Alkaloids are extremely important in tobacco (*Nicotiana tabacum*) leaf because they are a major factor in smoke quality, and provide a physiological stimulus which makes the use of tobacco products pleasurable. Nicotine and other alkaloids in *Nicotiana spp*. have been ascribed many functions, i.e

- detoxification products
- waste products
- nitrogen reserve
- regulatory substances for plant growth
- protection for the plant against insects and other herbivores

and consequently they appear to be important to the evolution of the genus as it is known today. However, as important as *Nicotiana* alkaloids are in eliciting animal responses from the utilization of tobacco, growth and development of the plant proceed normally in the absence of accumulation of alkaloids.

Nicotine levels in tobacco are affected by genetics, environmental conditions and cultural practices. Influence of environmental conditions and cultural practices on nicotine accumulation in tobacco has been the subject of many reviews (e.g. Bush and Saunders, 1977; Chaplin and Miner, 1980; Bush and Crowe, 1989). Genetic and physiologic aspects of tobacco alkaloid biosynthesis and accumulation were reviewed last by Leete (1980), Bush (1981), and Strunz and Findley (1985).

Nicotine and Related Alkaloids: Absorption, distribution, metabolism and excretion. Edited by J.W. Gorrod and J. Wahren. Published in 1993 by Chapman & Hall, London. ISBN 0 412 55740 1

In the 60-plus species of *Nicotiana,* most alkaloids are 3-pyridyl derivatives with nicotine the principal alkaloid in 50 to 60% of the species. Based on amounts of alkaloid accumulation in leaves of *Nicotiana spp.,* nicotine, nornicotine, anatabine and anabasine are the major alkaloids present in the genus. Nornicotine is the major alkaloid in 30 to 40% of the species. Anabasine is usually the major alkaloid in *N. acaulis, N. glauca, N. petuniodes* and *N. solanifolia.* Anatabine is usually not the principal alkaloid in any species but will be a relatively higher percentage of the total in *N. otophora, N. tomentosa* and *N. tomentosiformis.* Species that accumulate primarily nicotine tend to have higher total alkaloid concentration in their leaves. There are many minor alkaloids found in tobacco leaves which are derivatives of the major alkaloids and some of these are shown in Figures 1.1

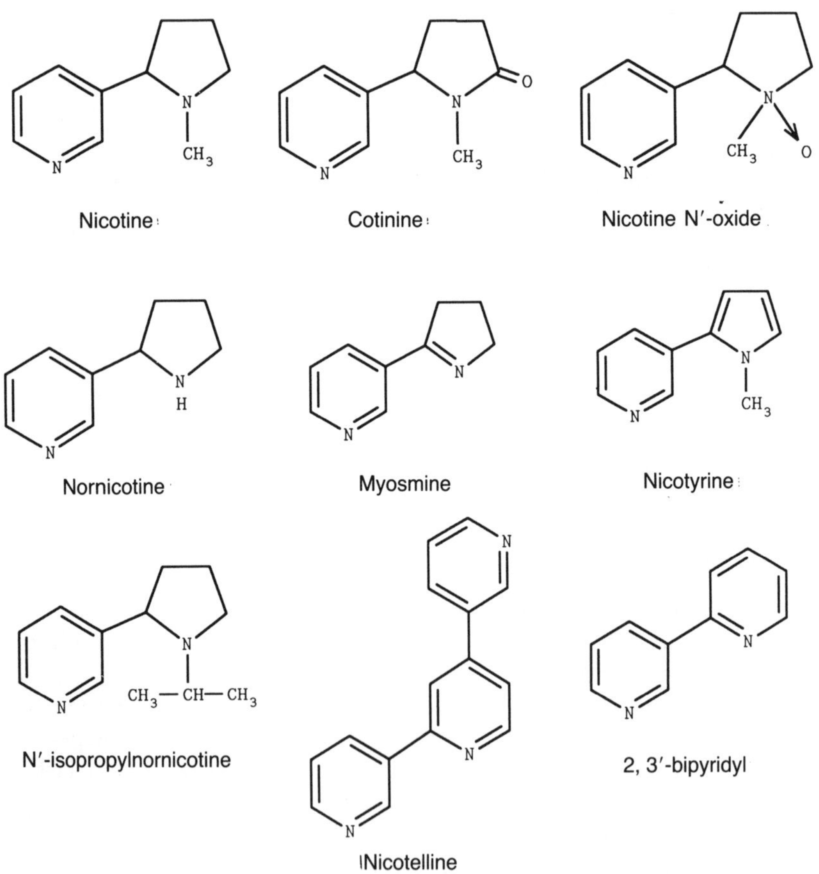

Figure 1.1 Structures of selected *Nicotiana* alkaloids.

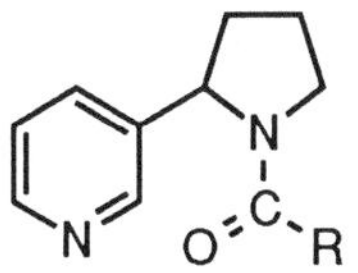

N′-acylnornicotines

R = —H
—CH_3
—CH_2CH_3
—$(CH_2)_2CH_3$
—$(CH_2)_4CH_3$
—$(CH_2)_6CH_3$
—$(CH_2)_4CH(OH)CH_2CH_3$
—$(CH_2)_5CH(OH)CH_3$
—CH_2—CH(OH)—$(CH_2)_8CH_3$
—CH_2—CH(OH)—$(CH_2)_9CH_3$
—CH_2—CH(OH)—$(CH_2)_{10}CH_3$
—CH_2—CH(OH)—$(CH_2)_{11}CH_3$
—CH_2—CH(OH)—$(CH_2)_{12}CH_3$
—$(CH_2)_{10}CH_3$
—$(CH_2)_{11}CH_3$

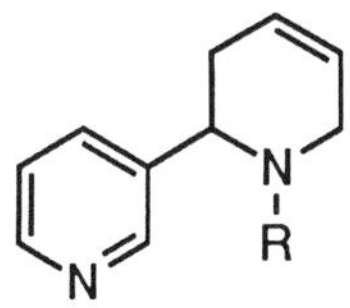

R = H, Anatabine
CH_3, Methylanatabine
COH, N′-formylanatabine

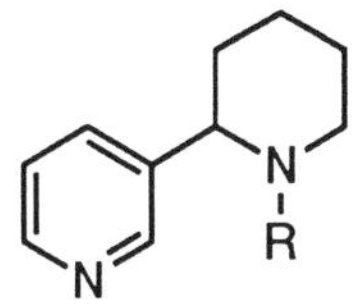

R = H, Anabasine
CH_3, Methylanabasine
COH, N′-formylanabasine

Figure 1.2 Derivatives of nornicotine, anabasine and anatabine.

and 1.2. Most of the minor alkaloids are present in less than 50 μg/g dry weight and many are present in nanogram amounts. Many of these alkaloids are apparently aberrant metabolic or minor catabolic products of the major alkaloids and some may even be artifacts formed during isolation from tobacco.

1.2 ALKALOID BIOSYNTHESIS

The precursors for the pyridine, pyrrolidine and piperidine rings of nicotine, nornicotine, anatabine and anabasine have been determined and biosynthetic schemes developed that are consistent with the data; however, many of the intermediates in the biosynthetic sequence are not known, nor have many of the enzymes required for the biosynthetic steps been fully characterized. Precursor and biosynthetic sequences are important in understanding alkaloid metabolism, but to manipulate leaf chemistry the metabolic regulation must be understood. Metabolic regulation of alkaloid accumulation should ideally be treated in both its qualitative and quantitative aspects, with consideration of alternative pathways and controls, as well as variations in levels of enzyme activity and metabolic effectors. The knowledge of the biosynthetic and degradative pathways of alkaloids in tobacco that is needed for studies of this kind is still incomplete. Accordingly, the most accepted routes for alkaloid biosynthesis will be presented and then the role of significant enzymes and potential metabolic regulation will be discussed.

1.2.1 NICOTINE

The pyridine ring of nicotine, nornicotine, anabasine and anatabine is formed from nicotinic acid. Quinolinic acid (pyridine-2,3-dicarboxylic acid) was an efficient immediate precursor of the nicotinic acid incorporated into the pyridine ring (Yang *et al.*, 1965). They also showed that glycerol was incorporated without randomization into C-4, C-5 and C-6 of the pyridine ring. 3-^{14}C-Aspartate or 3-^{14}C-malate fed to *N. rustica* resulted in the label being almost exclusively incorporated into the C-2 and C-3 positions of the pyridine ring (Jackanioz and Byerrum, 1966). All intermediates in this synthetic sequence have not been elucidated (Figure 1.3).

Substantial evidence indicates that the major biosynthetic route for the N-methylpyrrolidine ring of nicotine proceeds through putrescine (1,4-diaminobutane), N-methylputrescine and 4-methylaminobutanal (N-methyl-4-aminobutyraldehyde) to the N-methyl-Δ′-pyrrolinium salt (Figure 1.4). Dewey *et al.* (1955) and Leete (1955) fed 2-^{14}C-ornithine to tobacco and found nicotine to be symmetrically labelled at C-2′ and C-5′. Symmetrical incorporation of 2-^{14}C-ornithine could occur from a free

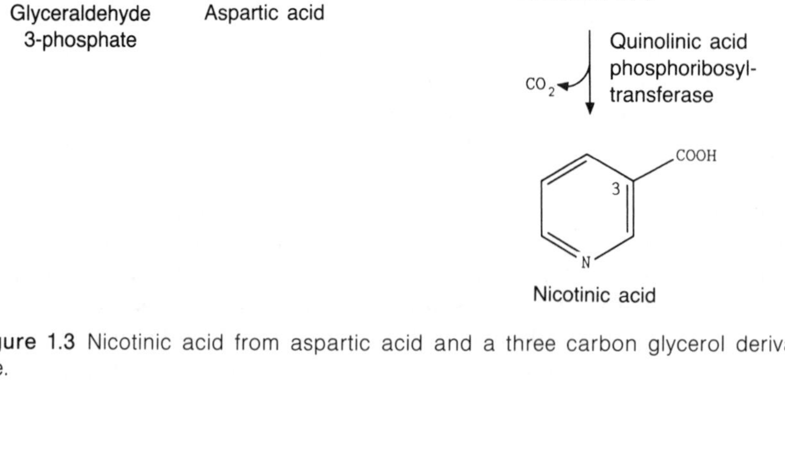

Figure 1.3 Nicotinic acid from aspartic acid and a three carbon glycerol derivative.

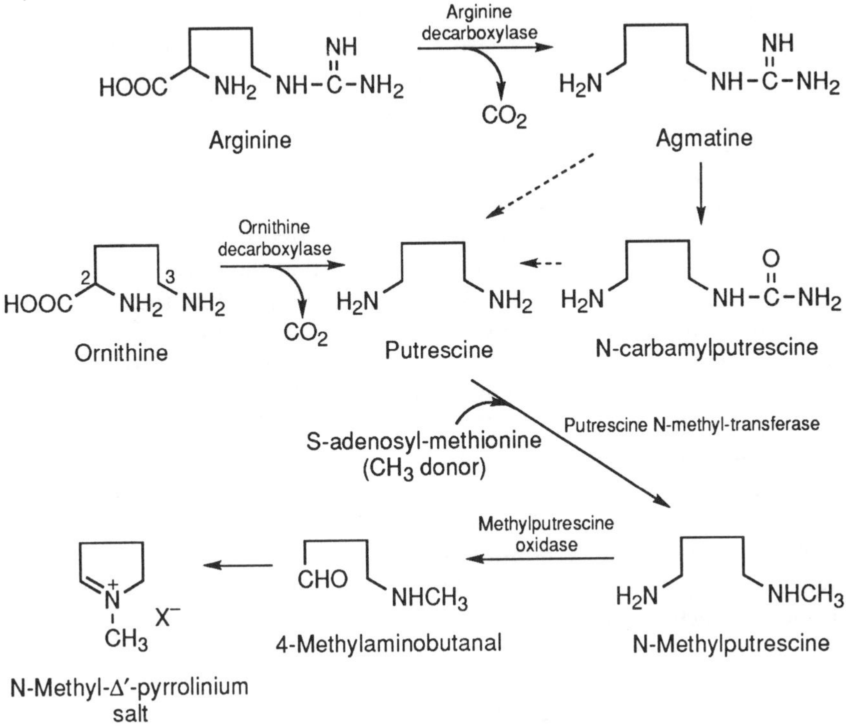

Figure 1.4 Synthesis of N-methyl-Δ′-pyrrolinium ion.

symmetrical intermediate such as putrescine. S-Adenosylmethionine was an effective *in vivo* methyl donor for putrescine N-methyltransferase measurements (Mizusaki *et al.*, 1971) and is most likely the *in vivo* methyl donor.

Yoshida and Mitake (1966) and Yoshida (1969a; 1969b) demonstrated that arginine, agmatine and N-carbamylputrescine, in sequence, were efficiently incorporated into putrescine. It has been generally accepted that this pathway is important in higher plants with the exception of the *Solanaceae* where decarboxylation of ornithine contributed significantly to putrescine formation (Leete, 1980). Berlin (1981) has shown in tobacco cell cultures that arginine and ornithine were equally efficient precursors of putrescine for cinnamoyl putrescine formation. Arginine decarboxylase and ornithine decarboxylase activities were nearly the same in cell cultures, and by use of inhibitors and substrates he concluded that putrescine was formed by initial decarboxylation of each substrate and not interconversion.

Yang (1984) isolated ornithine and arginine decarboxylase from root tissue of intact plants and found about four times as much activity of ornithine as arginine decarboxylase activity. Removal of the apical meristem (induction of nicotine accumulation in the plant) increased ornithine and arginine decarboxylase activity but after three days the activity of arginine decarboxylase was the same in intact and induced plants, whereas activity of ornithine decarboxylase remained higher for at least 23 days.

The incorporation of nicotinic acid into nicotine is better understood. Dawson *et al.* (1960a, 1960b) showed that nicotinic acid was incorporated into the pyridine ring of nicotine with loss of the carboxyl group (Figure 1.5). Also, Yang *et al.* (1965) showed that the pyrrolidine ring of nicotine was attached at the C-3 position of the pyridine ring, the position from which the carboxyl group was lost. Dawson *et al.* (1960a) also showed that the hydrogen at position C-6 of nicotinic acid was lost during nicotine formation. From the above observations Dawson and Osdene (1972) and Leete (1977) presented the hypothesis for nicotine formation that included the reduction of nicotinic acid to 3,6-dihydronicotinic acid.

Friesen and Leete (1990) proposed that the dihydronicotinic acid yielded a zwitterion (a β-iminium carboxylic acid) by proton transfer which readily decarboxylates to yield 1,2-dihydropyridine. 1,2-Dihydropyridine reacts with N-methyl-Δ′-pyrrolinium salt to yield 3,6-dihydronicotine. The 3,6-dihydronicotine is oxidized with loss of the hydrogen originally present at C-6 of nicotinic acid and retention of the hydrogen added in the reduction step to 3,6-dihydronicotinic acid (Figure 1.5). All nicotine in tobacco plants is the (−)-2′*S*-nicotine.

A minor biosynthetic pathway for nicotine from nornicotine has been described by Leete (1984). He fed (*RS*)-[2′-^{14}C] nornicotine to *N. tabacum* and recovered small amounts (0.045%) of the radioactivity in nicotine with all ^{14}C still at the 2′ position. The data indicate direct methylation of nornicotine to nicotine and Leete suggested putrescine methyltransferase may be able to catalyse this methylation although no data were presented to support this hypothesis.

1.2.2 ANATABINE

Both pyridine rings of anatabine are formed ultimately from the two methylene carbons and nitrogen of aspartate and two carbons from a glycerol derivative via nicotinic acid as outlined for nicotine (Leete, 1975, 1977, 1978; Leete and Slattery, 1976) (Figure 1.6). Incorporation of labelled precursors suggests the reduction of nicotinic acid to 3,6-dihydronicotinic acid and decarboxylation to form 1,2-dihydropyridine and 2,5-dihydropyridine. 1,2-Dihydropyridine reacts with the electrophile 2,5-dihydropyridine to yield 3,6-dihydroanatabine which is aromatized to (−)-2′*S*-anatabine (Leete and Muller, 1982). In this proposed biosynthesis the point of attachment between the two pyridine rings is

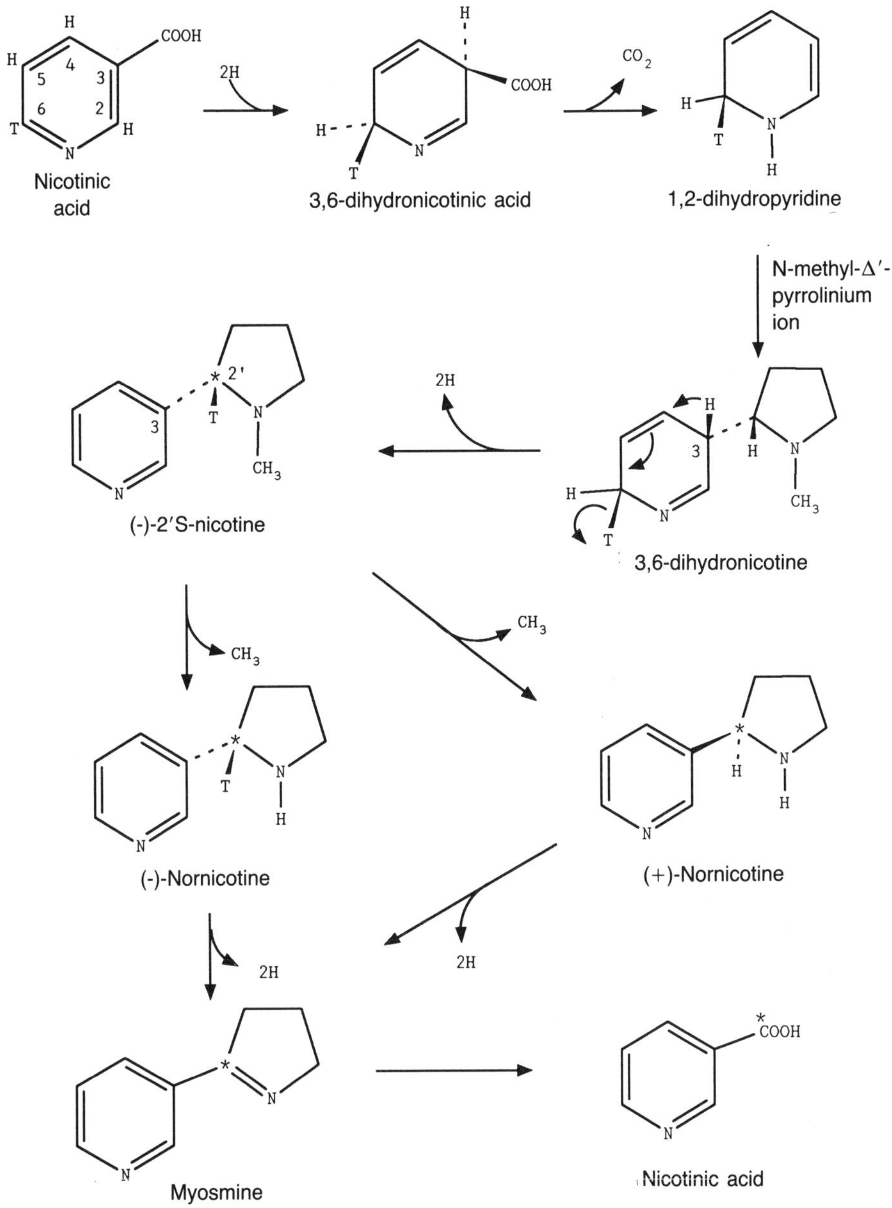

Figure 1.5 Formation of (−)-2′*S*-nicotine, nornicotine and myosmine from nicotinic acid and N-methyl-Δ′-pyrrolinium ion.

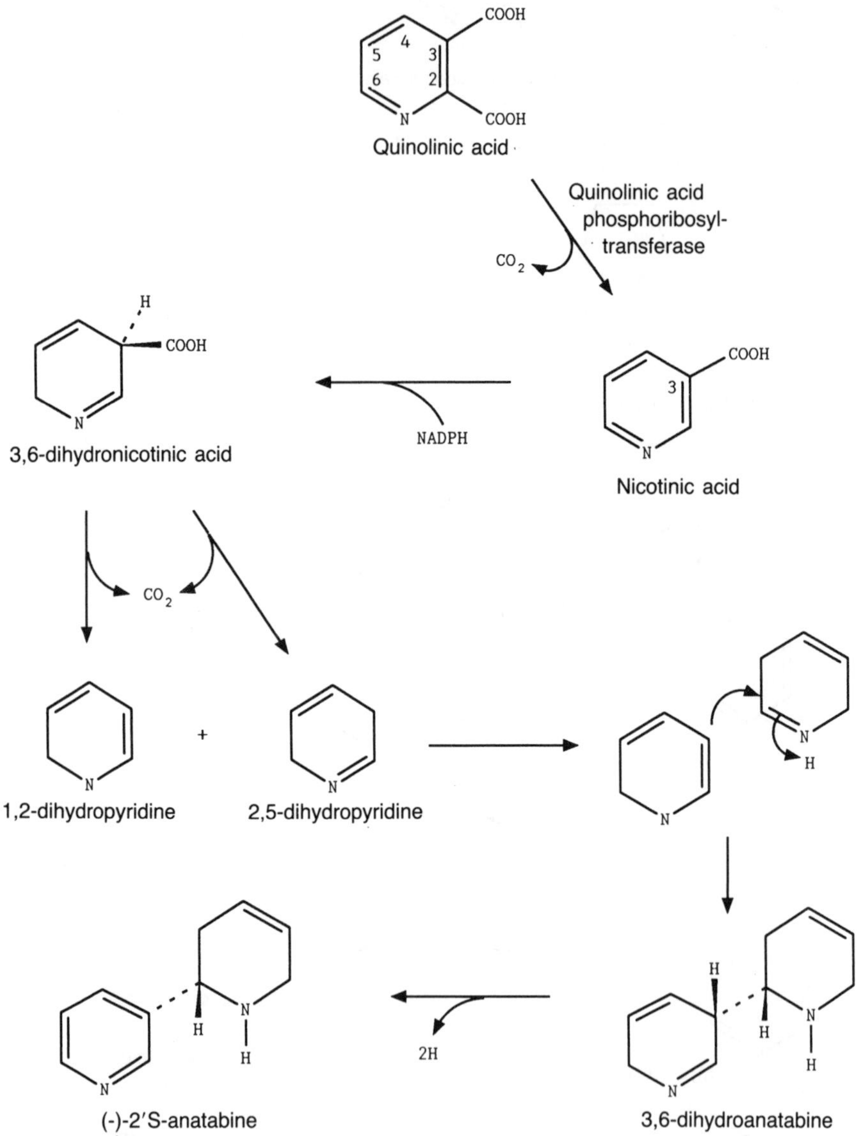

Figure 1.6 Anatabine formation from nicotinic acid.

the C-3, or the carbon from which decarboxylation occurred, to C-2 or C-6 of 2,5-dihydropyridine formed by the decarboxylation step. Even though anatabine and anabasine differ by only two hydrogens, there does not seem to be any interconversion between these two alkaloids.

1.2.3 ANABASINE

The pyridine ring of (−)-2′*S*-anabasine has been shown to originate from nicotinic acid with loss of the carboxyl group as proposed for the pyridine ring of nicotine (Solt *et al.*, 1960). The most accepted precursor for the piperidine ring is lysine via Δ′-piperidine with the nitrogen derived from the σ-nitrogen of lysine (Leete *et al.*, 1964). Incorporation studies with 2-^{14}C-lysine have shown that the radiolabel was incorporated into the C-2′ position of the piperidine ring of anabasine (Leete, 1958, 1980). This result is consistent with a bound cadaverine throughout steps with a potential symmetrical molecule, and is contrasted to the apparent free putrescine in N-methylpyrrolidine ring formation for nicotine. Leete (1977) proposed the same mechanism for anabasine formation from nicotinic acid and Δ′-piperidine as for nicotine formation from nicotinic acid and N-methyl-Δ′-pyrrolinium salt.

1.2.4 NORNICOTINE

Nornicotine seems to be formed irreversibly from nicotine in green leaves or during senescence (Dawson, 1951). Demethylation of (−)2′*S*-nicotine in tobacco yields (−)-2′*S*-nornicotine and (+)-2′*R*-nornicotine in nearly equal amounts. Data of Leete and Chedekel (1974) support a demethylation mechanism involving an iminium salt and tautomeric shift to obtain racemization of nornicotine. Their results demonstrate that the H atom at C2′ of (−)-2′*S*-nicotine is retained in formation of (−)-2′*S*-nornicotine, but is lost in formation of (+)-2′*R*-nornicotine (Figure 1.5). They also showed that the naturally occurring (−)-2′*S*-nicotine and (+)-2′*R*-nicotine were equally effective as substrate for the *in vivo* demethylation enzyme. Kisaki and Tamaki (1961) suggested that the amount of racemization of nornicotine is variable and may be influenced by environmental conditions and maturity or senescence of the plant. N-Demethylation of nicotine is a significant step in alkaloid metabolism in several *Nicotiana spp.* as it is the last step in production of nornicotine. Demethylation of nicotine to nornicotine has been described as being controlled by one of two genes. Characterization of the enzyme would provide potential plant modification for formation and accumulation of N′-nitrosonornicotine (NNN) and N′-acylnornicotines in tobacco products. The N-demethylation of nicotine is coupled with the oxidation of the methyl group to CO_2 (Stepka and Dewey, 1961) and was reported to be the primary source of nornicotine by Dawson (1945). A preliminary report by Chelvarajan *et al.* (1991) indicates that nicotine demethylase activity is associated with the microsomal fraction of the cell. Demethylation was dependent on the presence of reduced nucleotides. The activity of nicotine demethylase was over three times higher when NADPH was substituted for NADH. *In vitro* pH and temperature optima were 7.0 to 7.5 and 30°C, respectively. Higher concentrations of nornicotine in the assay decreased the rate of nicotine demethylation. Nornicotine was a more effective inhibitor of demethylation than an identical concentration of nicotine suggesting product inhibition rather than a general alkaloid effect. V_{max} and the apparent K_m for nicotine were estimated to be 11 pmoles/min and 118 μM, respectively. Reports indicate that many N-demethylases in higher plants are associated with cytochrome P450. Addition of tetcyclasis, a cytochrome P450 type enzyme inhibitor (Canivenc *et al.*, 1989), resulted in partial inhibition of demethylation of nicotine. Carbon monoxide also inhibits many P450 enzymes and the inhibition can often be reversed by light of 450 nm (Donaldson and Luster, 1991), but in this study CO did not inhibit nicotine demethylation. However, an antibody to Jerusalem artichoke NADPH cytochrome P450 reductase inhibited *in vitro* nicotine demethylation suggesting that the enzyme may be a P450 type enzyme (Benveniste *et al.*, 1986). These data are equivocal and further characterization is required.

In the growing plant, nornicotine may be dehydrogenated to myosmine (Kisaki and Tamaki, 1966; Leete and Chedekel, 1972). Leete (1984) fed ^{14}C-nornicotine to intact plants for 8 days and found 55% of the radioactivity in nornicotine and 15% in myosmine. After feeding ^{14}C-2′-myosmine to *N. glauca*, no labelled nicotine or nornicotine was found, suggesting that the dehydrogenation of nornicotine to myosmine was irreversible. However, activity was found in nicotinic acid and with most of the label in the carboxyl carbon (Figure 1.5).

1.3 REGULATION OF NICOTINE CONTENT IN TOBACCO PLANTS

1.3.1 GENETIC

Genetic control of alkaloid levels in tobacco has received much attention and will continue to be an area of intensive research. Identification of the gene systems responsible for tobacco alkaloid accumulation expands the possibility of success of genetic engineering programmes to control alkaloid production in plants or cell culture. Two principal genetic systems for regulating the level and kinds of alkaloids in tobaccos have been described.

The level of total alkaloids in tobacco is determined by genes at two loci (Legg *et al.*, 1969; Legg and Collins, 1971). These two loci are designated A and B. Commercial cultivars are considered homozygous dominant at the two loci (AABB) and low alkaloid genotypes homozygous recessive (aabb) at the two loci. Legg and Collins (1971) obtained the nine possible genotypes for the two-locus model and demonstrated that the relative dosage effect of the homozygotes at locus A were 2.4 times greater than the homozygotes at locus B. Dominant alleles at both the A and B alkaloid loci were associated with higher levels of total alkaloids, total N and total volatile nitrogenous bases, but lower levels of protein N, than were the recessive alleles. The decreased total N content associated with the recessive alleles compared with the dominant alleles was approximately equal to the decreased alkaloid N, and the differences in total volatile nitrogenous bases were mainly due to differences in the alkaloid N fraction. Significant genotypic differences were detected for several plant characteristics but no consistent ranking of genotypes was evident.

The other genetic system controls the conversion of nicotine to nornicotine in the plant. Nornicotine is produced by demethylation of nicotine (Dawson, 1952). There are two potential loci for nicotine demethylation in commercial tobacco, *N. tabacum*. One locus is found in the progenitor *N. tomentosiformis* genome and the other in the progenitor *N. sylvestris* genome (Mann *et al.*, 1964). The primary site of nicotine demethylation is the leaf; however, the demethylation gene from *N. tomentosiformis* causes nicotine to be converted to nornornicotine prior to senescence while the gene from *N. sylvestris* caused the conversion to occur during senescence (Wernsman and Matzinger, 1968). Plants with $c_1c_1c_2c_2$ genotypes are not capable of demethylating nicotine to nornicotine. Only one dominant allele at either locus is necessary for demethylation to occur. The C_1 locus from *N. tomentosiformis* appears to be active in the nicotine to nornicotine conversion in commercial cultivars but the high mutation rate at this locus is for nornicotine formation during senescence and not in the green plant as in the original *N. tomentosiformis* gene (Griffith *et al.*, 1955; Burk and Jeffrey, 1958; Mann *et al.*, 1964).

1.3.2 PHYSIOLOGIC

In the recent past the biochemist and physiologist have used these proposed pathways to propose enzyme systems and to provide mechanisms by which the biosynthesis actually occurs in the plant. To date, most of this research has been focused on understanding nicotine biosynthesis and the regulation of nicotine biosynthesis in the plant.

Mizusaki *et al.* (1973) reported activities of three enzymes in the proposed sequence to 4-methylaminobutanal formation (Figure 1.4). These authors found ornithine decarboxylase (ODC), putrescine N-methyltransferase (PMT), and N-methylputrescine oxidase (MPO) activities to be high in tobacco roots (primary site of alkaloid biosynthesis), but low or no detectable activities in tobacco leaves. Root enzyme activities increased 2- to 16-fold in 24 h following nicotine induction (plant decapitation). Saunders and Bush (1979) reported that PMT and MPO increased after induction of high alkaloid genotype plants but MPO did not increase after induction of low alkaloid genotype plants. Pudliner (1980) used citrate synthase as a check for general plant metabolic activity and found ODC and citrate synthase activities increased after induction of both high and low alkaloid genotypes, but there were no differences between the two genotypes.

PMT activity was proportional to leaf nicotine content in related and unrelated high and low nicotine accumulation genotypes and tissues (Yoshida, 1973; Saunders and Bush, 1979; Wagner *et al.*, 1986b). Activities of ODC and MPO were not proportional to nicotine content. Data from Saunders and Bush (1979) and Feth *et al.* (1986) strongly suggest PMT is the enzyme under stringent control for the pyrrolidine moiety of nicotine biosynthesis.

The role of putrescine supply and utilization may be a regulatory step in biosynthesis of the pyrrolidine moiety. A major branch point on the biosynthetic route to nicotine appears to exist at putrescine, where at least four pathways utilizing this diamine diverge (Figure 1.7). Two important alternative events seem to be the methylation (Mizusaki *et al.*, 1971) and N-aminopropylation (Baxter and Coscia, 1973) of putrescine as steps in the biosynthesis of nicotine and polyamines, respectively. The cinnamoyl putrescines constitute as much as 6% of the dry weight of cell cultures (Schiel *et al.*, 1984) and levels of γ-aminobutyrate may be as much as 80 mg per plant (Tso and McMurtrey, 1960).

Putrescine levels are normally low in tobacco, but elevated levels have been reported to be associated with several mineral deficiencies (Yoshida, 1967). ODC activity was ten times higher in tobacco callus than in roots of nicotine induced tobacco plants (Pudliner, 1980). High ODC activity and high polyamine levels are associated with rapid cell proliferation of tobacco callus but not high nicotine accumulation.

Tobacco callus contains low alkaloid levels compared to plant tissue. High ODC activity in callus suggests that putrescine formation is not limiting nicotine biosynthesis, but perhaps the utilization of putrescine and S-adenosylmethionine for polyamine biosynthesis is limiting availability of putrescine for nicotine biosynthesis. Mizusaki *et al.* (1973) did not detect PMT in callus. Since this enzyme specifically catalyses the formation of N-methylputrescine from putrescine and S-adeno-

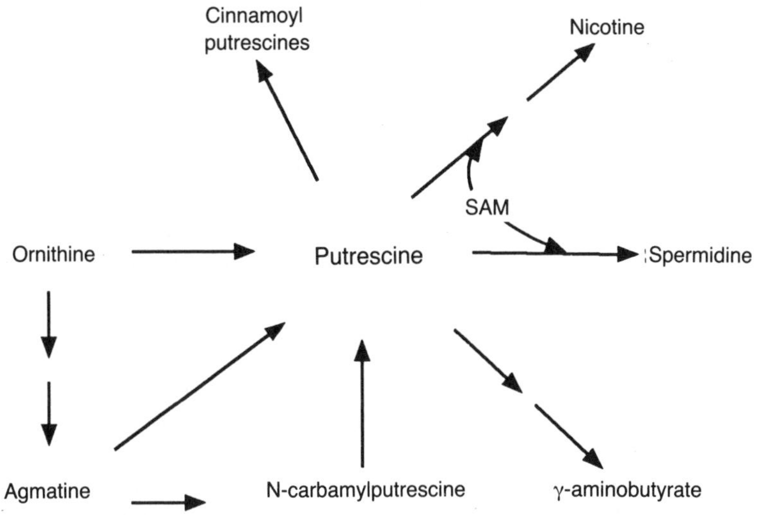

Figure 1.7 Involvement of putrescine in metabolic pathways of *Nicotiana spp.*

sylmethionine, it suggests that nicotine biosynthesis in callus is limited by methylation of putrescine.

Yang (1984) showed ODC activity in roots increased eight-fold within 24 h of plant induction and then after three days decreased to a level three times that in non-induced plants and remained at this level above the control plants for at least 23 days. Increased ODC activity and the subsequent putrescine formation could be a contributing factor in the observed increased nicotine accumulation following induction.

Madsen *et al.* (1984) reported twice the amount of polyamines in senesced leaves of a low alkaloid genotype than in leaves of the isogenic high alkaloid genotype. Work of Ravishankar and Mehta (1982) support the premise that nicotine production is dependent upon putrescine and ornithine availability. They found urea enhanced nicotine production and ornithine aminotransferase activity in callus while ornithine transcarbamylase activity was suppressed. This suggests greater availability of ornithine for nicotine biosynthesis. The inhibition of ornithine transcarbamylase may be due to putrescine formation via ornithine decarboxylase (Stalon *et al.*, 1972).

Examination of the substrate specificity of partially purified MPO of *N. tabacum* revealed that putrescine and cadaverine were both utilized 40% as readily by the enzyme as N-methylputrescine (Mizusaki *et al.*, 1972). Since oxidation of putrescine to 4-aminobutanal could lead to formation of nornicotine, some explanation – be it compartmentalization, differential degradation rates of the two alkaloids, or removal of the 4-aminobutanal by oxidation to 4-aminobutyric acid – is needed to account for the disproportionate differences in nicotine and nornicotine content in nonconverter tobacco genotypes. Subsequently, root protein fractions from nicotine and nornicotine accumulator genotypes of *N. tabacum* were found to oxidize putrescine and N-methylputrescine to the same relative extent, suggesting that a difference in substrate specificity is probably not a major factor in different alkaloid accumulation patterns (Saunders and Bush, 1979). Regulation of nicotine or nornicotine biosynthesis by MPO seems unlikely from available data, although Feth *et al.* (1986) suggest MPO is regulated in a concerted manner with PMT.

Another potential physiologic regulation of nicotine biosynthesis could be mediated through nicotinic acid availability. For biosynthesis of the pyridine ring, quinolinic acid (pyridine-2,3-dicarboxylic acid) is an efficient precursor (Yang *et al.*, 1965). High levels of quinolinic acid phosphoribosyltransferase (QPT) in roots but not leaves of intact, nearly mature *N. rustica* plants suggested that this enzyme was utilized in nicotine biosynthesis (Mann and Byerrum, 1974). In contrast, levels of nicotinic acid phosphoribosyltransferase and nicotinamide deaminase in the same tissue were within the range of activity seen in a few species that do not accumulate pyridine alkaloids. QPT activities in root tissue from four *N. tabacum* genotypes having different alkaloid accumulations were proportional to leaf nicotine levels (Saunders and Bush, 1979). Enzyme activity increased within 24 h after induction for all genotypes except the homozygous recessive low alkaloid genotype. Pudliner (1980) and Wagner *et al.* (1986a, 1986b) reported that QPT activity was greater in induced nicotine accumulating tissues and in genotypes with higher nicotine content than in non-induced tissues or low nicotine genotypes. The differential increase in QPT activity in these experiments was uniquely different as citrate synthase increased in both induced high and low alkaloid genotypes, but there was no difference between genotypes for citrate synthase activity (Pudliner, 1980).

The general conclusion from these data is that QPT is the main regulatory enzyme for nicotinic acid production in nicotine biosynthesis. In addition, Wagner *et al.* (1986a) demonstrated that the nicotinic acid pool increased by synthesis of nicotinamide adenine dinucleotide (NAD) and degradation via nicotinamide mononucleotide, and by a more direct route from nicotinic acid mononucleotide by a glycohydrolase (Figure 1.8). The K_m for the nicotinic acid mononucleotide glycohydrolase is high ($\sim$ 4 mmol/l) and probably prevents complete depletion of the nicotinamide mononucleotide pool to ensure continued NAD availability under conditions of rapid nicotine accumulation. We have not found that nicotinic acid mononucleotide glycohydrolase has any apparent regulatory function in nicotine or anabasine biosynthesis and accumulation in *N. tabacum* and *N. glauca* (Fannin and Bush, 1988).

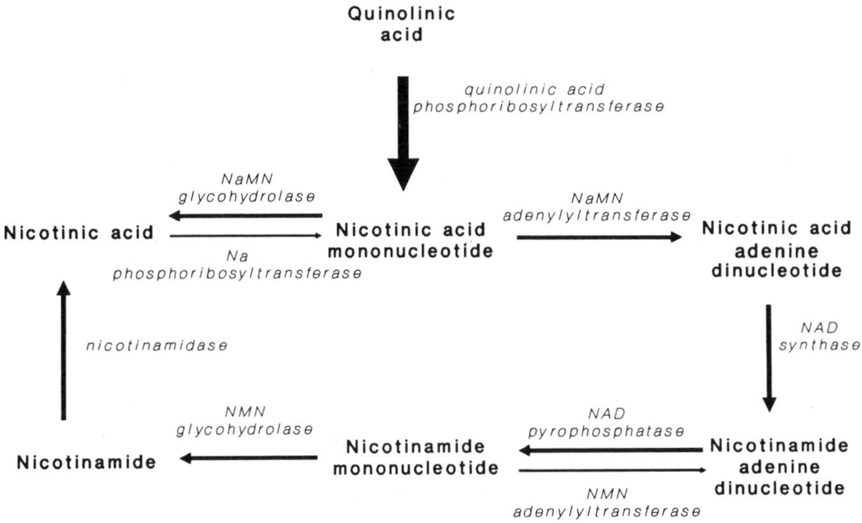

Figure 1.8 Pyridine nucleotide cycle and nicotinic acid availability for nicotine biosynthesis.

Wagner *et al.* (1986b) also concluded that NAD-pyrophosphatase, nicotinamide mononucleotide glycohydrolase and nicotinic acid phosphoribosyltransferase have a secondary regulatory function in nicotinic acid formation.

In commercial (AABB genotype) tobaccos, nicotine synthase may be rate limiting since addition of methylputrescine and nicotinic acid do not significantly increase the accumulation of nicotine as they do in low alkaloid tobaccos. Friesen and Leete (1990) detected nicotine synthase activity in a cell-free system; however, their report contained minimal *in vitro* characterization of the enzyme. The enzyme was absent during the first 100 h of seed germination. This absence of activity is consistent with the observations of nicotine changes in germinating tobacco seeds (Weeks and Bush, 1974). Friesen and Leete (1990) found nicotine synthase to be active at a narrow range near pH 7.0, stimulated by Mg^{2+}, required O_2 and was inhibited by NADH, FAD and ATP. Addition of NADPH to the reaction mixture did not change measured activity. Specificity of nicotine synthase is unknown but is important because if MPO oxidizes both putrescine and N-methylputrescine then enzyme specificity dictates that the N-methyl-Δ′-pyrrolinium ion is utilized and nicotine is formed rather than nornicotine. Anabasine and anatabine are formed by similar decarboxylation condensation reactions and it is not known if these reactions are catalysed by the same enzyme.

1.3.3 CELL AND TISSUE CULTURE SUPPLEMENTATION

Nicotine accumulates in callus and cell cultures of tobacco (Dawson, 1960; Speake *et al.*, 1964; Krikorian and Steward, 1969; Neumann and Muller, 1971) as well as intact plants. Nicotine concentrations are influenced by the concentrations of growth regulators used for culture maintenance but generally are much lower than those found in plants (Furuya *et al.*, 1967, 1971; Tabata *et al.*, 1971; Takahashi and Yamada, 1973; Tabata and Hiraoka, 1976; Miller *et al.*, 1983; Lockwood and Essa, 1984; Pinol *et al.*, 1984). Where the relationship between auxin level, growth rate and nicotine concentration has been established, nicotine accumulation is in general inversely proportional to the growth rate of proliferating cells (Takahashi and Yamada, 1973).

Ravishankar and Mehta (1982) obtained increased levels of nicotine (maximum level 0.1%) on both a culture and dry weight basis with increased urea concentration in the culture medium. The authors suggested that the increased nicotine levels may be the result of increased available ornithine for nicotine biosynthesis but this seems unlikely since added ornithine did not stimulate nicotine accumulation (Miller *et al.*, 1983; Lockwood and Essa, 1984). Precursors for nicotine biosynthesis supplemented to culture media generally reduced nicotine accumulation (Miller *et al.*, 1983). Even though callus from four isogenic burley tobacco genotypes differing in alkaloid content had a much lower nicotine level than the plants from which the cells were derived, the nicotine level in the callus was proportional to the level in the respective plant. Based on the results from experiments of supplementing nicotine precursors into the culture medium without increasing alkaloid levels in any of the genotypes they concluded that control of nicotine accumulation in callus was not in a biosynthetic block in pyridine or pyrrolidine ring formation. Lockwood and Essa (1984) also added nicotine precursors to the culture medium and measured decreased nicotine levels, but many precursors increased the level of anatabine. These authors reported the presence of myosmine, anabasine, anatabine anatalline and nicotelline in the cell suspension cultures.

Mantell *et al.* (1983) obtained 1 to 2% nicotine concentration in cell suspension cultures. They achieved these results by decreasing the growth regulator by 90%, increasing sucrose concentration to 50 g/l and reducing phosphate to 15 mg/l in the medium. They found that when sucrose was depleted from the medium nicotine accumulation stopped, but that when phosphate was depleted rapid nicotine accumulation occurred. Nicotine appeared to be synthesized in cells not undergoing cytokinesis and phosphate starved cells were arrested in the G_1 or to a lesser extent in G_2 stages of the cell cycle. These cells remained viable and apparently produced nicotine.

Leete *et al.* (1971) and Leete (1979) fed 5-fluoronicotinic acid to *N. tabacum* and found that 5-fluoronicotine and 5-fluoroanabasine were produced. Leete and Chedekel (1972) found that feeding N-methyl-Δ′-piperidine, which is not a normal precursor of anabasine, to *N. glauca* resulted in formation of N′-methylanabasine. Walton *et al.* (1988) reported that feeding cadaverine to hairy root cultures of *N. rustica* (*N. rustica* plants transformed with *Agrobacterium rhizogenes*) resulted in increased production of anabasine and inhibition of nicotine accumulation. We have obtained similar results by feeding cadaverine to intact plants of a low alkaloid *N. tabacum* genotype. From short-term diamine feeding of detached roots, demonstration of increased accumulation of alkaloid is more difficult. However, supplementing the feeding solution with nicotinic acid dramatically increased alkaloid accumulation. For example, when low alkaloid *N. tabacum* genotype roots were fed cadaverine alone for 7 h, anabasine concentration increased from 10 μg/g to about 20 μg/g tissue dry wt., but with addition of 2 mM nicotinic acid to the cadaverine feeding solution anabasine increased to about 70 μg/g. Supplemental nicotinic acid alone increased anatabine levels. In *N. glauca* nicotine synthesis can be stimulated by feeding N-methylputrescine plus nicotinic acid (up to eight times control level). Nicotine level increased only slightly with addition of only methylputrescine. Recently, Walton and McLauchlan (1990) reported that the diamine oxidase from *N. tabacum* hairy root cultures exhibited a high specificity for N-methylputrescine compared to cadaverine. In spite of this specificity, 1 mM cadaverine prevented the increased nicotine and nornicotine accumulation attributable to feeding 1 mM N-methylputrescine. They suggested that factors other than substrate specificity of MPO oxidase must be involved in the sensitivity of nicotine biosynthesis to perturbation. Our feeding studies with detached root cultures of *N. tabacum* and *N. glauca* fed N-methylputrescine and nicotinic acid indicate that MPO activity is not rate-limiting under these conditions and that penetration of exogenous diamine or availability of endogenous nicotinic acid is rate-limiting. Feeding of putrescine with or without nicotinic acid had little effect on nornicotine levels. Although putrescine can be oxidized to aminobutanal by tobacco root and leaf extracts containing MPO (Saunders and Bush, 1979), there is little, if any, evidence to support the idea that pyrrolinium is a normal precursor for nornicotine.

1.4 ALKALOID TRANSFORMATIONS DURING SENESCENCE AND CURING

It should be noted that, in this review, senescence is defined as being initiated at the time when the plant is decapitated (induction of rapid nicotine biosynthesis and accumulation). The plant is still accumulating dry matter, but mainly in the upper leaves of the plant (Burton *et al.*, 1983). Significant accumulation of the pyridine alkaloids is occurring in the upper leaves, but little or no alkaloid accumulation is occurring in the lower leaves. In addition to the accumulation of the four major pyridine alkaloids nicotine, anabasine, anatabine and nornicotine, there is an increase in minor pyridine alkaloids (Burton *et al.*, 1988; Andersen, *et al.* 1989). These include cotinine, 2,3′-bipyridyl, N-acylnornicotines, N-acylanatabines and N-acylanabasines (Figs. 1.1, 1.2). Even though there are only trace quantities of these minor alkaloids in mature leaves, the quantities increase significantly during air curing (Burton *et al.*, 1988).

1.4.1 OXIDIZED PYRIDINE ALKALOIDS

The two major oxidized alkaloids in tobacco are cotinine and 2,3′-bipyridyl. Cotinine is derived from nicotine by way of enzymatic or auto-oxidation (Enzell *et al.*, 1977). Nicotine is conveniently converted to cotinine through auto-oxidation in low yield (Burton *et al.*, 1987). The low level of cotinine in green tissue is most likely derived enzymatically rather than via auto-oxidation. Those tobaccos with relatively high content of nornicotine generally contain relatively high content of cotinine suggesting that the apparent P450 enzymatic demethylation of nicotine to nornicotine may somehow be related to cotinine formation as P450 systems are required for cotinine formation in mammals. As plant maturity increases and during air curing, there is greater accumulation of cotinine. There appears to be no effect of curing temperature on the accumulation of cotinine (Burton *et al.*, 1988). 2,3′-Bipyridyl is formed by oxidation of anatabine. This compound was first identified by Frankenburg *et al.* (1952) as a fermentation product along with 3-acetylpyridine. Even though 2,3′-bipyridyl accumulates during fermentation it is also present in low concentrations in green leaf (approx. 2.0 ppm). Although there are detectable amounts of 2,3′-bipyridyl in mature leaf, it is only approximately 0.001% of the level of anatabine. There is also a direct relationship between plant maturity and the level of 2,3′-bipyridyl (Burton *et al.*, 1988). Curing environment influences the accumulation of this alkaloid. Curing tobacco at 32°C/83% relative humidity results in a six-fold increase of 2,3′-bipyridyl in comparison with curing tobacco at 24°C/70% relative humidity. Curing tobacco at high relative humidity increases the moisture content of the cured lamina and enhances the fermentation process. These conditions result in the increased accumulation of 2,3′-bipyridyl. Other oxidized alkaloids present in green and cured tobacco include nicotyrine, 3-acetylpyridine, nicotinic acid, nicotinamide and the nicotine-N-oxides. These oxidation products also increase during curing and have been reviewed (Enzell *et al.*, 1977).

1.4.2 N-ACYLATED PYRIDINE ALKALOIDS

A unique class of pyridine alkaloids that has recently been studied is the acylated derivatives of the secondary amine alkaloids nornicotine, anatabine and anabasine. The first acylated alkaloids identified in tobacco and shown in Figure 1.2 were n-hexanoyl- and n-octanoylnornicotine (Bolt, 1972). Warfield *et al.* (1972) identified N′-formyl- and N′-acetylnornicotine, Matsushima *et al.* (1983) isolated N′-butanoylnornicotine, Miyano *et al.* (1979) isolated N′-formylanatabine and N′-formylanabasine, and Matsushima *et al.* (1983) identified N′-(6-hydroxyoctanoyl)- and N′-(7-hydroxyoctanoyl)-nornicotine in cured tobaccos. The above acylated pyridine alkaloids are distributed within the leaf matrix. Severson *et al.* (1988) isolated and identified a series of C_{12} to C_{16} hydroxyacyl- and C_{12} and C_{13} acyl-nornicotines from the leaf surface of *N. repanda, N. stocktonii* and *N. nesophila.* The acylated compounds that contained a hydroxyl group at the 3 position of the acid side chain had greatest biological activity in insect, plant and microbial bioassays. The major component of the hydroxyacylnornicotines [iso-N′-(3-hydroxy-12-methyltetradecanoyl)]-nornicotine is biosynthesized in the trichomes on the surface of the leaf (Zador and Jones, 1986) and not in the leaf matrix as the simpler acylated pyridine alkaloids. N′-Formylnornicotine may be formed via formylation of nornicotine or oxidation of the N-methyl group of nicotine. This latter mechanism was proposed by Leete (1977) to explain conversion of nicotine to nornicotine. However, Burton *et al.* (1988) concluded that the accumulation profiles of N′-formylnornicotine, N′-acetylnornicotine and N′-formylanatabine were similar, suggesting that these compounds are formed via acylation of the secondary amine alkaloid.

Matsushima *et al.* (1983) first reported changes in levels of acylated nornicotine alkaloids. Except for 6-hydroxyoctanoyl- and 7-hydroxyoctanoylnornicotine, they reported decreased acylated nornicotines during curing. Burton *et al.* (1988) investigated the changes in the acylated alkaloids during senescence and air curing of burley tobacco. In contrast to Matsushima's study, there was a significant increase of the acylated alkaloids throughout air curing of burley tobacco. This observation was substantiated in a parallel study reported by Andersen *et al.* (1989).

Curing environment and plant maturity influence accumulation of these minor alkaloids. Tobacco harvested one week after decapitating (nicotine induction) contained 20% of the acylated alkaloids that were in the tobacco harvested four weeks after decapitation (Burton *et al.*, 1988). Tobacco harvested seven weeks after decapitation did not contain significantly greater quantities of these alkaloids than tobacco from the 4-week induction treatment. Not only does leaf maturity affect accumulation of the acylated alkaloids, but curing environment also influences the accumulation of these alkaloids. Curing tobacco at elevated temperatures (32°C vs. 24°C) resulted in greater accumulation of the acylated alkaloids, regardless of plant maturity. Increases of these alkaloids generally occurred during the later stages of curing. This would indicate their accumulation occurred after the cell membranes lost integrity. Comparisons of acylated nornicotine data from air and flue cured tobacco also show curing environment influences the accumulation of these constituents (Burton *et al.*, 1987). Flue curing tobacco resulted in reduced accumulation of these acylated derivatives. It should be noted that curing time interval for flue curing is approximately 7 days at above ambient temperatures, whereas air curing requires approximately 6 to 8 weeks for complete curing at ambient temperatures. The difference in curing environments would allow greater accumulation of acylated alkaloids to occur in the air curing environment.

The levels of acylated secondary alkaloids also vary in different tobacco genotypes. For example, the concentration of 6- and 7-hydroxy-octanoylnornicotines have been reported to accumulate in flue type tobacco in significant quantities (Djordjevic *et al.*, 1990). In one cultivar the levels of these hydroxylated derivatives were the predominant acylated nornicotine. Burley tobacco cultivars contained detectable quantities of the hydroxylated derivatives but they were less than 1 ppm (unpublished data of authors). In general, the level of acylated alkaloids parallels the concentration of the secondary alkaloid content of the tobacco (Burton *et al.*, 1988; Djordjevic *et al.*, 1990).

In summary, the levels of the acylated secondary alkaloids are influenced by curing environment, plant maturity, genetics and production practices. Even though the concentrations of these derivatives are small in comparison to the total alkaloid concentration in tobacco their accumulation may be of significance to flavour perception and reduction of the secondary amine alkaloids.

1.4.3 TOBACCO SPECIFIC NITROSAMINES

Interest in tobacco specific N′-nitrosamines (TSNA) has been a result of the report that some of these nitrosamines induced malignant tumours in mice, rats and hamsters (Boyland *et al.*, 1964a, 1964b; Hecht *et al.*, 1980). The nitrosamines of greatest interest have been N′-nitrosonornicotine (NNN), N′-nitrosoanatabine (NAT), N′-nitrosoanabasine (NAB), and 4-(methylnitrosamino)-1-(3-pyridyl)-1-butanone (NNK) (Figure 1.9). It should be noted that only NNN and NNK are reported to have significant tumorigenic activity. There have been many studies on the

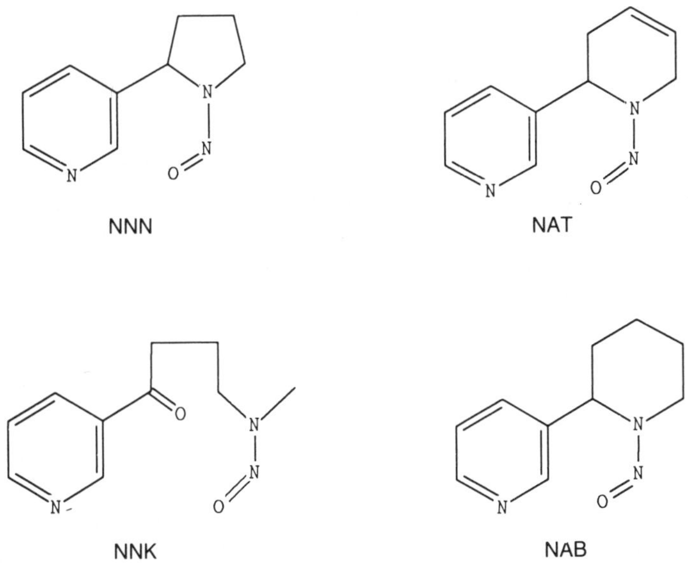

Figure 1.9 Major N′-nitrosamines derived from *Nicotiana* alkaloids.

TSNA content and formation in tobacco yet conclusions from these studies in several instances are conflicting. NAT and NAB are formed directly from their precursor alkaloids anatabine and anabasine by reaction with nitrite. Concentrations of anabasine and anatabine are more than adequate for the amount of nitrosamine formed, since concentration of the alkaloids is at least 100 times greater than the reaction products. Also, the relative concentrations of NAT and NAB reflect the levels of anatabine and anabasine in the leaf (Djordjevic *et al.*, 1989).

Proposed precursors and sources of NNN have not been as straightforward. An initial study by Hecht *et al.* (1978) indicated nicotine as well as nornicotine were precursors of NNN. From radiolabelling they proposed nicotine was the major precursor of NNN. Caldwell *et al.* (1991) proposed nornicotine was the major precursor for NNN. This was based on data obtained from kinetic studies on the formation of NNN from nicotine and nornicotine. Studies by Djordjevic *et al.* (1989) and Andersen *et al.* (1989) showed the correlation coefficient between NNN and nornicotine was higher than between NNN and nicotine. Because of the complex nature of the tobacco matrix, additional studies on nitrosation of alkaloids in tobacco are needed.

NNK is derived from nicotine since it contains the methylamine group. It is formed by oxidation, ring opening between the C1′ and C2′ of the pyrrolidine ring, and nitrosation. The mechanism for this reaction has not been determined, therefore it is not possible to discuss the key intermediates required for its formation. Treatment of nicotine and aqueous sodium nitrite with dilute hydrochloric acid (pH 5) will result in the accumulation of small quantities of NNK along with cotinine (Hecht *et al.*, 1978;). These observations show NNK does not require an enzymatic process for its formation.

Hoffmann *et al.* (1980) reported that 46% of the NNN in tobacco was transferred to the smoke and the remaining pyrosynthesized. However, more recently Fischer *et al.* (1990) concluded that pyrosynthesis of NNN does not occur. No definitive experiment has been conducted to compare tobacco types for amount of TSNA accumulation. However, from various reports of the amounts of TSNA in tobacco, it is generally concluded that dark tobacco types have greatest TSNA concentration followed in descending order by burley, flue cured and oriental tobacco (Fischer *et al.*, 1989; Burton *et al.*, 1990). These differences in TSNA accumulation probably do not reflect any genetic difference, but a culture difference of nitrogen fertilizer application and subsequent nitrate accumulation as a source of nitrite formation for nitrosation. There have been independent reports that the levels of TSNA in tobacco are directly related to alkaloids, nitrate, or nitrite content of the lamina. Djordjevic *et al.* (1989) and Andersen *et al.* (1989) found there was a direct relationship between alkaloids and TSNA in cured tobacco. Low levels of alkaloids in tobacco will result in reduced levels of TSNA, but production of low alkaloid tobacco is not practical since these tobaccos will not meet consumer acceptance. One would predict the positive relationship between alkaloids and TSNA in tobacco since pyridine alkaloids are direct precursors for TSNA.

Brunnemann *et al.* (1983) reported a positive and significant correlation between TSNA and nitrate in cigarette tobacco. In a nitrogen fertilization study, MacKown *et al.* (1984) reported no significant relationship between nitrate and TSNA even though there was a ten-fold difference in leaf nitrate. Fischer *et al.* (1989) reported there was a positive correlation between TSNA and nitrate from tobacco varieties and blended cigarette tobaccos. Therefore, nitrate level in tobaccos may be an indicator of TSNA accumulation. These observations suggest the reduction of nitrate to nitrite may be the rate limiting step for TSNA accumulation.

Since nitrite is an immediate precursor of TSNA and since the alkaloids are present in two to four orders of magnitude higher concentration than nitrite, the obvious conclusion is that the major limiting factor for TSNA accumulation is nitrite. One could argue that nitrite is a transient reactive intermediate and there would be no correlation with TSNA accumulation, but this is not the case. In a study by Burton *et al.* (1989) on the curing of burley tobacco in a controlled environment of

32°C and 83% relative humidity, there was a significant increase of TSNA and nitrite in the plant tissue. In one instance the level of TSNA approached 1000 μg/g and nitrite N was 1.0 mg/g. This represented a 100-fold increase in TSNA and 1200-fold increase in nitrite over tobacco cured at ambient conditions. These temperatures and relative humidities certainly will support microbial activity in and on the tobacco matrix. It was assumed the nitrate reductase activity was from bacteria or fungi and not from residual nitrate reductase activity remaining in the plant tissue. Relationships between high levels of nitrite and TSNA have also been reported for smokeless tobaccos that were stored at elevated moisture content and 32°C (Andersen *et al.*, 1989). This again supports nitrate reductase activity being responsible for nitrite accumulation.

The elevated TSNA and nitrite concentrations obtained from tobaccos stored at high moisture content are not representative of the levels obtained under normal air curing conditions. Burton *et al.* (1990) conducted a two-year study with three tobacco genotypes and several different tissue samples to determine the relationship between TSNA precursors and TSNA accumulation. They reported positive and significant correlations between the low endogenous levels of nitrite in tobacco and accumulation of individual TSNA. This positive correlation certainly is an over-simplification of the nitrosation reactions that occur in the tobacco matrix. For the reactions to occur the reactants must be in proximity to each other and most of the TSNA accumulation occurs after the cell membrane has become leaky. The average pH of tobacco is 6, which is not the ideal condition for optimum nitrosation. In addition the moisture content is below optimum for the reaction to occur. However, even though optimum conditions do not prevail there is nitrosation of the tobacco alkaloids and accumulation of TSNA.

It was proposed that higher concentrations of TSNA in the leaf would be related to the concentration of the rate limiting precursor (Burton *et al.*, 1992) and the distribution of these constituents within the leaf was determined. Leaves from an air cured dark tobacco were segmented into 4 cm × 7 cm portions and the midvein was divided into 7 cm lengths. Surface plots were calculated from the analytical data obtained for these constituents in each leaf segment. Alkaloid levels were greatest on the periphery of the leaf and lowest along the midvein. Nornicotine is shown as a representative accumulation pattern of each of the alkaloids (Figure 1.10a). TSNA levels were lowest at the tip of the leaf and highest in the midvein at the base of the leaf (Figure 1.10b). Nitrate concentration was greatest along the midvein. Nitrite concentrations were highest in the midvein at the base of the leaf and more closely paralleled the profile of TSNA. When the TSNA and nitrite data are presented as percentage distribution in the leaf, the patterns are almost identical (Figure 1.10c and 1.10d). Both TSNA and nitrite are greatest in the midvein at the base of the leaf. These observations suggest that the nitrosation of alkaloids in tobacco leaf is dependent on the capacity for nitrite production within the system. Alkaloids and nitrate are essential for nitrosamine formation, but if nitrite is not produced, no nitrosamines will be formed.

This explanation for nitrosation of the alkaloids is somewhat speculative at this time. We are proposing that in stalk-cut air cured tobacco the midvein attached to the stalk is the last part of the leaf to dry. Maintaining moisture over a prolonged time may allow microbial activity to occur. Some of these microbes reduce nitrate (Parsons *et al.*, 1986) and could explain the increased accumulation of nitrite at the base of the leaf. Preliminary results from primed air cured segmented leaf which loses moisture more uniformly support this speculation. The conditions required for the accumulation of TSNA have not been totally clarified but it is apparent that disruption of plant cell membranes and apparent microbial reduction of nitrate to nitrite is required for nitrosa-

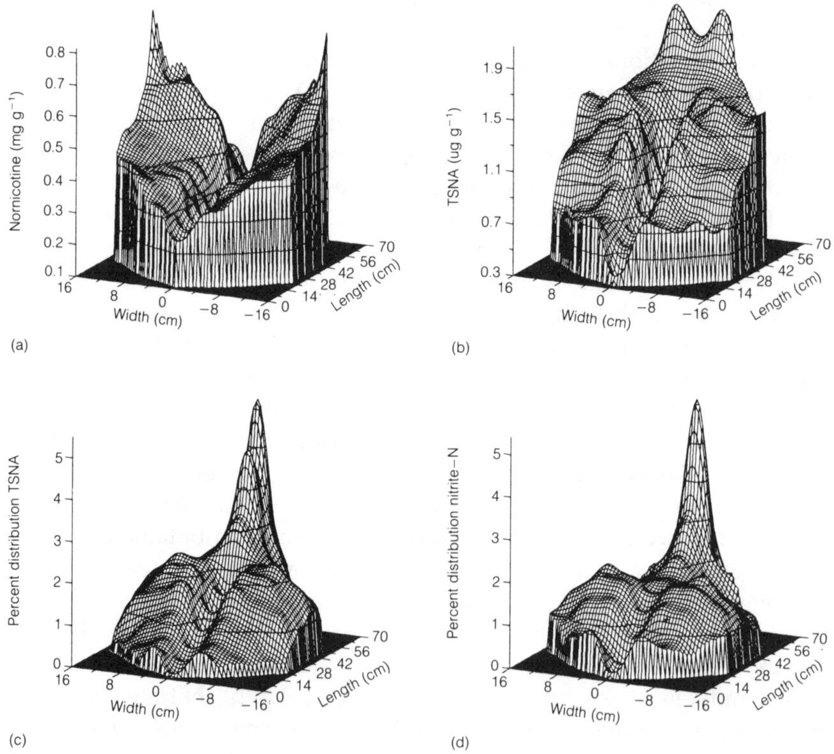

Figure 1.10 Distribution of (a) nornicotine, (b) tobacco specific nitrosamines and percentage distribution of (c) nitrite and (d) tobacco specific nitrosamines in cured leaf. X-axis coordinates 0, 0 are the tip of the leaf with midrib samples on the line from coordinates 0, 0 to 0, 70.

tion of the alkaloids to occur. From present knowledge, it will be possible to produce tobacco with predictable levels of TSNA.

ACKNOWLEDGEMENT

The authors are indebted to support from US Department of Agriculture – Agricultural Research Service under Co-operative Agreements Nos 58–6430–2–120 and 58–6430–1–121 and the Kentucky Tobacco Research Board project 5–41096.

REFERENCES

Andersen, R.A., Fleming, P.D., Burton, H.R. *et al.* (1989) N′-Acyl- and N′-nitroso-pyridine alkaloids in alkaloid lines of burley tobacco during growth and air-curing. *J. Agric. Food Chem.*, **37**, 44–50.

Baxter, C. and Coscia, C.J. (1973) *In vitro* synthesis of spermidine in the higher plant, *Vinca rosea*. *Biochem. Biophys. Res. Comm.*, **54**, 147–54.

Benveniste, I., Gabriac, B. and Durst, F. (1986) Purification and characterization of the NADPH-cytochrome P450 (cytochrome c) reductase from higher-plant microsomal fraction. *Biochem. J.*, **235**, 365–73.

Berlin, J. (1981) Formation of putrescine and cinnamoyl putrescines in tobacco cell cultures. *Phytochem.*, **20**, 53–5.

Bolt, A.J. (1972) 1′-Hexanaylnornicotine and 1′-octanoylnornicotine from tobacco. *Phytochem.*, **11**, 2341–3.

Boyland, E., Roe, F.J.C., Gorrod, J.W. *et al* (1964a) The carcinogenicity of nitrosoanabasine, a possible constitutent of tobacco smoke. *Br. J. Cancer*, **XVIII**, 265–70.

Boyland, E., Roe, F.J.C. and Gorrod J.W. (1964b) Induction of pulmonary tumours in mice by nitrosonornicotine, a possible constituent of tobacco smoke. *Nature*, **202**, 1126.

Brunnemann, K.D., Masaryk, J., Hoffmann, D. (1983) Role of tobacco stems in the formation of N-nitrosamines in tobacco and mainstream and sidestream smoke. *J. Agric. Food Chem.*, **31**, 1221–4.

Burk, L.G. and Jeffrey, R.N. (1958) A study of the inheritance of alkaloid quality in tobacco. *Tobacco Sci.*, **2**, 139–41.

Burton, H.R., Andersen, R.A., Fleming, P.D. *et al.* (1988) Changes in chemical composition of burley tobacco during senescence and curing. 2. Acylated pyridine alkaloids. *J. Agric. Food Chem.*, **36**, 579–84.

Burton, H.R., Bush, L.P. and Djordjevic, M.V. (1989) Influence of temperature and humidity on the accumulation of tobacco-specific nitrosamines in stored burley tobacco. *J. Agric. Food Chem.*, **37**, 1372–7.

Burton, H.R., Bush, L.P. and Hamilton, J.L. (1983) Effect of curing on the chemical composition of burley tobacco. *Recent Adv. Tob. Sci.*, **9**, 91–153.

Burton, H.R., Djordjevic, M., Childs, G. *et al.* (1987) On the formation of acylated pyridine alkaloids in tobacco. *Tob. Chemists' Res. Conf.*, **41**, 39.

Burton, H.R., Dye, N.K., and Bush, L.P. (1990) Accumulation of nitrites and tobacco-specific nitrosamines from different tobacco types during curing. *CORESTA Symposium*, Cooperation Center for Scientific Research Relative to Tobacco, Paris, APTS02, p. 97

Burton, H.R., Dye, N.K., and Bush, L.P. (1992) Distribution of tobacco constituents in tobacco leaf tissue. 1. Tobacco-specific nitrosamines, nitrate, nitrite and alkaloids. *J. Agric. Food Chem.*, **40**, 1050–5

Bush, L.P. (1981) Physiology and biochemistry of tobacco alkaloids. *Recent Adv. Tobacco Science*, **7**, 75–106.

Bush, L.P. and Crowe, M.W. (1989) *Nicotiana* alkaloids, in *Toxicants of Plant Origin*, (Ed. P.R. Cheeke) CRC Press, Inc., Boca Raton, pp. 87–107.

Bush, L.P. and Saunders, J.W. (1977) Accumulation, manipulation and regulation of nicotine content in tobacco, in *Recent Advances in the Chemical Composition of Tobacco and Tobacco Smoke*, Proc. Am. Chem. Soc. Symp. (Ed. J.L. McKenzie), Pierce, Ashokie, N.C., pp, 389–425.

Caldwell, W.S., Greene, J.M., Plowchalk, D.R. *et al.* (1991) The nitrosation of nicotine: A kinetic study. *Chem. Res. Tox.*, **4**, 513–6.

Canivenc, M.-C., Cagnac, B., Cabanne, F. *et al.* (1989) Induced changes of chlorotoluron metabolism in wheat cell suspension cultures. *Plant Physiol. Biochem.*, **27**, 193–201.

Chaplin, J.F. and Miner, G.S. (1980) Production factors affecting chemical components of the tobacco leaf. *Recent Adv. Tobacco Science*, **6**, 3–63.

Chelvarajan, R.L., Fannin, F.F., Bush, L.P. (1991) *In vitro* demethylation of nicotine. *Tob. Chemists' Res. Conf.*, **45**, 18.

Dawson, R.F. (1945) On the biogenesis of nornicotine and anabasine. *J. Am. Chem. Soc.*, **67**, 503–4.

Dawson, R.F. (1951) Alkaloid biogenesis. III. Specificity of the nicotine–nornicotine conversion. *J. Am. Chem. Soc.*, **73**, 4218–21.

Dawson, R.F. (1952) Alkaloid biogenesis: Nicotine demethylation in excised leaves of *Nicotiana glutinosa*. *Am. J. Bot.*, **39**, 250–3.

Dawson, R.F. (1960) Biosynthesis of the *Nicotiana* alkaloids. *Am. Scientist*, **48**, 321–40.

Dawson, R.F., Christman, D.R., D'Adamo, A. *et al.* (1960a) The biosynthesis of nicotine from isotopically labelled nicotinic acids. *J. Am. Chem. Soc.*, **82**, 2628–33.

Dawson, R.F., Christman, D.R., Solt, M.L. *et al.* (1960b) The biosynthesis of nicotine from nicotinic acid. Chemical and radiochemical yields. *Arch. Biochem. Biophys.*, **91**, 144–50.

Dawson, R.F. and Osdene, T.S. (1972). A speculative view of tobacco alkaloid biosynthesis, in *Recent Adv. Phytochem.* (Eds V.C. Runecles and T.C. Tso), **5**, 317–38.

Dewey, L.J., Byerrum, R.U. and Ball, C.D. (1955) The biosynthesis of the pyrrolidine ring of nicotine. *Biochem. Biophys. Acta*, **18**, 141–2.

Djordjevic, M.V., Bush, L.P., Gay, S.L. *et al.* (1990) Accumulation and distribution of acylated nornicotine derivatives in flue-cured tobacco alkaloid lines. *J. Agric. Food Chem.*, **38**, 347–50.

Djordjevic, M.V., Gay, S.L., Bush, L.P. *et al.* (1989) Tobacco-specific nitrosamine accumulation and distribution in flue-cured tobacco isolines. *J. Agric. Food Chem.*, **37**, 752–6.

Donaldson, R.P. and Luster, D.G. (1991) Multiple forms of plant cytochrome P450. *Plant Physiol.*, **96**, 669–74.

Enzell, C.R., Wahlberg, I. and Aasen A.J. (1977) Isoprenoids and alkaloids of tobacco. *Prog. Chem. Org. Nat. Prod.*, **34**, 1–79.

Fannin, F.F. and Bush, L.P. (1988) Enzymes for alkaloid biosynthesis in *N. tabacum, N. glauca* and their hybrid. *Tob. Chemists' Res. Conf.*, **42**, 20.

Feth, F., Wagner, R. and Wagner, K.G. (1986) Regulation in tobacco callus of enzyme activities of the nicotine pathway. I. The route ornithine to methylpyrroline. *Planta*, **168**, 402–7.

Fischer, S., Spiegelhalder, B., Eisenbarth, J. *et al.* (1990) Investigations on the origin of tobacco-specific nitrosamines in mainstream smoke of cigarettes. *Carcinogenesis*, **11**, 723–30.

Fischer, S., Spiegelhalder, B. and Preussmann, R. (1989) Preformed tobacco-specific nitrosamines in tobacco – role of nitrate and influence of tobacco type. *Carcinogenesis*, **10**, 1511–7.

Frankenburg, W.G., Gottscho, A.M., Mayaud, E.W. *et al.* (1952) The chemistry of tobacco fermentation. I. Conversion of the alkaloids. A. The formaion of 3-pyridylmethylketone and of 2,3'dipyridyl. *J. Am. Chem. Soc.*, **74**, 4309–14.

Friesen, J.B. and Leete, E. (1990) Nicotine synthase – An enzyme from *Nicotiana* species which catalyses the formation of (*S*)-nicotine from nicotinic acid and 1-methyl-Δ'-pyrrolinium chloride. *Tetrahedron Lett.*, **31**, 6295–8.

Furuya, T., Kojima, H. and Syono, K. (1967) Regulation of nicotine synthesis in tobacco callus tissue. *Chem. Pharm. Bull.*, **15**, 901–3.

Furuya, T., Kojima, H. and Syono, K. (1971) Regulation of nicotine synthesis by auxin in tobacco callus cultures. *Phytochem.*, **10**, 1529–32.

Griffith, R.B., Valleu, W.D. and Stokes, G.W. (1955) Determination and inheritance of nicotine to nornicotine conversion in tobacco. *Science*, **121**, 343–4.

Hecht, S.S., Chen C.B., Hirota, N. *et al.* (1978) Tobacco-specific nitrosamines: Formation from nicotine *in vitro* and during curing and carcinogenicity in strain A mice. *J. Nat. Cancer Inst.*, **60**, 819–24.

Hecht, S.S., Chen, C.B., Ohmori, T. *et al.* (1980) Comparative carcinogenicity in F-344 rats of the tobacco-specific nitrosamine, N-nitrosonornicotine and 4-(methyl-N-nitrosamino)-1-(3-pyridyl)-1-butanone. *Cancer Res.*, **40**, 298–302.

Hoffmann, D., Adams, J.D., Piade, J.J. *et al.* (1980) Chemical studies on tobacco smoke. LXVII. Analysis of volatile and tobacco-specific nitrosamines in tobacco products, in *N-Nitroso Compounds: Analysis, Formation and Occurrence* (ed. E.A. Walker, M. Castegnaro, L. Griciute, *et al.*) IRAC Scientific Publication No. **31**, International Agency for Research on Cancer, Lyon, France, pp. 507–515.

Jackanioz, T.M. and Byerrum, R.U. (1966) Incorporation of aspartate and malate into the pyridine ring of nicotine. *J. Biol. Chem.* **241**, 1246–9.

Kisaki, T., and Tamaki, E. (1961) Phytochemical studies on the tobacco alkaloids. I. Optical rotatory power of nornicotine. *Arch. Biochem. Biophys.*, **92**, 351–5.

Kisaki, T. and Tamaki, E. (1966) Phytochemical studies on the tobacco alkaloids. X. Degradation of the tobacco alkaloids and their optical rotatory changes in tobacco plants. *Phytochem.*, **5**, 293–300.

Krikorian, A.D., and Steward, F.C. (1969) Biochemical differentiation: The biosynthetic potentialities of growing and quiescent tissue, in *Plant Physiology* (Ed. F.C. Steward), Academic Press, **5B**, 227–326.

Leete, E. (1955) Biogenesis of nicotine. *Chem. Ind.* (London) 537.

Leete, E. (1958) The biogenesis of the *Nicotiana* alkaloids. VI. The piperidine ring of anabasine. *J. Am. Chem. Soc.*, **80**, 4393–4.

Leete, E. (1975) Biosynthesis of anatabine and anabasine in *Nicotiana glutinosa. J. Chem. Soc. Chem. Commun.* 9–10.

Leete, E. (1977) Biosynthesis and metabolism of the tobacco alkaloids, in *Recent Advances in the Chemical Composition of Tobacco and Tobacco Smoke,* Proc. Am. Chem. Soc. Symp. (Ed. J.L. McKenzie), Pierce, Ashokie, N.C., pp. 365–88.

Leete, E. (1978) Stereochemistry of the reduction of nicotinic acid when it serves as a precursor of anatabine. *J. Chem. Soc. Chem. Commun.*, 610–11.

Leete, E. (1979) Aberrant biosynthesis of 5-fluoroanabasine from 5-fluoro-[5,6-$^{14}C^{13}C_2$]nicotinic acid, established by means of carbon-13 nuclear magnetic resonance. *J. Org. Chem.*, **44**, 165–8.

Leete, E. (1980) Alkaloids derived from ornithine, lysine, and nicotinic acid, in *Secondary Plant Products, Encycl. Plant Physiology, New Series* (eds E.A. Bell and B.V. Charlwood), **8**, 65–91.

Leete, E. (1984) The methylation of nornicotine to nicotine, a minor biosynthetic pathway in *Nicotiana tabacum. Beit. Tabak.*, **12**, 113–6.

Leete, E., Bodem, G.E. and Manuel, M.F. (1971) Formation of 5-fluoronicotine from 5-fluoronicotinic acid in *Nicotiana tabacum. Phytochem.*, **10**, 2687–92.

Leete, E. and Chedekel, M.R. (1972) The aberrant formation of (−)-N-methylanabasine from N-methyl-Δ′-piperideinium chloride in *Nicotiana tabacum* and *N. glauca. Phytochem.*, **11**, 2751–6.

Leete, E. and Chedekel, M.R. (1974) Metabolism of nicotine in *Nicotiana glauca. Phytochem.*, **13**, 1853–9.

Leete, E., Gros, E.G. and Gilbertson, T.J. (1964) Biosynthesis of anabasine, origin of the nitrogen of the piperidine ring. *J. Am. Chem. Soc.*, **86**, 3907–8.

Leete, E. and Muller, M.E. (1982) Biomimetic synthesis of anatabine from 2,5-dihydropyridine produced by the oxidative decarboxylation of baikiain. *J. Am. Chem. Soc.*, **104**, 6440–4.

Leete, E. and Slattery, S.A. (1976) Incorporation of [2-^{14}C]- and [6-^{14}C] nicotinic acid into the tobacco alkaloids. Biosynthesis of anatabine and α,β-dipyridyl. *J. Am. Chem. Soc.*, **98**, 6326–30.

Legg, P.D., Chaplin, J.F. and Collins, G.B. (1969) Inheritance of percent total alkaloids in *Nicotiana tabacum* L. Populations derived from crosses of low alkaloid lines with burley and flue-cured varieties. *J. Hered.*, **60**, 213–17.

Legg, P.D. and Collins, G.B. (1971) Inheritance of percent total alkaloids in *Nicotiana tabacum* L. II. Genetic effects of two loci in Burley 21 × LA Burley 21 populations. *Can. J. Genet. Cytol.*, **13**, 287–91.

Lockwood, G.B. and Essa, A.K. (1984) The effect of varying hormonal and precursor supplementation on levels of nicotine and related alkaloids in cell cultures of *Nicotiana tabacum. Plant Cell Reports*, **3**, 109–11.

MacKown, C.T., Eivagi, F., Sims, J.L. *et al.* (1984) Tobacco-specific N-nitrosamines. Effect of burley isolines and nitrogen fertility management. *J. Agric. Food Chem.*, **32**, 1269–72.

Madsen, J.P., Bush, L.P. and Gay, S.L. (1984) Effect of curing on polyamine content of *Nicotiana tabacum* L. genotypes with different alkaloid levels. *J. Agric. Food Chem.*, **33**, 1182–5.

Mann, D.F. and Byerrum, R.U. (1974) Activation of the *de novo* pathway for pyridine nucleotide biosynthesis prior to ricinine biosynthesis in castor beans. *Plant Physiol.*, **53**, 603–9.

Mann, T.J., Weybrew, J.A., Matzinger, D.F. *et al.* (1964) Inheritance of the conversion of nicotine to nornicotine in varieties of *Nicotiana tabacum* L. and related amphiploids. *Crop Sci.*, **4**, 349–53.

Mantell, S.H., Pearson, D.W., Hazell, L.P. *et al.* (1983) The effect of initial phosphate and sucrose levels on nicotine accumulation in batch suspension cultures of *Nicotiana tabacum* L. *Plant Cell Reports*, **2**, 73–7.

Matsushima, S., Ohsumi, T. and Sugawara, S. (1983) Composition of trace alkaloids in tobacco leaf lamina. *Agric. Biol. Chem.*, **47**, 507–10.

Miller, R.D., Collins, G.B. and Davis, D.L. (1983) Effects of nicotine precursors on nicotine content in callus cultures of burley tobacco alkaloid lines. *Crop Sci.*, **23**, 561–5.

Miyano, M., Matsushita, H., Yasumatsu, N. *et al.* (1979) New minor alkaloids in burley tobacco (*Nicotiana tabacum*). *Agric. Biol. Chem.*, **43**, 1607–8.

Mizusaki, S., Tanabe, Y., Noguchi, M. *et al.* (1971) Phytochemical studies on tobacco alkaloids. XIV. The occurrence and properties of putrescine N-methyltransferase in tobacco roots. *Plant Cell Physiol.*, **12**, 633–40.

Mizusaki, S., Tanabe, Y., Noguchi, M. *et al.* (1972) N-Methylputrescine oxidase from tobacco roots. *Phytochem.*, **11**, 2757–62.

Mizusaki, S., Tanabe, Y., Noguchi, M. *et al.* (1973) Changes in the activities of ornithine decarboxylase, putrescine N-methyltransferase and N-methylputrescine oxidase in tobacco roots in relation to nicotine biosynthesis. *Plant Cell Physiol.*, **14**, 103–10.

Neumann, D. and Muller, E. (1971) Contributions to the physiology of alkaloids. V. Alkaloid production in callus and suspension cultures of *Nicotiana tabacum* L. *Biochem. Physiol. Pflanzen.*, **162**, 503–13.

Parsons, L.L., Smith, M.S., Hamilton, J.L. *et al.* (1986) Nitrate reduction during curing and processing of burley tobacco. *Tob. Sci.*, **30**, 100–3.

Pinol, M.T., Palazon, J. and Lerrano, M. (1984) Growth and nicotine content of tobacco callus culture without organogenesis. *Plant Sci. Letters*, **35**, 219–23.

Pudliner, H.J. (1980) Nicotine accumulation in tobacco and tobacco callus. Master Sci. Thesis, Univ. Kentucky, Lexington, KY.

Ravishanker, G.A. and Mehta, A.L. (1982) Regulation of nicotine biogenesis. III. Biochemical basis of increased nicotine biogenesis by urea in tissue cultures of tobacco. *Can. J. Bot.*, **60**, 2371–4.

Saunders, J.W. and Bush, L.P. (1979) Nicotine biosynthetic enzyme activities in *Nicotiana tabacum* L. genotypes with different alkaloid levels. *Plant Physiol.*, **64**, 236–40.

Schiel, O., Jarchow-Redecker, K., Gerd-Walter, P. *et al.* (1984) Increased formation of cinnamoyl putrescines by fedbatch fermentation of cell suspension cultures of *Nicotiana tabacum. Plant Cell Reports*, **3**, 18–20.

Severson, R.F., Arrendale, R.F., Cutler, H.G. *et al.* (1988) Chemistry and biological activity of acylnornicotines from *Nicotiana respondae*, in *Am. Chem. Soc. Sym. Series* No. **380** (ed H.G. Cutler), Am. Chem. Soc., pp. 335–62.

Solt, M.L., Dawson, R.F. and Christman, D.R. (1960) Biosynthesis of anabasine and of nicotine by excised root cultures of *Nicotiana glauca. Plant Physiol.*, **35**, 887–94.

Speake, T., McCloskey, P. and Smith W.K. (1964) Isolation of nicotine from cell cultures of *Nicotiana tabacum. Nature*, **201**, 614–15.

Stalon, V., Lamos, F., Pierard, A. *et al.* (1972) Regulation of catabolic ornithine carbamoyltransferase of *Pseudomonas flourescens*: A comparison with the anabolic transferase and with a mutationally modified catabolic transferase. *Eur. J. Biochem.*, **29**, 25–35.

Stepka, W. and Dewey, L.J. (1961) Conversion of nicotine to nornicotine in harvested tobacco: Fate of the methyl group. *Plant Physiol.*, **36**, 592–7.

Strunz, G.M. and Findlay, J.A. (1985) Pyridine and piperidine alkaloids, in *The Alkaloids* (ed. A. Brossi), Academic Press, **26**, 89–183.

Tabata M. and Hiraoka, N. (1976) Variation of alkaloid production in *Nicotiana rustica* callus cultures. *Physiol. Plant.*, **38**, 19–23.

Tabata, M., Yamamoto, H., Hiraoka, N. *et al.* (1971) Regulation of nicotine production in tobacco tissue culture by plant growth regulators. *Phytochem.*, **19**, 723–9.

Takahashi, M. and Yamada, Y. (1973) Regulation of nicotine production by auxins in tobacco cultured cells *in vitro. Agric Biol. Chem.*, **37**, 1755–7.

Tso, T.C. and McMurtrey, J.E., jun. (1960) Mineral deficiency and organic constituents in tobacco plants. II. Amino acids. *Plant Physiol.*, **35**, 865–70.

Wagner, R., Feth, F. and Wagner, K.G. (1986a) Regulation in tobacco callus of enzyme activities of the nicotine pathway. II. The pyridine–nucleotide cycle. *Planta*, **168**, 408–13.

Wagner, R., Feth, F. and Wagner, K.G. (1986b) The regulation of enzyme activities of the nicotine pathway in tobacco. *Physiol. Plant*, **68**, 667–72.

Walton, N.J. and McLauchlan, W.R. (1990) Diamine oxidation and alkaloid production in transformed root cultures of *Nicotiana tabacum. Phytochem.*, **29**, 1455–7.

Walton, N., Robins, R.J. and Rhodes, M.J.C. (1988) Perturbation of alkaloid production by cadaverine in hairy root cultures of *Nicotiana rustica. Plant Sci.*, **54**, 125–31.

Warfield, A.H., Galloway, W.D. and Kallianos, A.G. (1972) Some new alkaloids from burley tobacco. *Phytochem.*, **11**, 3371–5.

Weeks, W.W. and Bush, L.P. (1974) Alkaloid changes in tobacco seeds during germination. *Plant Physiol.*, **53**, 73–5.

Wernsman, E.A. and Matzinger, D.F. (1968) Time and site of nicotine conversion in tobacco. *Tob. Sci.*, **12**, 226–8.

Yang, K.S., Gholson, R.K. and Waller, G.R. (1965) Nicotine biosynthesis. *J. Am. Chem. Soc.*, **87**, 4184–8.

Yang, Y. (1984) Putrescine biosynthetic enzymes in tobacco roots – purification, properties and function. Master Sci. Thesis, Univ. Kentucky, Lexington, KY.

Yoshida, D. (1967) Effects of nutrient deficiencies on the biosynthesis of nicotine in tobacco plants. *Soil Sci. Plant Nutr.*, **13**, 107–12.

Yoshida, D. (1969a) Effect of agmatine on putrescine formation in tobacco plants. *Plant Cell Physiol.*, **10**, 923–4.

Yoshida, D. (1969b) Formation of putrescine from ornithine and arginine in tobacco plants. *Plant Cell Physiol.*, **10**, 393–7.

Yoshida, D. (1973) Mechanisms of alkaloid accumulations in the varieties of tobacco plants containing lower alkaloid. *Bull. Hatano Tob. Expt. Stn.*, **73**, 245–57.

Yoshida, D. and Mitake, T. (1966) Agmatine and N–carbamylputrescine as intermediates in the formation of nicotine by tobacco plants. *Plant Cell Physiol.*, **7**, 301–5.

Zador, E. and Jones, D. (1986) Biosynthesis of a novel nicotinoid alkaloid in the trichomes of *Nicotiana stocktonii. Plant Physiol.*, **82**, 479–84.

The mammalian metabolism of nicotine: an overview

2

J.W. Gorrod

2.1 INTRODUCTION

The widespread use of tobacco products is an international pastime obviously enjoyed by a very large number of people. Some of this pleasure may well be associated with the pharmacological effects of alkaloids present in tobacco and transferred via tobacco smoke to the lung and the buccal and nasal epithelium where they are absorbed. In addition to the pleasure people gain from tobacco use it is undoubtedly associated with an increased health risk. Recently it has been shown that tobacco consumption may be associated with beneficial effects in Parkinson's disease and may give protective effects in inflammatory bowel syndrome. Nicotine* is the major alkaloid in commercial tobacco and whilst it may not be directly associated with a chronic health risk it is clearly a substance to which the human population at large is routinely and continuously exposed, albeit at extremely low doses via environmental tobacco smoke.

Additional exposure may come about due to the use of nicotine in horticulture or agriculture or through the biosynthesis of the alkaloid by plants and vegetables used as normal foodstuffs. Therefore it is important to understand the biotransformation of this alkaloid in humans, the enzymes involved and the factors controlling and influencing these processes. This chapter is intended to give an overview of the current situation regarding nicotine metabolism in mammals.

*Throughout this chapter nicotine means (*S*)-(−)-nicotine unless the (*R*)-(+)- isomer is specifically mentioned.

Nicotine and Related Alkaloids: Absorption, distribution, metabolism and excretion. Edited by J.W. Gorrod and J. Wahren. Published in 1993 by Chapman & Hall, London. ISBN 0 412 55740 1

Nicotine (3′-pyridyl-2-N-methylpyrrolidine) is a tobacco alkaloid base with two nitrogenous centres of different pKa and hybridization which (Figure 2.1(a)) would be expected to undergo both carbon and nitrogen oxidation in biological systems. In addition, the presence of a lone pair of electrons on each nitrogen atom would allow their utilization in quaternization reactions.

2.2 PRIMARY CARBON OXIDATION PRODUCTS OF NICOTINE

One of the primary metabolites of nicotine is cotinine (Figure 2.1(d)). This compound was originally isolated as a urinary metabolite in the dog but has now been shown to be a significant metabolite in nearly every species examined, including rat, rabbit, mouse, guinea pig, hamster, monkey and man. This metabolite, which is formed by hepatic preparations *in vitro,* is particularly interesting in that the oxygen present in the carbonyl group is not derived from molecular oxygen as occurs with most xenobiotics oxidized by mixed function oxidases (Gillette, 1963), but from water. This led Murphy (1973) to suggest that cotinine was formed via a $\Delta^{1'(5')}$ iminium ion intermediate (Figure 2.1(b)). This idea gained support when Murphy showed that the presence of cyanide in microsomal incubates of nicotine led to incorporation of the cyanide to produce 5′-cyanonicotine. The mechanism of formation of nicotine $\Delta^{1'(5')}$ iminium ion has not been fully elucidated and the possibility exists that it is formed from the α-carbinolamine (Figure 2.1(c)), a metabolite originally proposed by Hucker *et al.* (1960). An alternative mechanism would involve initial 5′-hydrogen or electron abstraction (Guengerich and MacDonald, 1990; Peterson *et al.* 1987; Peterson and Castagnoli, 1988) by a cytochrome P450 isozyme.

The second stage in the formation of cotinine from nicotine undoubtedly involves the oxidation of nicotine $\Delta^{1'(5')}$ iminium ion by cytosolic aldehyde oxidase (Brandange and Lindblom, 1979; Gorrod and Hibberd, 1982; Hill *et al.* 1972). An analogous series of reactions in which an iminium ion (Figure 2.1(e)) is formed by oxidation of the N-methyl group and is in equilibrium with the unstable carbinolamine (Figure 2.1(f)), would lead to nornicotine (Figure 2.1(g)). Such reactions have been proposed (Nguyen *et al.*, 1979). The N-demethylation of nicotine to nornicotine has been reported both *in vitro* (Papadopoulis, 1964) and *in vivo* (McKennis *et al.*, 1962b), although other groups have failed to find this compound as a metabolite. This ambiguity may in part be due to difficulty in the analysis of nornicotine (Beckett *et al.*, 1972). An alternative pathway involving nicotine $\Delta^{1'(2')}$ iminium ion (Figure 2.2(b)), has not been recognized in the *in vitro* biological oxidation of nicotine although it is formed chemically (Sanders *et al.*, 1975; Brandange *et al.*, 1983). In a very recent report, Neurath *et al.* (1992) reported the presence

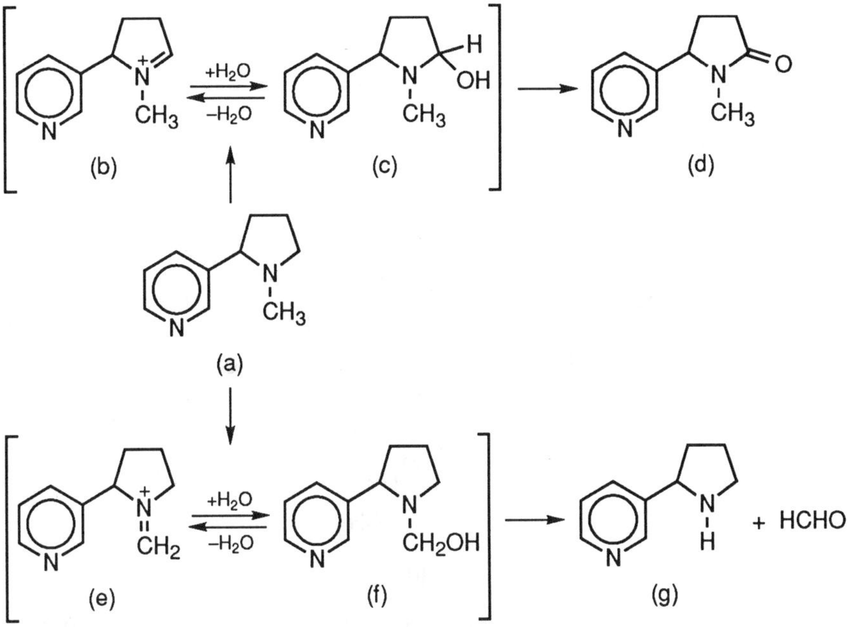

Figure 2.1 Primary carbon oxidation of nicotine.

of both the $\Delta^{1'(2')}$- and $\Delta^{1'(5')}$-nicotine iminium ions in the urine of smokers by use of the cyanide trapping technique. If the 2′-hydroxy compound (Figure 2.2(c)) is formed during the metabolism of nicotine, then it may be expected to result in ring opening, as α-carbinolamines derived from basic tertiary amines are very unstable giving rise to a secondary amine and an aldehyde (McMahon, 1966). In this case because the carbinol function is part of a cyclic sytem an aldehyde could not be formed and the corresponding ketone (Figure 2.2(d)) should arise. This compound has not been recognized as a metabolite of nicotine although the analogous route from cotinine has been recognized (see later). The finding of 2′-hydroxy nicotine (as the iminium ion) may account for the formation by Neurath *et al.* (1992) of nicotine-2′, 3′-enamine and nicotyrine, both of which have been reported as metabolites of nicotine.

As shown by Murphy (1973), nicotine iminium ions are very reactive moieties towards cyanide, and Gorrod and Jenner (1975) argued that this reactivity may enable 'nicotine' to react with other nucleophiles including cellular macromolecules and by this mechanism exert both pharmacological and toxicological effects. Since that time Hibberd and Gorrod (1981) have shown that nicotine 1′(5′)-iminium ions can react with glutathione, and Castagnoli and co-workers (Shigenaga *et al.*, 1988)

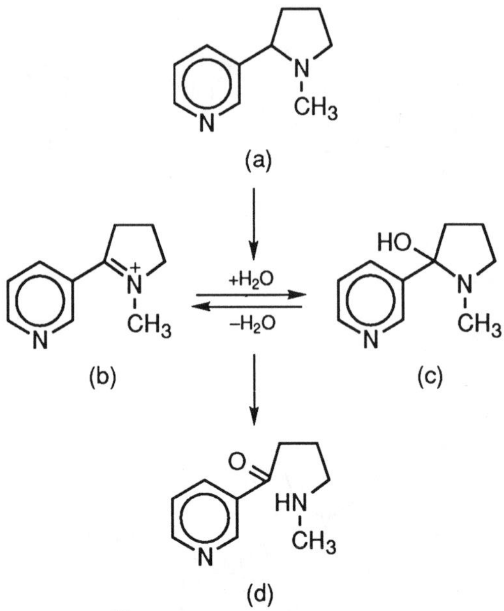

Figure 2.2 The possible involvement of nicotine $\Delta 1' \rightarrow (2')$ iminium ion in nicotine metabolism.

have provided evidence that this is the mechanism involved in the covalent binding observed when tritium labelled nicotine is incubated with fortified hepatic microsomes. In the light of the known reactivity of iminium ions it is perhaps surprising that Neurath *et al.* (1992) were able to detect their presence in smokers' urine. This may indicate that they can be formed locally by bladder epithelial cells or transported as conjugates of the carbinol and released extrahepatically. Either way this exciting observation has implications for the pharmacology and toxicology of tobacco alkaloids.

2.3 THE METABOLISM OF COTININE

Cotinine, like nicotine, can undergo both carbon and nitrogen oxidation. However, in this case N-oxidation occurs at the pyridine nitrogen (Dagne and Castagnoli, 1972) to give Figure 2.3(f). This is due to the delocalisation of the lone pair of electrons in the pyrrolidone ring by the carbonyl group, which means that the pyridine nitrogen of cotinine, unlike nicotine, is the most basic centre and as predicted by the pKa hypothesis (Gorrod, 1973) is susceptible to oxidation. This change in the site of nitrogen oxidation in the two compounds is reflected in the enzymology of the processes which are related to differences in both pKa and the hybridization states of the susceptible nitrogen atoms. In cotinine N-oxidation a cytochrome P450 isozyme is involved whereas nicotine-1′N-oxide is formed by a flavin containing amine oxidase as discussed later (Hibberd and Gorrod, 1985).

It is now recognized that most of the further metabolites of nicotine arise via cotinine; indeed it has been estimated that more than 70% of nicotine metabolites are probably formed via this route. The primary metabolic oxidation products of cotinine are shown in Figure 2.3.

N-Demethylcotinine (norcotinine) (Figure 2.3(c)) has been reported to be a urinary metabolite of nicotine in smokers and subjects receiving nicotine and cotinine (McKennis *et al.*, 1959). N-Demethylcotinine occurs as a urinary metabolite of cotinine in various species (Morselli *et al.* 1967), and is formed *in vitro* by hepatic, pulmonary and kidney tissue from several species (Gorrod and Jenner, 1975 and references therein). It would be expected that N-demethylation of cotinine would proceed via the corresponding hydroxymethyl intermediate (Figure 2.3(b)). This intermediate may be stabilized by the low pKa of the pyrrolidone nitrogen (Gorrod and Temple, 1976). If this is the case, this route may also play a role in the covalent binding of 'nicotine' observed during its metabolism as an enhanced binding of 4-aminoazobenzene to macromolecules has been found in the presence of formaldehyde (Roberts and Warwick, 1964), a situation where the N-hydroxymethyl compound is in equilibrium with the amine and formaldehyde.

Hydroxylation of cotinine occurs on the pyrrolidone ring at the 3′-position (4′- position in early nomenclature) to give both the *cis* and *trans* isomers. These compounds have been found as metabolites of nicotine and in the urine of smokers (Neurath *et al.* 1988; Vonken *et al.*, 1990);

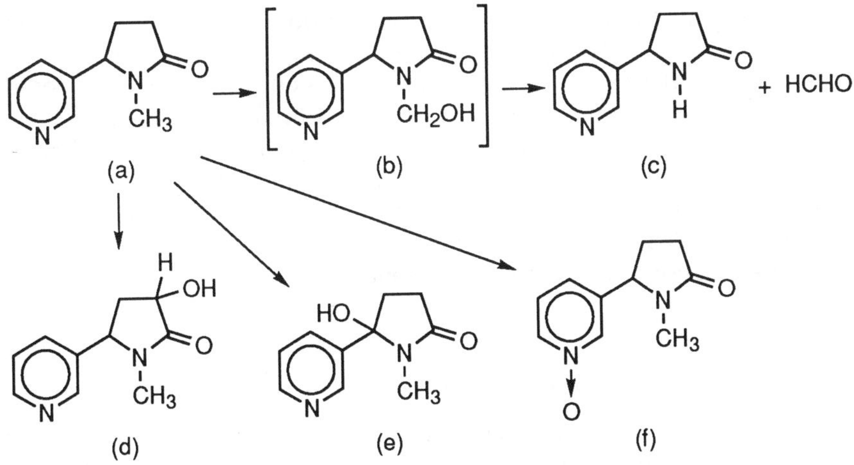

Figure 2.3 Primary C- and N-oxidation products of cotinine.

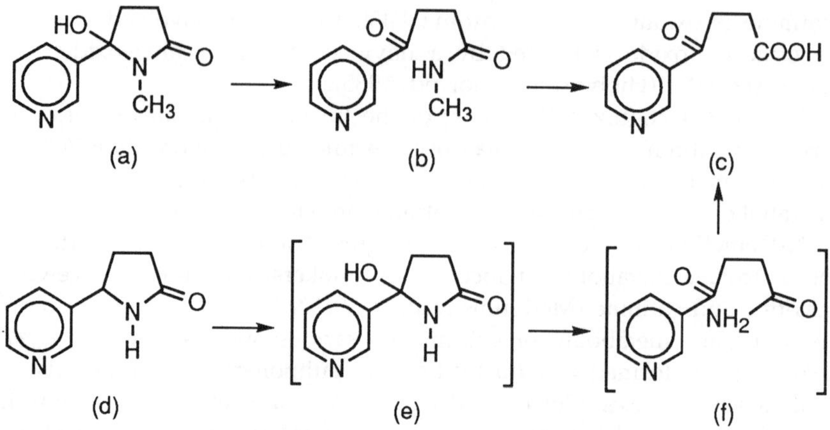

Figure 2.4 Formation of 4-(3-pyridyl)-4-oxo-butyric acid.

more recently we have detected their presence in guinea pig hepatic microsomal incubates of cotinine (Li *et al.*, 1992). *In vivo* and *in vitro* *trans*-3′-hydroxy cotinine is the major isomer formed which may indicate either stereoselective oxidation by one isoenzyme of cytochrome P450 or individual isoenzymes each effecting a stereospecific oxidation. These compounds are excreted as glucuronide conjugates in man (Curvall *et al.*, 1991). During the course of the *in vivo* studies on the metabolism of cotinine in monkeys Dagne *et al.* (1974) observed the excretion of a compound which they thought was 4-(3′-pyridyl)-4-oxo-N-methylbutyramide as McKennis *et al.* (1962a) had previously reported this as a metabolite of nicotine. Later, Nguyen *et al.* (1981) showed that this keto amide was a tautomer of 5′-hydroxycotinine (Figure 2.4(a)) which existed predominantly in the cyclic form. 5′-Hydroxycotinine (Figure 2.4(a)) and norcotinine (Figure 2.4(d)) have been implicated as intermediates *en route* to 4-(3-pyridyl)-4-oxo-butyric acid (Figure 2.4(c)). It has been proposed that 5′-hydroxycotinine and 5′-hydroxynorcotinine undergo ring cleavage (see above) to the corresponding carbonyl compounds (Figures 2.4(b) and 2.4(e)) which are then deaminated. In the case of the cotinine pathway (Figures 2.4(a) to 2.4(c)) clearly an N-methylamino group has to be lost, but whether cleavage of the amide bond occurs prior to or after N-demethylation is not known at present.

2.4 THE METABOLISM OF NORNICOTINE

The metabolism of nornicotine (Figure 2.5(a)) also involves both carbon and nitrogen centres; oxidation of the pyrrolidine nitrogen occurs to give the nitrone (Figure 2.5(c)) which probably arises by further oxida-

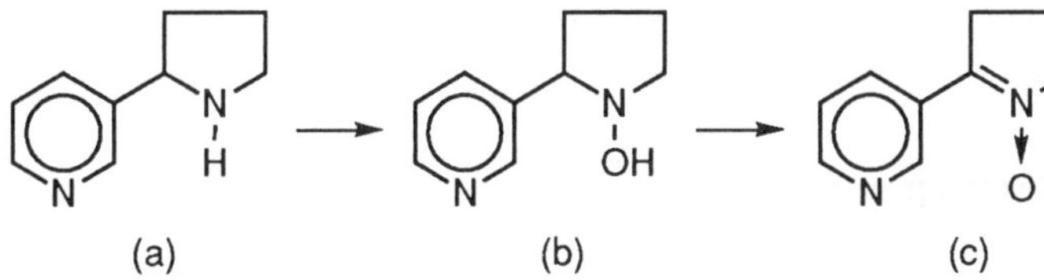

Figure 2.5 The N-oxidation pathway of nornicotine.

tion of N-hydroxynornicotine (Figure 2.5(b)) (Aislaitner *et al.*, 1992). The carbon oxidation of nornicotine (Figure 2.6(a)) has been demonstrated both *in vivo* (Wada *et al.*, 1961) and *in vitro* (Papadopoulos, 1964; Aislaitner *et al.*, 1992) and norcotinine (Figure 2.6(d)) recognized as a major metabolite. It is possible that this compound is formed via an analogous pathway (Figure 2.6(a) to 2.6(d)) to that involved in the metabolism of nicotine to cotinine. The formation of the carbinolamine (Figure 2.6(c)) would allow the production of the ring opened aldehyde (Figure 2.6(e)) which could cyclize to norcotinine via the acid (Figure 2.6(f)). The carbinolamine (Figure 2.6(c)) would be expected to be in equilibrium with the compound shown in Figure 2.6(b). This molecule would also be expected to react with nucleophiles as analogous reactions to those of tertiary alicyclic amines have been observed when secondary cyclic amines are incubated with microsomes in the presence of

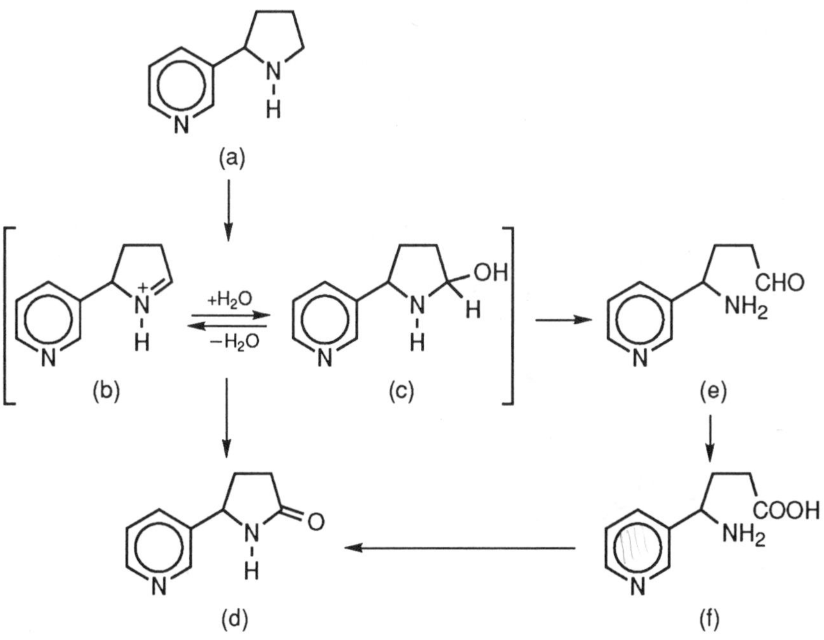

Figure 2.6 Alternative pathways to norcotinine.

cyanide (Murphy, personal communication; Gorrod and Aislaitner, unpublished observations).

2.5 THE METABOLISM OF 4-(3-PYRIDYL)-4-OXO-BUTYRIC ACID

As mentioned earlier, the formation of norcotinine, by either of the pathways, affords an alternative route to 4-(3-pyridyl)-4-oxo-butyric acid (Figure 2.7(a)). Clearly, this is a key intermediate in the metabolism of nicotine as it channels the previously described pathways ultimately producing 3-pyridylacetic acid (Figure 2.7(e)). This seems to be the major end product of nicotine degradation and might be expected to undergo conjugation with glucuronic acid and amino acids prior to excretion. 4-(3′-Pyridyl)-4-oxo-butyric acid is reduced to 4-(3′-pyridyl)-4-hydroxybutyric acid (Figure 2.7(b)), which can exist as two enantiomers. This compound can then undergo cyclization to produce the lactone (Figure 2.7(f)) although this lactone can be reversibly hydrolyzed back to the hydroxy acid.

The hydroxy acid (Figure 2.7(b)) also undergoes a dehydration reaction to produce 4-(3′-pyridyl)butenoic acid shown in Figure 2.7(c) which in turn is converted to 4-(3-pyridyl)-butyric acid (Figure 2.7(d)) which, presumably, enters the fatty acid degradation cycle where it loses a two-carbon fragment to form 3-pyridylacetic acid (Figure 2.7(e)). The detection of the above compounds and the elucidation of their role in nicotine metabolism was principally the work of McKennis and colleagues over many years (see Gorrod and Jenner, 1975 and McKennis, 1965), however much remains to be done.

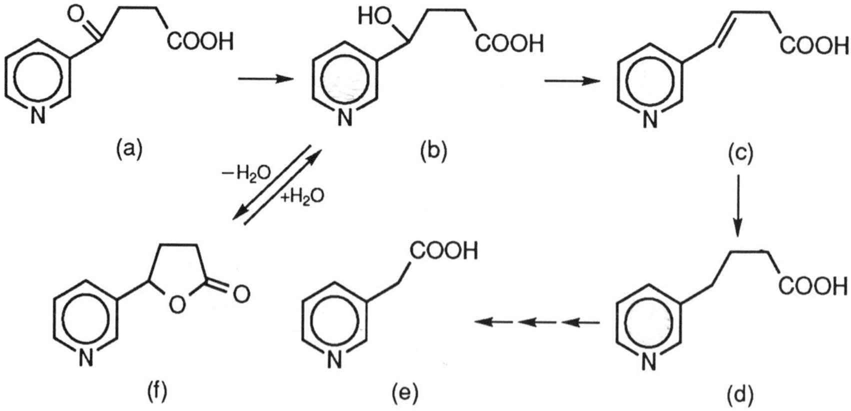

Figure 2.7 Route to 3-pyridylacetic acid.

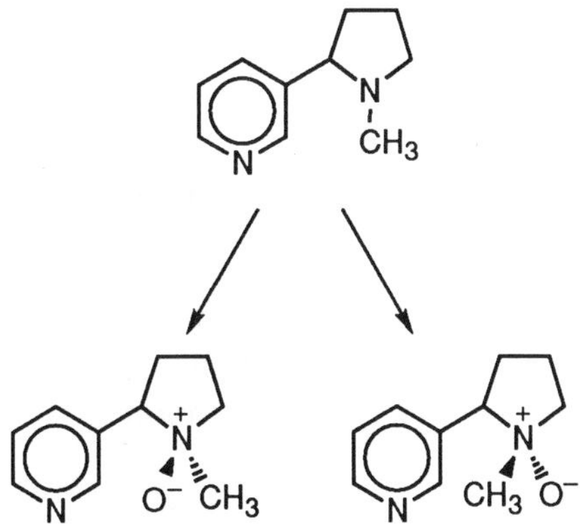

Figure 2.8 Major N-oxidation reactions of nicotine.

2.6 THE N-OXIDATION OF NICOTINE

Whilst many nicotine N-oxides are possible and have been produced by chemical synthesis, the only one which is significant in mammalian systems seems to be nicotine-1′-N-oxide (Figure 2.8). Nicotine-1′-N-oxide has been observed as a urinary metabolite of nicotine in many species including man (Beckett *et al.*, 1971). Booth and Boyland (1970) showed that this metabolite exists as both *cis* and *trans* isomers. Nicotine-1′-N-oxide is predominately, if not exclusively, formed by a flavin containing amine oxidase (Hibberd and Gorrod, 1985) which is very active in guinea pig (Booth and Boyland, 1971). It should be remembered that N-oxidation is a reversible reaction and the reduction of nicotine-1′-N-oxide has been observed in experimental animals (Dajani *et al.*, 1975) and man (Beckett *et al.*, 1970).

2.7 METHYLATION REACTIONS OF NICOTINE AND COTININE

The methylation reactions of nicotine and cotinine were early recognized as being involved in nicotine metabolism, (McKennis *et al.*, 1963). The enzymology of the processes involved has been investigated by Crooks and colleagues, who have shown a remarkable specificity of the N-methyl transferase (Cundy *et al.*, 1984). These reactions will be described in detail in chapter five.

2.8 CONJUGATION REACTIONS

The failure to account for the majority of nicotine administered to man or animals excreted as unchanged material or metabolites led to the search for the 'missing' material. Curvall *et al.* (1991) found that, after incubation of urine samples from smokers with β-glucuronidase, nicotine, cotinine and 3′-hydroxy cotinine were released. This has led to attempts to detect, synthesize and characterize these new compounds. This work is now coming to fruition (see Crooks, this volume). As nicotine and cotinine do not have a hydroxyl functional group it is assumed that these new metabolites of nicotine must be quaternary compounds. As quaternization can occur on either the pyridyl or pyrrolidine nitrogen at least two structural possibilities exist for nicotine. However, the delocalization of electrons by the carbonyl group in continine and 3-hydroxycotinine would direct quaternization exclusively to the pyridine nitrogen.

2.9 CONCLUSIONS

From the above overview it is clear that the metabolism of nicotine is extremely complex and shares with few other xenobiotics the number and diversity of structure of the known metabolites. The full metabolic map of nicotine is still not complete and considerable effort is still being made to fill the gaps. In addition, the recognition that during its metabolism nicotine generates several reactive metabolites, may lead to a greater understanding of the pharmacology of nicotine. Additionally, it must be recorded that despite our understanding of the chemical nature and properties of nicotine metabolites, the enzymology involved in the individual steps has only been elucidated in a few cases.

Finally, it is impossible in the space available to cite all the references related to this subject and readers are referred to other reviews (Gorrod and Jenner, 1975; Nagayama, 1988; Kyerematen and Vesell, 1991; Benowitz *et al.*, 1990; Castagnoli *et al.*, 1991) for more details. The latest views on these pathways, their enzymology and the factors affecting nicotine biotransformation in animals and man will be dealt with in the succeeding chapters.

ACKNOWLEDGEMENT

Studies carried out in my laboratories are currently supported by grants from Fabriques de Tabac Réunies SA, Neuchâtel, Switzerland.

REFERENCES

Aislaitner, G., Li, Y., and Gorrod, J.W., (1992) *In vitro* studies on (−)-(*S*)-nornicotine. *Medical Sci. Res.*, **20**, 897–9.

Beckett, A.H., Gorrod, J.W., and Jenner, P., (1970) Absorption of (−)-nicotine-1′-N-oxide in man and its reduction in the gastrointestinal tract. *J. Pharm. Pharmacol.*, **22**, 722–3.

Beckett, A.H., Gorrod, J.W., and Jenner, P., (1971) The analysis of nicotine-1'-N-oxide, in the presence of nicotine and cotinine, and its application to the study of *in vivo* nicotine metabolism in man. *J. Pharm. Pharmacol.*, **23**, 55S–61S

Beckett, A.H., Gorrod, J.W., and Jenner, P. (1972) A possible relation between pKa, and lipid solubility and the amounts excreted in urine of some tobacco alkaloids given to man. *J. Pharm. Pharmacol.*, **24**, 115–20.

Benowitz, N.L., Porchet, H. and Jacob, P. (1990) Pharmacokinetics, metabolism and pharmacodynamics of nicotine, in *Nicotine Psychopharmacology: Molecular, Cellular and Behavioural Aspects* (Eds S. Wonnacott, M.A.H. Russell and I.P. Stolerman), Oxford University Press, New York, pp. 112–57.

Booth, J., and Boyland, E. (1971) Enzymic oxidation of (−)-nicotine by guinea pig tissues *in vitro*. *Biochem. Pharmacol.*, **20**, 407–15.

Booth, J., and Boyland, E., (1970) The metabolism of nicotine into two optically-active stereoisomers of nicotine-1'-N-oxide by animal tissue *in vitro* and by cigarette smokers. *Biochem. Pharmacol.*, **19**, 733–42.

Brandänge, S., Lindblom, L., Pilotti, A. *et al.* (1983) Ring chain tautomerism of pseudooxynicotine and some other iminium compounds. *Acta Chem. Scand.* **B37**, 617–22.

Brandänge, S., and Lindblom, L. (1979) The enzyme 'aldehyde oxidase' is an iminium oxidase. Reaction with nicotine $\Delta^{1'(5')}$-iminium ion. *Biochem. Biophys. Res. Commun.*, **91**, 991–6.

Castagnoli, N., Shigenaga, M., Carlson, T. *et al.* (1991) The *in vitro* metabolic fate of (*S*)-nicotine, in *Effects of Nicotine on Biological Systems* (Eds F. Adlkofer and K. Thurau), Birkhäuser Verlag, Basel, pp. 25–34.

Cundy, K.C., Godin, C.S., and Crooks, P., (1984) Evidence of stereospecificity in the *in vivo* methylation of [^{14}C]-(±)-nicotine in the guinea pig. *Drug. Metab. Disposit.*, **12**, 755–9.

Curvall, M., Kazemi-Vala, E., and Englund, G. (1991) Conjugation pathways in nicotine metabolism, in *Effects of Nicotine on Biological Systems* (eds F. Adlkofer and K. Thurau), Birkhauser Verlag, Basel, pp. 69–75.

Dagne, E., and Castagnoli, N. (1972) Cotinine N-oxide, a new metabolite of nicotine. *J. Med. Chem.*, **15**, 840–41.

Dagne, E., Gruenke, L., and Castagnoli, N. (1974) Deuterium isotope effects in the *in vivo* metabolism of cotinine. *J. Med. Chem.*, **17**, 1330–33.

Dajani, R.M., Gorrod, J.W., and Beckett, A.H., (1975) *In vitro* hepatic and extra-hepatic reduction of (−)-nicotine-1'-N-oxide in rats. *Biochem. Pharmacol.*, **24**, 109–117.

Gillette, J.R., (1963) Metabolism of drugs and other foreign compounds by enzymatic mechanisms. *Progress in Drug Res.*, **6**, 17–73.

Gorrod, J.W., (1973) Differentiation of various types of biological oxidation of nitrogen in organic compounds. *Chem. Biol. Interact.*, **7**, 289–303.

Gorrod, J.W., and Hibberd, A.R., (1982) The metabolism of nicotine $\Delta^{1'(5')}$-iminium ion *in vivo* and *in vitro*. *Europ. J. Drug Metab. Pharmacokinet.* **7**, 293–8.

Gorrod, J.W., and Jenner, P.J., (1975) The metabolism of tobacco alkaloids. *Essays in Toxicology*, **6**, 35–78.

Gorrod, J.W., and Temple, D.J. (1976) The formation of an N-hydroxymethyl intermediate in the N-demethylation of N-methylcarbazole *in vivo* and *in vitro*. *Xenobiotica* **6**, 265–74.

Guengerich, F.P., and MacDonald, T.L., (1990) Mechanisms of cytochrome P450 catalysis. *FASEB Journal* **4**, 2453–9.

Hibberd, A.R., and Gorrod, J.W., (1981) Nicotine 1′ (5′)-iminium ion: a reactive intermediate in nicotine metabolism. *Advances Exptl. Biol & Med.*, **136B**, 1121–31.

Hibberd, A.R., and Gorrod, J.W., (1985) Comparative N-oxidation of nicotine and cotinine by hepatic microsomes, in *Biological Oxidation of Nitrogen in Organic Molecules, Chemistry Toxicology and Pharmacology* (ed. J.W. Gorrod and L.A. Damani), Ellis Horwood, Chichester, UK, pp. 246–50.

Hill, D.L., Laster, W.R., and Struck, R.F., (1972) Enzymatic metabolism of cyclophosphamide and nicotine and production of a toxic cyclophosphamide metabolite. *Canc. Res.* **32**, 658–65.

Hucker, H.B., Gillette, J.R., and Brodie, B.B., (1960) Enzymatic pathway for the formation of cotinine, a major metabolite of nicotine in rabbit liver. *J. Pharmacol. Exp. Ther.*, **129**, 94–100.

Kyerematen, G.A., and Vesell, E.S., (1991) Metabolism of Nicotine *Drug Metabolism Revs.*, **23**, 3–41.

Li, Y., Li, N.Y., Aislaitner, G., *et al.*, (1992) *In vitro* metabolism of cotinine. *Medical Sci. Res.*, **20**, 903–4.

McKennis, H., Turnball, L.B., and Bowman, E.R., (1963) N-Methylation of nicotine and cotinine *in vivo*. *J. Biol. Chem.*, **238**, 719–23.

McKennis, H., Turnbull, L.B., Bowman, E.R. *et al.*, (1959) Demethylation of cotinine *in vivo*. *J. Amer. Chem. Soc.*, **81**, 3951–4.

McKennis, H., Turnbull, L.B., Bowman, E.R. *et al.*, (1962a) The corrected structure of a ketonamide arising from the metabolism of (−)-nicotine. *J. Amer. Chem. Soc.*, **84**, 4598–9.

McKennis, H, Turnbull, L.B., Schwartz, S.L., *et al.*, (1962b) Demethylation in the metabolism of (−)-nicotine. *J. Biol. Chem.*, **237**, 541–5.

McKennis, H., (1965) Disposition and Fate of Nicotine in Animals, in *Tobacco Alkaloids and related compounds.* (ed. U.S. von Euler), Pergamon Press, Oxford, pp. 53–74.

McMahon, R.L., (1966) Microsomal dealkylation of drugs; substrate specificity and mechanism. *J. Pharm Sci.*, **55**, 457–66.

Morselli, P.L., Ong, H.H., Bowman, E.R., *et al.*, (1967) Metabolism of (±)-cotinine-[2-^{14}C] in the rat. *J. Med. Chem.*, **10**, 1033–6.

Murphy, P.J., (1973) Enzymatic oxidation of nicotine to nicotine-$\Delta^{1'(5')}$-iminium ion. *J. Biol. Chem.*, **248**, 2796–2800.

Nakayama, H. (1988) Nicotine metabolism in mammals. *Drug Met., and Drug Inter.*, **6**, 95–122.

Neurath, G.B., Dunger, M. and Orth, D., (1992) Detection and determination of tautomers of 5-hydroxynicotine and 2-hydroxynicotine in smokers' urine. *Med. Sci. Research*, **20**, 853–8.

Neurath, G.B., Dinger, M., Krenz, O. *et al.*, (1988) *trans*-3′-Hydroxycotinine – a main metabolite of nicotine in smokers. *Klin Wockenschrift*, **66**, 2–4.

Nguyen, T.L., Gruenke, L.D., and Castagnoli, N. (1979) Metabolic oxidation of nicotine to reactive intermediates. *J. Med. Chem.*, **22**, 259–63.

Nguyen, T.L., Dagne, E. Gruenke, L., *et al.*, (1981) The tautomeric structure of 5-hydroxynicotine; a secondary mammalian metabolite of nicotine. *J. Org. Chem.*, **46**, 758–60.

Papadopoulos, N., (1964) Formation of nornicotine and other metabolites from nicotine *in vitro* and *in vivo*. *Can. J. Biochem.*, **42**, 435–42.

Peterson, L.A., and Castagnoli, N. (1988) Regio- and stereo-chemical studies on the α-carbon oxidation of (*S*)-nicotine by cytochrome P450 model systems. *J. Med. Chem.*, **31**, 637–40.

Peterson, L.A., Trevor, A., and Castagnoli, N. (1987) Stereochemical studies on the cytochrome P450 catalysed oxidation of (*S*)-nicotine to the (*S*)-nicotine-$\Delta^{1'(5')}$-iminium species *J. Med Chem.*, **30**, 249–54.

Roberts, J.J., and Warwick, G.P., (1964) Reactions of ^{3}H-labelled Butter Yellow *in vivo* and N-hydroxymethylaminoazobenzene *in vitro*. *Biochem. J.*, **93**, 18P.

Sanders, E.B., De Bardeleben, J.F., and Osdene, T.S., (1975) Nicotine Chemistry: 5′-Cyanonicotine. *J. Org. Chem.*, **40**, 2848–9.

Shigenaga, M.K., Trevor, A.J., and Castagnoli, N. (1988) Metabolism-dependent covalent binding of (*S*)-[5-^{3}H]-nicotine to liver and lung microsomal macromolecules. *Drug Metabolism Dispos.*, **16**, 397–402.

Voncken, P., Rustemeirer, K., and Schepers, G., (1990) Identification of *cis*-3′-hydroxycotinine as a urinary nicotine metabolite. *Xenobiotica*, **20**, 1353–6.

Wada, E., Bowman, E.R., Turnbull, L.B., *et al.*, (1961) Norcotinine (desmethylcotinine) as a urinary metabolite of nornicotine. *J. Med. Pharm. Chem.*, **4**, 21–30.

The roles of cytochrome P450 in nicotine metabolism

3

H. Nakayama, T. Kita, T. Nakashima, S. Imaoka and Y. Funae

3.1 INTRODUCTION

The conversion of nicotine to cotinine is a major pathway of nicotine metabolism and cytochrome P450 catalyses the first step of this pathway (Nakayama, 1988). To date, numerous studies have shown that cytochrome P450 is composed of many molecular forms and their expression is regulated by many factors such as drug administration, age, sex and pathophysiological conditions (Ryan and Levin, 1990; Soucek and Gut, 1992; Funae and Imaoka, 1992).

Approximately five years ago, nicotine metabolism in mammals was reviewed, and it was emphasized that knowledge regarding cytochrome P450s participating in nicotine metabolism must be obtained (Nakayama, 1988). The results of extensive research performed in recent years suggest that only limited forms of cytochrome P450 have an important role in nicotine metabolism. Furthermore, considerable knowledge about cytochrome P450s makes it possible to consider the properties of cytochrome P450 dependent nicotine metabolism and its regulation. In this chapter, mechanisms involved in the conversion of nicotine into cotinine are described first, and then present knowledge about cytochrome P450 dependent nicotine metabolism and its regulation is dealt with.

Nicotine and Related Alkaloids: Absorption, distribution, metabolism and excretion. Edited by J.W. Gorrod and J. Wahren. Published in 1993 by Chapman & Hall, London. ISBN 0 412 55740 1

3.2 MECHANISMS OF COTININE FORMATION

Hucker *et al.* (1960) first pointed out the two-step reactions by which nicotine is metabolized to cotinine. In their study, nicotine was first oxidized to 5′-hydroxynicotine by rabbit hepatic microsomal enzyme and then converted to cotinine by a soluble enzyme (Figure 3.1). Since 5′-hydroxynicotine is unstable, however, the intermediate was not quantitated and these conversion mechanisms were not fully characterized at that time. Hill *et al.* (1972) found that purified aldehyde oxidase catalysed the formation of cotinine from the intermediate which was derived from nicotine in the presence of hepatic microsomes, O_2 and NADPH. These authors postulated that this intermediate was the corre-

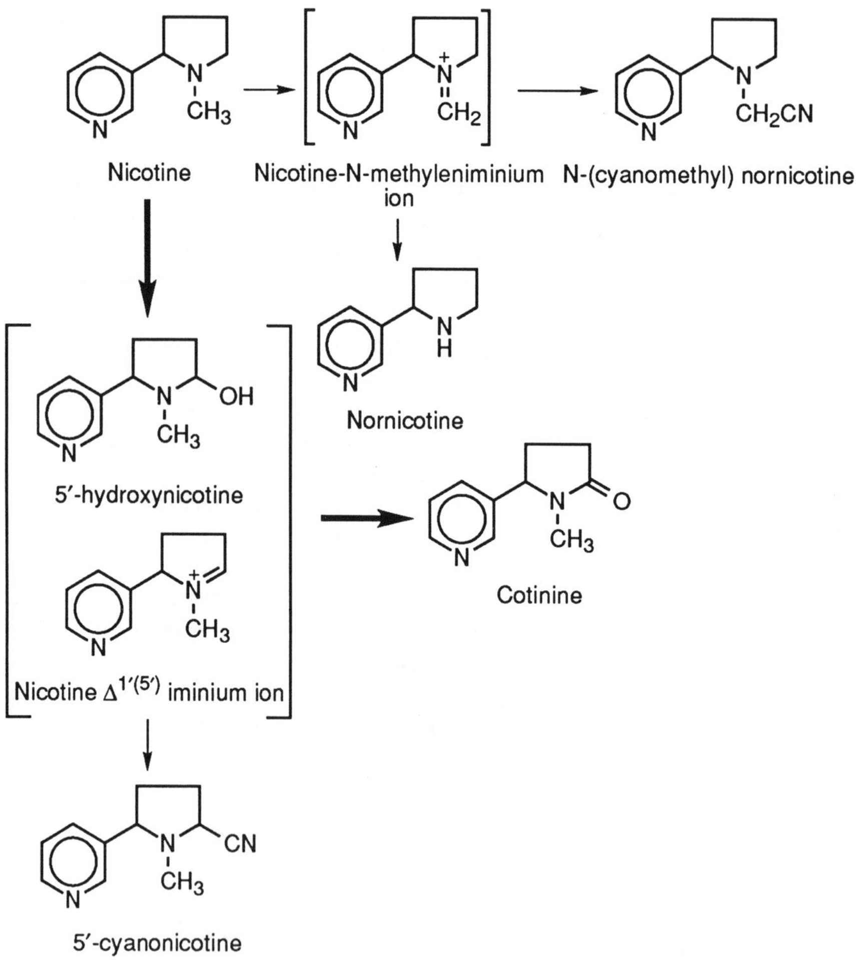

Figure 3.1 Metabolism of nicotine to cotinine and nornicotine.

sponding aldehyde which was derived non-enzymatically from 5′-hydroxynicotine. Cyanide, an inhibitor of aldehyde oxidase, inhibited the conversion of nicotine to cotinine. Murphy (1973) investigated this cyanide inhibition and found that 5′-cyanonicotine was formed from nicotine in the presence of KCN, NADPH, O_2 and rabbit hepatic microsomes and the increase in cyanonicotine formation paralleled the decrease in cotinine formation. Murphy postulated that nicotine $\Delta^{1'(5')}$-iminium ion served as the precursor of 5′-cyanonicotine. In addition to 5′-cyanonicotine, N-(cyanomethyl)nornicotine was detected as a minor metabolite after nicotine was metabolized in the presence of rabbit hepatic microsomes, NADPH, O_2 and NaCN (Nguyen *et al.*, 1979; Peterson and Castagnoli, 1988). Chemical synthesis of nicotine $\Delta^{1'(5')}$-iminium ion (Brändange and Lindblom, 1979a; Hibberd and Gorrod, 1981) allowed experiments to investigate its role in cotinine formation. Brändange and Lindblom (1979b) precisely investigated the conversion of the iminium ion to cotinine by aldehyde oxidase and showed that the iminium ion but not 5′-hydroxynicotine is a substrate for aldehyde oxidase. The Km value of aldehyde oxidase for the iminium ion was very low (Km = 2 μM at pH 7.4). Using the iminium ion synthesized, soluble fractions prepared from guinea pig, rat, mouse and hamster livers, were each shown to catalyse efficiently the conversion of the iminium ion to cotinine (Hibberd and Gorrod, 1981; Gorrod and Hibberd, 1982). Furthermore, intraperitoneal administration of the iminium ion to animals resulted in the formation of cotinine as a urinary metabolite (Gorrod and Hibberd, 1982). To quantitate nicotine $\Delta^{1'(5')}$-iminium ion after [^{3}H]nicotine oxidation, the iminium ion has been separated from nicotine and other metabolites by high-performance liquid chromatography (HPLC) and its radioactivity determined (Peterson and Castagnoli, 1988; Williams *et al.*, 1990a, 1990b). A monoclonal antibody against the iminium ion has also been used to determine the iminium ion (Obach and Vunakis, 1990). Gas chromatography (GC) and HPLC have been used for the determination of 5′-cyanonicotine (Murphy, 1973; McCoy *et al.*, 1986a).

By GC–electron impact mass spectra analysis of deuterium-labelled products obtained from specifically deuterium labelled nicotine, Castagnoli and his co-workers have shown that the conversion of nicotine to 5′-cyanonicotine proceeds regio- and stereoselectively (Nguyen *et al.*, 1979; Peterson *et al.*, 1987; Peterson and Castagnoli, 1988). In this case, the conversion of nicotine $\Delta^{1'(5')}$-iminium ion to cotinine occurs with the diastereoselective loss of the pro(E)proton *trans* to the pyridine ring. When nicotine oxidation is performed with other haemoproteins (horseradish peroxidase, methaemoglobin and chloroperoxidase), oxidation of the 5′- carbon atom occurs predominantly, but the 2′- carbon atom is also oxidized to some extent (Peterson and Castagnoli, 1988). Of these three haemoproteins, only horseradish peroxidase catalyses an oxidative attack at the N-methyl group. Peterson and Castagnoli, (1988) proposed that the regio- and stereoselective oxidation of nicotine is due to a highly ordered nicotine–cytochrome P450 complex and nicotine $\Delta^{1'(5')}$-iminium ion and nicotine-N-methylene-iminium ion are produced via an iminium radical intermediate. In their model, nicotine binds at the active site with the pyridine ring extended away from the porphyrin ring system.

Gorrod *et al.* (1971) first reported that the metabolic N- and C-oxidations of nicotine were mediated by different processes. Using hepatic microsomes of guinea pigs, antibodies against NADPH-cytochrome P450 reductase and several effectors against cytochrome P450 and flavin-containing monooxygenase (FMO), we showed that cytochrome P450s participate in cotinine formation and FMO catalyses nicotine-1′-oxide formation (Nakayama *et al.* 1987). Recently, in reconstituted experiments with P450s purified from rabbits and rats, several forms of cytochrome P450 have been reported to catalyse the conversion of nicotine to nicotine $\Delta^{1'(5')}$-iminium ion and other metabolites (McCoy *et al.*, 1989; Williams *et al.*, 1990a, 1990b). Cytochrome P450s purified from rabbit livers have been found to catalyse the formation of the iminium ion and nornicotine (McCoy *et al.*, 1989). In this case, the iminium ion was formed between 1.4 to 4 times more extensively than nornicotine. A rabbit lung cytochrome P450 has been reported to catalyse the formation of cotinine much more efficiently than that of nornicotine (Williams *et al.*, 1990a). Rabbit nasal P450 NMa produced nicotine iminium ion as a major product, but significant amounts of nornicotine and nicotine-1′-oxide were also produced (Williams *et al.*, 1990b). Interestingly, only nicotine iminium ion has been reported to be produced by human cytochrome P450s (Flammang *et al.*, 1992). Similar results were obtained by the use of human hepatic microsomes (Peterson *et al.*, 1987).

3.3 FORMS OF CYTOCHROME P450 PARTICIPATING IN HEPATIC MICROSOMAL NICOTINE OXIDATION

3.3.1 RAT LIVER

Numerous studies have shown that drug administration leads to enhancement of not only its own metabolism but also the metabolism of other drugs, which is due to overlapping specificities of substrates and inducers of cytochrome P450. Many kinds of cytochrome P450 are induced by a large number of drugs, xenobiotics and carcinogens, most of which are substrates for cytochrome P450s (Ryan and Levin, 1990; Soucek and Gut, 1992; Funae and Imaoka, 1992). To better understand nicotine metabolism and its regulation, therefore, it is necessary to identify specific cytochrome P450s which participate in nicotine oxidation. Direct evidence for cytochrome P450s participating in nicotine oxidation has been obtained from experiments with reconstituted purified cytochrome P450s and the inhibition of nicotine metabolism by specific antibodies against cytochrome P450s. This chapter deals with hepatic cytochrome P450s; extrahepatic nicotine metabolism is described elsewhere in this book.

Participation of phenobarbital (PB) inducible cytochrome P450 in hepatic nicotine metabolism was first shown in a reconstituted system with PB inducible P450s purified from rats and guinea pigs (Nakayama *et al.*, 1982b; Nakayama *et al.*, 1985). In addition, antibodies against PB inducible P450 purified from guinea pigs inhibited microsomal nicotine oxidation in guinea pigs (Nakayama *et al.*, 1985). Advances in purification techniques have shown that PB inducible cytochrome P450s can be further separated into several forms, (Ryan and Levin, 1990; Soucek and Gut 1992; Funae and Imaoka, 1992). Four forms of PB inducible cytochrome P450 (P450 2B1, 2B2, 2C6 and 3A2) have been purified from rat liver. Table 3.1 shows the expression of four PB inducible cytochrome P450s in hepatic microsomes of PB treated and untreated rats (Imaoka *et al.*, 1989). P450 2B1 and 2B2 are present in hepatic microsomes of untreated male rats at very low levels but are strongly induced by PB treatment. P450 2B1 is most abundant in liver microsomes of PB treated rats. P450 2C6 and 3A2 are expressed constitutively in hepatic microsomes of untreated male rats, and are induced two- to three-fold with PB. Table 3.1 shows nicotine oxidation activities of these PB inducible cytochrome P450s in a reconstituted system. Three groups have employed different assay methods for nicotine oxidation activity. Williams *et al.* (1990a, 1990b) determined nicotine $\Delta^{1'(5')}$-iminium ion with HPLC after nicotine oxidation. Hammond *et al.* (1991) have performed nicotine oxidation in the presence of cytosol and the cotinine formed has been determined by use of a monoclonal antibody against cotinine. We determined nicotine oxidation activity in a reconstituted system by disappearance of nicotine (Nakayama *et al.*, 1992). The assay mixture contained 0.1 M potassium phosphate buffer (pH 7.4), 10 μg dilauroylphosphatidylcholine, 0.5 mM NADPH, 0.5 mM nicotine, 0.3 unit NADPH-cytochrome P450 reductase and 40–50 pmol of purified cytochrome P450 in a total volume of 0.1 ml. Quinoline was used as an internal standard and the incubation was performed for 30 min at 37°C. After incubation, nicotine was extracted with chloroform and then determined by use of GC with a nitrogen/phosphorus detector. Three groups have shown that P450 2B1 has nicotine oxidative activity. Hammond *et al.* (1991) and ourselves detected the particularly high nicotine oxidation activity of P450 2B1. In contrast, P450 2C6 had no detectable nicotine oxidation activity. P450 2B2 shares a 97% sequence

Table 3.1 Expression of PB-inducible P450s and its nicotine oxidation activity in a reconstituted system

Expression in rat liver[a]

	Expression (pmol/mg protein)[b]	
P450	Untreated	PB-treated
2B1	2 ± 1	785 ± 51
2B2	5 ± 3	329 ± 28
2C6	196 ± 31	499 ± 34
3A2	150 ± 44	367 ± 71
Total	669 ± 72	1513 ± 385

P450-dependent nicotine oxidation activity

	Activity (nmol/min/nmol P450)		
P450	Nakayama *et al.*	Hammond *et al.*	Williams *et al.*
2B1	5.31 ± 0.18 (3)	15.6	1.02 (iminium ion) 0.17 (N′-oxide)
2B2	1.44 ± 0.50 (6)	–	–
2C6	ND (2)	ND	–
3A2	(ND)	–	–

[a]Values are expressed as means ± S.D. of pmol of P450/mg of microsomal protein. (Data presented are from Imaoka *et al.*, 1989.)
[b]Values are expressed as mean ± S.D.; numbers in parentheses represent number of experiments; ND = not detectable. (Data presented are from Williams *et al.*, 1990a, Hammond *et al.*, 1991, and Nakayama *et al.* 1993.)

homology with P450 2B1 (Funae and Imaoka, 1992) yet had low nicotine oxidation activity. Reconstitution with P450 3A2 has not been performed but antibody against P450 3A2 did not significantly inhibit microsomal nicotine oxidation activity in liver of PB treated rats. From these results, it is concluded that of four PB inducible cytochrome P450s, P450 2B1 plays the most important role in the hepatic microsomal nicotine oxidation in PB treated rats.

Table 3.2 shows oxidation activities towards nicotine of another 11 cytochrome P450 forms tested in the reconstituted system. Of these 11 cytochrome P450 forms tested, P450 2C11 had the highest nicotine oxidation activity and P450 1A2 and 2D1 showed low activity towards nicotine. Based on experiments with inhibitors and antibodies against PB inducible P450 and NADPH-cytochrome P450 reductase, the involvement of constitutive forms of cytochrome P450 in nicotine oxidation of untreated animals was previously emphasized (Nakayama *et al.*, 1982a, 1986). This is supported by the above reconstitutive experiments.

Table 3.2 Nicotine oxidation activity of purified cytochrome P450s in a reconstituted system[a]

P450	Activity (nmol/min/nmol P450)
1A1	ND (4)
1A2	1.68 ± 0.39(4)
2A1	ND (2)
2A2	ND (2)
2C7	ND (4)
2C11	3.45 ± 0.48(4)
2C12	ND (4)
2C13	ND (2)
2D1	1.26 ± 0.41(4)
2E1	ND (2)
4A1	ND (2)

[a]Cytochrome P450s were purified from hepatic microsomes of rat. Values are expressed as mean ± S.D.; numbers in parentheses represent numbers of experiments; ND = not detectable.

P450 2C11 is a major male specific cytochrome P450 and its expression is developmentally regulated (Kamataki *et al.* 1981; Kamataki *et al.* 1983; Kamataki *et al.* 1985; Waxman *et al.* 1985; Imaoka *et al.* 1991). In recent experiments, the level of P450 2C11 was high at 14–52 weeks of age and then gradually decreased in male rats (Imaoka *et al.*, 1991). In contrast, a major female specific cytochrome P450, P450 2C12, had no detectable nicotine oxidative activity. Other male specific cytochrome P450s, P450 2A2 and 2C13, also had no detectable nicotine oxidative activity. The activity of P450 2B1 was higher than that of P450 2C11 but its level is very low in hepatic microsomes of untreated rats. In addition, P450 1A2 having low nicotine oxidation activity is also present at a very low level in hepatic microsomes of untreated rats at all ages (Imaoka *et al.*, 1991). These results suggest that most of the cytochrome P450 dependent nicotine oxidation is catalysed by P450 2C11 in hepatic microsomes of untreated male rats.

Treatment of rats with 3-methylcholanthrene (MC) or β-naphthoflavone (β-NF) induces two forms of cytochrome P450 which are involved in metabolic activation of benzo(a)pyrene which is a carcinogen (Ryan and Levin, 1990; Funae and Imaoka, 1992). Since tobacco contains benzo(a)pyrene, metabolic relations between benzo(a)pyrene and nicotine have been studied. P450 1A1 and 1A2 are typical MC inducible cytochrome P450s. The level of P450 1A1 is higher than that of P450 1A2 in liver microsomes of MC treated rat (Funae and Imaoka, 1992).

P450 1A2 had low nicotine oxidation activity, whereas P450 1A1 showed no detectable catalytic activity towards nicotine. These findings are consistent with *in vivo* and microsomal experiments, which are discussed in section 3.4.2.

These reconstituted experiments with many kinds of cytochrome P450s provide another interesting feature of cytochrome P450 dependent nicotine oxidation. The relative catalytic activities of 15 rat cytochrome P450s towards nicotine are very similar to the benzphetamine N-demethylation activities of these cytochrome P450s (Funae and Imaoka, 1992). Benzphetamine metabolism has been extensively studied and shown to be catalysed by rat hepatic cytochrome P450 forms. It is now necessary to compare the cytochrome P450 catalysed oxidation of nicotine with other tertiary amines from a stereochemical point of view.

3.3.2 SPECIES DIFFERENCES IN HEPATIC CYTOCHROME P450 DEPENDENT NICOTINE OXIDATION

Table 3.3 shows species differences in cytochrome P450 dependent nicotine oxidation. McCoy and his co-workers have investigated the participation of many cytochrome P450 forms in rabbit and human nicotine oxidation (McCoy *et al.*, 1989; Flammang *et al.*, 1992). The nicotine oxidation activities of six cytochrome P450 forms purified from rabbit hepatic microsomes were determined in a reconstituted system and P450 2B4 and 2C3 were found to have high nicotine oxidation activity (McCoy *et al.*, 1989). P450 2B4 is a typical PB inducible form having high benzphetamine N-demethylase activity (Ryan and Levin, 1990; Funae and Imaoka, 1992). P450 2C3 is a constitutive form and efficiently catalyses 6β- and 16α-hydroxylation of testosterone (Ryan and Levin, 1990; Funae and Imaoka, 1992). P450 2C11 also catalyses efficiently testosterone 16α-hydroxylation (Ryan and Levin, 1990; Funae and Imaoka 1992). Of 12 human cytochrome P450 forms tested, P450 2B6 had the highest nicotine oxidase activity and P450 2E1 and 2C9 showed intermediate values. P450 2B6 is a PB inducible form (Soucek and Gut, 1992). Human cytochrome P450 2C9 shows a high degree of sequence homology with rat P450 2C11 (Soucek and Gut, 1992). More recently, McCracken *et al.* (1992) have shown that human P450 2B6 and 2D6 have high nicotine oxidation activity whereas the catalytic activity of P450 2E1 towards nicotine is not detectable. These results concerning P450 2D6 and 2E1 are in disagreement with the findings by Flammang *et al.* (1992). What the cause of this discrepancy is, remains uncertain. The current studies on nicotine metabolism by cytochrome P450s are still in progress. Nevertheless, cytochrome P450 forms catalysing nicotine oxidation in rat liver are similar to those in rabbit and human livers

Table 3.3 Comparison of nicotine oxidation activity of multiple P450 forms in rat, rabbit and human

Rat[a]	Rabbit[b]	Human[c]
1A1	△1A1	
△1A2	△1A2	1A2
2A1		
2A2		△2A6
◎2B1	◎2B4	◎2B6
△2B2		
2C6		
2C7		△2C8
◎2C11	◎2C3	○2C9
2C12		
2C13		
△2D1		2D6
2E1	△2E1	○2E1
		△2F1
3A2		3A3
		3A4
	△3A6	3A5
4A1		△4B1

◎, ○ and △ represent high, intermediate and low values of nicotine oxidation activity.
[a]Nakayama *et al.*, 1992; [b]McCoy *et al.*, 1989; [c]Flammang *et al.*, 1992.

and only limited forms of cytochrome P450 have an important role in nicotine metabolism.

There still remain some ambiguities as to the affinity of cytochrome P450s for nicotine. McCoy and co-workers have performed nicotine oxidation in the presence of KCN to detect 5′-cyanonicotine with HPLC and have shown much lower affinity of cytochrome P450s for nicotine than the affinity reported by other groups. For example, the apparent Km values of cytochrome P450s in microsomes of hamsters and rabbits for 5′-cyanonicotine formation were shown to be 6.3 mM and in the range of 5–8 mM, respectively (McCoy *et al.* 1986a, 1986b, 1989). Treatment of hamsters with PB did not change the Km value. Furthermore, a recent report showed the apparent Km of human P450 2B6 for nicotine was 2.8 mM (Flammang *et al.*, 1992). In contrast, Hammond *et al.* (1991) determined the Km of rat hepatic P450 2B1 for cotinine formation in the presence of cytosol and found it to be 5–7 μM. Williams *et al.* (1990a, 1990b) determined the Km of rabbit lung P450 2 and rabbit nasal P450 NMa for nicotine oxidation in the reconstituted system and showed them to be 68 and 35 μM, respectively. Thus, the Km values found by McCoy and his co-workers are two to three orders higher than those established by other groups. This may depend not only on the forms of cytochrome P450 examined but also on the methods used. The Km value of the specific cytochrome P450s have not been compared in the presence or absence of KCN. Williams *et al.* (1990a), however, have pointed out the cyanide inhibition of total nicotine oxidation. To perform kinetic studies, this problem must be kept in mind.

3.4 REGULATION OF CYTOCHROME P450 DEPENDENT NICOTINE METABOLISM

3.4.1 EFFECTS OF NICOTINE TREATMENT ON CYTOCHROME P450 DEPENDENT NICOTINE METABOLISM

Administration of certain drugs often results in the selective elevation of some forms of cytochrome P450 in hepatic microsomes. Many reports have dealt with effects of nicotine treatment on nicotine metabolism. For details the reader is referred to reviews on these problems (Gorrod and Jenner, 1975; Nakayama, 1988). To elevate the tissue concentration of nicotine, it has often been injected repeatedly throughout the day or administered in drinking water. Despite extensive investigation, however, no pronounced induction of nicotine metabolism with nicotine treatment has been found. To understand this problem, at least three factors must be considered. First, nicotine may not be an inducer of the cytochrome P450s catalysing nicotine metabolism, or at best a poor inducer. The effects of nicotine on other drug metabolic pathways are further complicated (Gorrod and Jenner, 1975; Nakayama, 1988) and the effects of nicotine treatment on specific cytochrome P450 have not been reported to date. Secondly, injection of large doses of nicotine results in toxic effects and the production of stress. In contrast, large doses of MC and PB have been administered to animals. Furthermore, it is necessary to take into consideration the rapid metabolism and excretion of nicotine (Benowitz, 1990). Thirdly, nicotine causes a number of pharmacological effects involving the central and peripheral nervous system (Fuxe, 1990). For example, injection of only 0.5 mg/kg of nicotine to rats led to marked behavioural effects; the ambulatory activities were decreased in the initial period and then increased gradually (Kita *et al.*, 1986). Furthermore, repeated injections of nicotine for several days result in tolerance of the initial depressant effects, whereas the increasing effects observed in the later period are enhanced. Nicotine causes the release of many kinds of neuropeptides and neurotransmitters in the brain and consequently corticosteroid is released from the adrenal cortex (Fuxe, 1990). These complicated effects of nicotine may make it difficult to characterize nicotine as a potent inducer. Although the initial attempts to induce nicotine metabolism were unsuccessful, recent studies on participation of cytochrome P450s in nicotine metabolism make it possible to consider regulation of cytochrome P450 dependent nicotine metabolism at molecular levels.

3.4.2 CYTOCHROME P450s CATALYSING NICOTINE OXIDATION AND ITS REGULATION

In the search for correlations between cytochrome P450 and changes in nicotine oxidation activity with drug treatment, experiments with PB and MC or β-NF have been extensively performed. The kinetics of nicotine elimination in perfused liver prepared from PB and β-NF treated rats have shown that PB but not β-NF induces nicotine metabolism (Foth *et al.*, 1990, 1991). Similar results were obtained by using hepatocytes prepared from PB or β-NF treated rats (Kyerematen *et al.*, 1990). In addition, treatment of rats with PB increased microsomal nicotine oxidation activity per unit protein in liver, whereas MC treatment decreased the activity per unit cytochrome P450 (Nakayama *et al.*, 1982a). These results are consistent with the reconstitution experiments shown above. To more clearly account for these findings, however, not only the increase in PB and MC inducible P450s but also changes in P450 2C11 must be taken into consideration. MC and PB treatment decrease the P450 2C11 level (Funae and Imaoka, 1992), which suggests that the contribution of P450 2C11 to hepatic microsomal nicotine oxidation decreases with MC and PB. This decrease of the contribution of P450 2C11 may be in part compensated by elevation of P450 1A2 in hepatic microsomes of MC treated rats. Indeed, MC treatment did not change the specific activity of hepatic microsomal nicotine oxidation activity (Nakayama *et al.*, 1982a). In contrast, PB treatment leads to an elevation of P450 2B1 dependent nicotine oxidation as shown above. The elevation of P450 2B1 dependent nicotine oxidation probably exceeds decreases of the P450 2C11 dependent reaction.

Adir *et al.* (1980) have investigated the effects of ethanol treatment on cotinine formation in rats. Based on a pharmacokinetic analysis of cotinine formation, the apparent volume of distribution of cotinine and its rate of formation were found to increase significantly. In contrast, ethanol administration to hamsters led to no increase in the rate of 5′-cyanonicotine formation (McCoy *et al.* 1986b). P450 2E1 is an ethanol inducible cytochrome P450 and its nicotine oxidation activity has been shown to exhibit species differences as shown in Table 3.3. However, it is rather difficult to explain the effect of ethanol administration on nicotine metabolism in terms of P450 2E1. To account for *in vivo* experiments, further studies at a molecular level are necessary. Although there have been several reports on the effects of smoking on nicotine metabolism (Beckett *et al.*, 1971; Beckett and Triggs, 1967; Kyerematen *et al.*, 1982), the specific chemicals in tobacco which induce microsomal nicotine oxidation have not been established. Studies on the effects of tobacco smoking on cytochrome P450 forms may be useful for resolving this problem.

Previous reports concerning age and sex differences in nicotine metabolism have already been reviewed (Gorrod and Jenner, 1975; Nakayama, 1988). Using recent radiometric HPLC techniques, Kyerematen *et al.* (1988) have precisely investigated sex differences in nicotine metabolism in the rat and have shown that male rats metabolize nicotine faster than female rats. Furthermore, castration of male rats results in a decrease of nicotine metabolism and the effect of castration is reversed by testosterone treatment. Male-specific P450 2C11 is also induced by testosterone and its expression is developmentally regulated (Kamataki *et al.*, 1983, 1985; Waxman *et al.*, 1985; Imaoka *et al.*, 1991). These age and sex differences in nicotine metabolism show a close correlation between nicotine metabolism and P450 2C11. At present, which cytochrome P450s are important in hepatic microsomal nicotine oxidation in untreated female rats has not been determined.

Starvation and diabetes increase the levels of P450 2E1, 3A2 and 4A2, but repress the expression of P450 2C11 and 2C13 (Schenkman *et al.*, 1989; Imaoka *et al.*, 1990). Under these pathophysiological conditions, therefore, hepatic microsomal nicotine oxidation activity may decrease. Although the effects of changes in pathophysiological conditions on nicotine metabolism have not been elucidated, advances in cytochrome P450 studies will permit more insight into cytochrome P450 dependent nicotine metabolism under several pathophysiological conditions.

ACKNOWLEDGEMENTS

The authors wish to thank Miss Kazuko Sakata for preparation of this manuscript. This work was supported in part by the Smoking Research Foundation, Japan.

REFERENCES

Adir, J., Wildfeuer, W. and Miller, R.P. (1980) Effect of ethanol pretreatment on the pharmacokinetics of nicotine in rats. *J. Pharmacol. Exp. Ther.*, **212(2)**, 274–9.

Beckett, A.H. and Triggs, E.J. (1967) Enzyme induction in man caused by smoking. *Nature*, **216**, 587.

Beckett, A.H., Gorrod, J.W. and Jenner, P. (1971) The effect of smoking on nicotine metabolism *in vitro* in man. *J. Pharm. Pharmacol.*, **23**, 62S–7S.

Benowitz, N. L (1990) Pharmacokinetic considerations in understanding nicotine dependence, in *The Biology of Nicotine Dependence* (Eds Bock, G. and Marsh, J.), John Wiley & Sons, Chichester, UK, pp. 186–209.

Brandänge, S. and Lindblom, L. (1979a) Synthesis, structure and stability of nicotine $\Delta^{1'(5')}$ iminium ion, an intermediary metabolite of nicotine. *Acta Chem. Scand.*, **B 33**, 187–91.

Brandänge, S. and Lindblom, L. (1979b) The enzyme 'aldehyde oxidase' is an iminium oxidase. Reaction with nicotine $\Delta^{1'(5')}$ iminium ion. *Biochem. Biophys. Res. Comm.*, **91(3)**, 991–6.

Flammang, A.M., Gelboin, H.V., Aoyama, T., *et al.* (1992) Nicotine metabolism by cDNA-expressed human cytochrome P450s. *Biochem. Arch.*, **8**, 1–8.

Foth, H., Walther, U.I. and Kahl, G.F. (1990) Increased hepatic nicotine elimination after phenobarbital induction in the conscious rat. *Toxicol. Appl. Pharmacol.*, **105(3)**, 382–92.

Foth, H., Looschen, H., Neurath, H. *et al.* (1991) Nicotine metabolism in isolated perfused lung and liver of phenobarbital- and benzoflavone-treated rats. *Arch. Toxicol.*, **65**, 68–72.

Funae, Y. and Imaoka, S. (1992) P450 in rodents, in *Handbook of Experimental Pharmacology 105*, (Eds Schenkman, J.B. and Greim, H.), Heidelberg/Springer, 221–38.

Fuxe, K., Agnati, L.F., Jansson, A., *et al.* (1990) Regulation of endocrine function by the nicotinic cholinergic receptor, in *The Biology of Nicotine Dependence*, (Eds Bock, G. and Marsh, J.), John Wiley & Sons, Chichester, UK, pp. 113–27.

Gorrod, J. W., Jenner, P., Keysell, G. *et al.* (1971) Selective inhibition of alternative pathways of nicotine metabolism *in vitro*. *Chem. Biol. Interactions*, **3**, 269–70.

Gorrod, J.W. and Jenner, P. (1975) The metabolism of tobacco alkaloids. *Essays in Toxicology*, **6**, 35–78.

Gorrod, J.W. and Hibberd, A.R. (1982) The metabolism of nicotine-$\Delta^{1'(5')}$-iminium ion, *in vivo* and *in vitro*. *Eur. J. Drug Metabol. Pharmacokinetics*, **7(4)**, 293–8.

Hammond, D.K., Bjercke, R.J., Langone, J.J. *et al.* (1991) Metabolism of nicotine by rat liver cytochromes P450. *Drug Metab. Disp.*, **19(4)**, 804–7.

Hibberd, A.R. and Gorrod, J.W. (1981) Nicotine $\Delta^{1'(5')}$ iminium ion: a reactive intermediate in nicotine metabolism. *Advances in Experimental Medicine and Biology*. **136 PtB**, 1121–31.

Hill, D.L., Laster, jun., W.R. and Struck, R.F. (1972) Enzymatic metabolism of cyclophosphamide and nicotine and production of a toxic cyclophosphamide metabolite. *Cancer Res.*, **32(4)**, 658–65.

Hucker, H.B., Gillette, J.R. and Brodie, B.B. (1960) Enzymatic pathway for the formation of cotinine, a major metabolite of nicotine in rabbit liver. *J. Pharmacol. Exp. Ther.*, **129(1)**, 94–100.

Imaoka, S., Terano, Y. and Funae, Y. (1989) Expression of four phenobarbital-inducible cytochrome P450s in liver, kidney, and lung of rats. *J. Biochem.*, **105(6)**, 939–45.

Imaoka, S., Terano, Y. and Funae, Y. (1990) Changes in the amount of cytochrome P450s in rat hepatic microsomes with starvation. *Arch. Biochem. Biophys.*, **278(1)**, 168–78.

Imaoka, S., Fujita, S. and Funae, Y. (1991) Age-dependent expression of cytochrome P450s in rat liver. *Biochim. Biophys. Acta*, **1097(3)**, 187–92.

Kamataki, T., Maeda, K., Yamazoe, Y., *et al.* (1981) Partial purification and characterization of cytochrome P450 responsible for the occurrence of sex difference in drug metabolism in the rat. *Biochem. Biophys. Res. Comms,* **103(1),** 1–7.

Kamataki, T., Maeda, K., Yamazoe, Y., *et al.* (1983) Sex difference of cytochrome P450 in the rat. *Arch. Biochem. Biophys.,* **225(2),** 758–70.

Kamataki, T., Maeda, K., Shimada, M. *et al.* (1985) Age-related alteration in the activities of drug-metabolizing enzymes and contents of sex-specific forms of cytochrome P450 in liver microsomes from male and female rats. *J. Pharmacol. Exp. Ther.,* **233(1)** 222–8.

Kita, T., Nakashima, T. and Kurogochi, Y. (1986) Circadian variation of nicotine-induced ambulatory activity in rats. *Jap. J. Pharmacol.,* **41,** 55–60.

Kyerematen, G.A., Damiano, M.D., Dvorchik, B.H. *et al.* (1982) Smoking-induced changes in nicotine disposition: Application of a new HPLC assay for nicotine and its metabolites. *Clin. Pharmacol. Ther.,* **32(6),** 769–80.

Kyerematen, G.A., Owens, G.F., Chattopadhyay, B., *et al.* (1988) Sexual dimorphism of nicotine metabolism and distribution in the rat. *Drug Metab. Disp.,* **16(6),** 823–8.

Kyerematen, G.A., Morgan, M., Warner, G., *et al.* (1990) Metabolism of nicotine by hepatocytes. *Biochem. Pharmacol.,* **40(8),** 1747–56.

McCoy, G.D., Howard, P.C. and DeMarco, G.J. (1986a) Characterization of hamster liver nicotine metabolism. 1. Relative rates of microsomal C and N oxidation. *Biochem. Pharmacol.,* **35(16),** 2767–73.

McCoy, G.D. and DeMarco, G.J. (1986b) Characterization of hamster liver nicotine metabolism. 2. Differential effects of ethanol or phenobarbital pretreatment on microsomal N and C oxidation. *Biochem. Pharmacol.,* **35(24),** 4590–2.

McCoy, G.D., DeMarco, G.J. and Koop, D.R. (1989) Microsomal nicotine metabolism: A comparison of relative activities of six purified rabbit cytochrome P450 isozymes. *Biochem. Pharmacol.,* **38(7),** 1185–8.

McCracken, N.W., Cholerton, S. and Idle, J.R. (1992) Cotinine formation by cDNA-expressed human cytochromes P450. *Medical Sci. Res.,* **20(33),** 877–8.

Murphy, P.J. (1973) Enzymatic oxidation of nicotine to nicotine $\Delta^{1'(5')}$ iminium ion. *J. Biol. Chem.,* **248(8),** 2796–800.

Nakayama, H., Nakashima, T. and Kurogochi, Y. (1982a) Heterogeneity of hepatic nicotine oxidase. *Biochim. Biophys. Acta,* **715(2),** 254–7.

Nakayama, H., Nakashima, T. and Kurogochi, Y. (1982b) Participation of cytochrome P450 in nicotine oxidation. *Biochem. Biophys. Res. Comm.,* **108(1),** 200–5.

Nakayama, H., Nakashima, T. and Kurogochi, Y. (1985) Cytochrome P450-dependent nicotine oxidation by liver microsomes of guinea pigs. *Biochem. Pharmacol.,* **34(13),** 2281–6.

Nakayama, H., Nakashima, T. and Kurogochi, Y. (1986) Participation of microsomal electron transport systems in nicotine metabolism by livers of guinea pigs. *Biochem. Pharmacol.,* **35(23),** 4343–5.

Nakayama, H., Fujihara, S., Nakashima, T. *et al.* (1987) Formation of two major nicotine metabolites in livers of guinea pigs. *Biochem. Pharmacol.,* **36(24),** 4313–7.

Nakayama, H. (1988) Nicotine metabolism in mammals. *Drug Metabolism and Drug Interactions,* **6(2),** 95–122.

Nakayama, H., Okuda, H., Nakashima, T. *et al.* (1993) Nicotine metabolism by rat hepatic cytochrome P450s. *Biochem. Pharmacol.*, in press.

Nguyen, T-L., Gruenke, L.D. and Castagnoli, jun., N. (1979) Metabolic oxidation of nicotine to chemically reactive intermediates. *J. Med. Chem.*, **22(3)**, 259–63.

Obach, R.S. and Vunakis, H.V. (1990) Radioimmunoassay of nicotine-$\Delta^{1'(5')}$-iminium ion, an intermediate formed during the metabolism of nicotine to cotinine. *Drug Metab. Disp.*, **18(4)**, 508–13.

Peterson, L.A., Trevor, A. and Castagnoli, jun., N. (1987) Stereochemical studies on the cytochrome P450 catalysed oxidation of (*S*)-nicotine to the (*S*)-nicotine $\Delta^{1'(5')}$-iminium species, *J. Med. Chem.*, **30(2)**, 249–54.

Peterson, L.A. and Castagnoli, jun., N. (1988) Regio- and stereochemical studies on the α-carbon oxidation of (*S*)-nicotine by cytochrome P450 model systems, *J. Med. Chem.*, **31(3)**, 637–40.

Ryan, D.E. and Levin, W. (1990) Purification and characterization of hepatic microsomal cytochrome P450. *Pharmacol. Ther.*, **45**, 153–239.

Schenkman, J.B., Thummel, K.E. and Favreau, L.V. (1989) Physiological and pathophysiological alterations in rat hepatic cytochrome P450. *Drug Metabolism Reviews*, **20**, 557–84.

Soucek, P. and Gut, I. (1992) Cytochrome P450 in rats. *Xenobiotica*, **22(1)**, 83–103.

Waxman, D.J., Dannan, G.A. and Guengerich, F.P. (1985) Regulation of rat hepatic cytochrome P450: Age-dependent expression, hormonal imprinting, and xenobiotic inducibility of sex-specific isoenzymes. *Biochemistry*, **24(16)**, 4409–17.

Williams, D.E., Shigenaga, M.K. and Castagnoli, jun. N. (1990a) The role of cytochrome P450 and flavin-containing monooxygenase in the metabolism of (*S*)-nicotine by rabbit lung. *Drug Metab. Disp.*, **18(4)**, 418–28.

Williams, D.E., Ding, X. and Coon, M.J. (1990b) Rabbit nasal cytochrome P450 NMa has high activity as a nicotine oxidase. *Biochem. Biophys. Res. Comm.*, **166(2)**, 945–52.

Nicotine metabolism beyond cotinine

4

G.B. Neurath

4.1 INTRODUCTION

Many fundamental investigations on the metabolism of nicotine were performed in the years between 1957–1972, when McKennis and colleagues identified a great variety of nicotine metabolites after separation by chromatographic techniques and comparison with synthetic samples, which his group had prepared for this purpose. In this work several secondary metabolites of cotinine were already prominent. Products from the further degradation and hydroxylation of the pyrrolidone ring of cotinine were observed in the urine of smokers and in the urine of dogs, rats, mice, monkeys and humans receiving doses of cotinine. The formation of cotinine and its further metabolism was at this time recognized as being very important in the disposition of nicotine (Bowman *et al.*, 1959, 1964, Bowman and McKennis, 1962; McKennis *et al.*, 1961, Morselli *et al.*, 1967; McKennis, 1965).

Among the metabolites of cotinine, 4-(3′-pyridyl)-4-oxo-N-methylbutyramide, 4-(3′-pyridyl)-4-methylaminobutyric acid, 4-(3′-pyridyl)-4-oxobutyric acid, 4-(3′-pyridyl)-4-hydroxybutyric acid and 5-(3′-pyridyl)-tetrahydrofuranone-2 (the lactone of cotinine) were identified in the urine after administration of either nicotine, cotinine or one of the cited intermediates to one of the species i.e. dog, rat or rabbit (McKennis *et al.*, 1957, 1958, 1959, 1961, 1964a, 1964b; Hucker *et al.*, 1960; Schwartz and McKennis, 1963). Further, *S*-(−)-norcotinine was identified in the urine of nicotine treated dogs (McKennis *et al.*, 1959). Another pyridine compound isolated from dog urine and capable of being acetylated was

Nicotine and Related Alkaloids: Absorption, distribution, metabolism and excretion. Edited by J.W. Gorrod and J. Wahren. Published in 1993 by Chapman & Hall, London. ISBN 0 412 55740 1

tentatively assigned the structure of 3-hydroxycotinine. The metabolic methylation of cotinine, with retention of the absolute configuration, to *S*-(−)-cotinine methonium ion was also observed in this period (McKennis *et al.*, 1963).

For the majority of the isolated metabolites, methods with sufficient sensitivity for their quantitation were not available at that time. Separation and detection techniques permitting reliable determinations of the variety of concomitant pyridine derivatives were still missing at that time. In addition the ease of reversibility of cyclic and ring open structures as for instance in the case of cotinine and 4-(3′-pyridyl)-4-methylaminobutyric acid may have further hampered early progress.

Progressive oxidation in the course of metabolic transformations leads to substances of ever increasing polarity. The extraordinary water solubility of many of the metabolites derived from nicotine and in particular from cotinine causes extreme difficulty in their extraction from urine. The ordinary procedure of 'salting out' is not sufficient for general applicability. Saturation of the samples with potassium carbonate or even sodium hydroxide are successfully used in some of these cases. For instance Shulgin *et al.* (1987) have mentioned the extraordinary polarity of cotinine-N-oxide as hindering the development of extraction-based analysis. These authors have applied sodium hydroxide to shift the partition coefficient to the acetonitrile used as a solvent in their study. Another example is that the preponderance of hydroxycotinine has been overlooked for more than two decades caused by the extreme water solubility of this metabolite. A further problem is the lack of volatility which may have caused some metabolites to have escaped detection by gas chromatography. This most potent separation technique for quantitation and identification in combination with mass spectrometry is not applicable to very polar and non-volatile substances without derivatization.

For many years, studies on nicotine pharmacokinetics were almost exclusively restricted to the determination of nicotine and cotinine, i.e. those compounds which were easily extractable by organic solvents and to which gas chromatography as the most convenient method was directly applicable without derivatization, and sometimes nicotine-1′-N-oxide following reduction to the parent base.

In recent years growing interest in pharmacokinetics as a scientific tool for understanding the effects of drugs in therapy and toxicology, evoked growing efforts for the elucidation of metabolic pathways and contributed to the enhancement of the evolution of sophisticated and more sensitive analytical methods.

For nicotine, as one of the most widespread stimulants, several research groups in the USA, the UK, Sweden and Germany re-synthesized the various metabolites identified in the early work of McKennis and his co-workers. Better extraction procedures and gas chromatographic methods with nitrogen specific and other sensitive detection or more selective assays like molecular ion monitoring mass spectrometry for the quantitation of these compounds were developed. These – high-performance liquid chromatography and radioimmunoassays (Langone *et al.*, 1973) – are currently the techniques which are most widely utilized. As representative, the gas chromatographic determination of nicotine and cotinine (Feyerabend *et al.*, 1975; Curvall *et al.*, 1982), the high-performance liquid chromatographic determination of the quaternary nicotine ions (Cundy and Crooks, 1984) may be mentioned. The importance of the use of closely related compounds as internal standards for the metabolites was recognized (Jacob *et al.*, 1981). The mass spectra of mammalian nicotine metabolites and related compounds were studied for further comparison into the nature of and for the quantitation of the numerous compounds (Pilotti *et al.*, 1976). Derivatization in order to achieve better volatility and sensitivity by electron capture detection was also used (Neurath *et al.*, 1987) in order to detect and quantitate nicotine metabolites. The aim of these efforts was partly to develop techniques for the simultaneous determination of nicotine and as many as possible of its metabolites in biological samples. A radiometric high-performance liquid chromatographic assay for nicotine and 12 of its metabolites after intra-arterial administration of 2-^{14}C-pyrrolidine labelled nicotine to rats, has been reported (Kyerematen *et al.*, 1987). But great problems still exist, to find extraction, separation and detection techniques combined in one method applicable to the whole spectrum of different properties of the great variety of nicotine metabolites.

4.2 OXIDATION OF NICOTINE

α-Oxidation in the 5′-position clearly causes the most abundant metabolites of nicotine in the human and in all animal species studied so far. Originally detected by McKennis *et al.* (1957) in urine of anaesthetized dogs after administration of nicotine by slow intravenous infusion, *S*-(−)-cotinine, though age dependant (Klein and Gorrod, 1978), has been considered the main metabolite of nicotine for many years (McKennis, 1965; Gorrod and Jenner, 1975; Gorrod and Hibberd, 1982). It has been obtained following the administration of nicotine to the rabbit, the rat, the mouse and in humans. The amount of cotinine isolated from pooled urine of male smokers clearly exceeded that part which could be attributed to preformed cotinine transferred from tobacco.

Cotinine is not a primary microsomal metabolite, but is formed by the oxidation of an initial microsomal product.

The oxidation of nicotine has been studied in synthetic and micro-

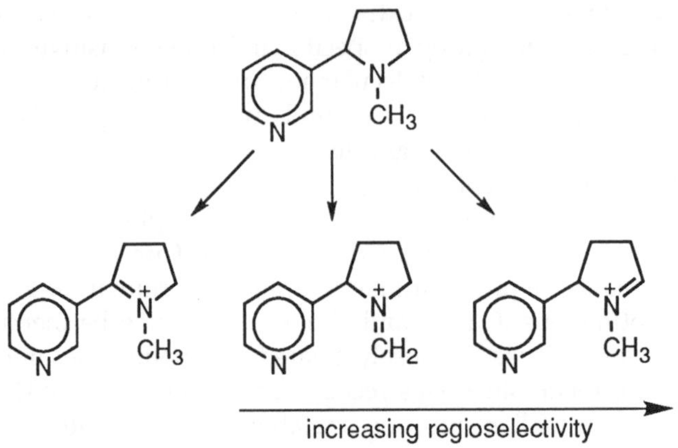

Figure 4.1 Nicotine-iminium ions formed by electron transfer to cytochrome P450. (From Peterson and Castagnoli, 1987.)

biological models since 1973. As an intermediate in the cytochrome P450 catalysed degradation of nicotine, nicotine-$\Delta^{1',5'}$ iminium ion was identified by the formation of 5′-cyanonicotine with the strong nucleophile cyanide (Murphy, 1973). Supporting results for this concept were presented by the isolation of cotinine after intraperitoneal administration of nicotine-$\Delta^{1',5'}$ iminium ions to rats or guinea pigs (Gorrod and Hibberd, 1982). The iminium ion appeared more likely to be primarily formed in nicotine metabolism than its unstable tautomer 5′-hydroxynicotine (Hucker *et al.*, 1960), which exists in equilibrium with it (Brandänge and Lindblom, 1979b). The iminium ion was the only form observed in freshly prepared acidic or neutral solutions of synthetic nicotine-$\Delta^{1',5'}$ iminium perchlorate, whereas the carbinol, 5′-hydroxynicotine, was the only form observed in strongly alkaline solutions (Brandänge and Lindblom, 1979a). The content of the carbinol tautomer at the pH of cytoplasm (7.4) was estimated by these authors to be approximately 25%. Recent studies on microsomal preparations asserted a regiospecificity on the heme in favour of oxidation of the pro-E-proton at the 5′-position of the nicotine molecule (Peterson *et al.*, 1987).

In recent years the theory has generally been put forward that iminium species play a widespread role in biological systems as a charge transfer agent (Kovacic, 1984). Experimental proof of the occurrence of this first intermediate in nicotine metabolism in smokers' biofluids, however, was still missing. Recently the presence of the nicotine-$\Delta^{1',5'}$ iminium ion, tautomer with 5′-hydroxynicotine, and nicotine-$\Delta^{1',2'}$ iminium ion were observed in smokers' urine samples by the mass spectrometric identification of 5′-cyanonicotine and 2′-cyanonicotine after the treatment of urine samples with potassium cyanide. Reaction

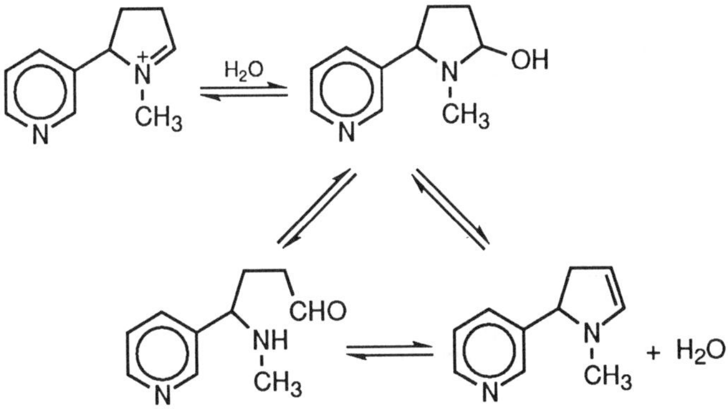

Figure 4.2 Nicotine-Δ1′,5′-iminium ion and tautomers.

with propionic acid chloride revealed the presence of a hydroxynicotine in all urine samples of 30 smokers (Neurath *et al.*, 1992).

It may be an attractive speculation to consider conjugation reactions of the carbinol form of the 5′-hydroxy intermediate. On the action of hydrolytic enzymes on those postulated conjugates, nicotine and cotinine may be formed by disproportionation or, the latter, by immediate oxidation.

The other iminium ions demonstrated in Figure 4.1 are known to compete with nicotine-$\Delta^{1',5'}$ iminium ion in the first oxidation step of nicotine. The product of the methylene-iminium ion as an intermediate, identified in rabbit postmitochondrial supernatant fractions as the corresponding 1′-N-cyanomethyl-nornicotine by Nguyen *et al.* (1979), would be nornicotine (McKennis *et al.*, 1963; Papadopoulos and Kintzias, 1963). The compound is definitely formed in humans as a metabolite, but the yield is very low. In good agreement it was found at ratios of 0.3% after intravenous administration of deuterium-labelled nicotine (Jacob and Benowitz, 1991) and at 0.5% of the dose after infusion of the purified alkaloid (Neurath *et al.*, 1991). The occurrence of deuterium-labelled nornicotine and its urinary concentration–time curve after infusion of pure nicotine to male subjects indicated the formation of small amounts of nornicotine as a metabolite, but it accounts for only a small proportion of the total amount of metabolites excreted. It is not clear, however, whether this may be caused by a low rate of formation or by rapid further metabolism, for instance to norcotinine. The third possible product of α-hydroxylation, 2′-hydroxynicotine (pseudooxynicotine), originating from the nicotine-$\Delta^{1',2'}$ iminium ion, was found by Wada and Yamasaki (1954) as a metabolite in the degradation of nicotine by soil bacteria. Like its 5′-hydroxy isomer this compound exists as four tautomers (Brandänge *et al.*, 1983) causing many analytical problems in

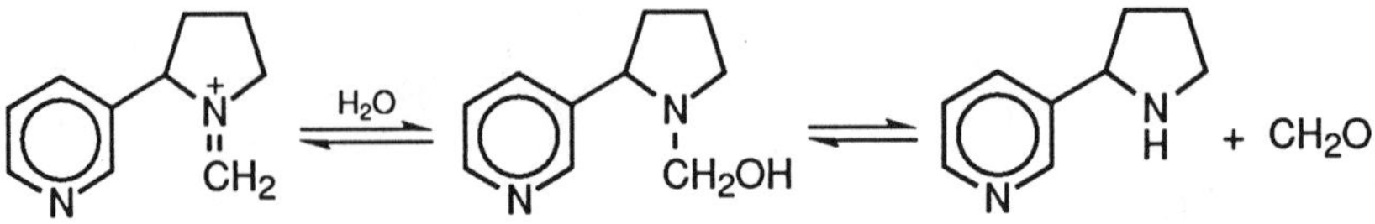

Figure 4.3 Nicotine-methylene-iminium ion and tautomers.

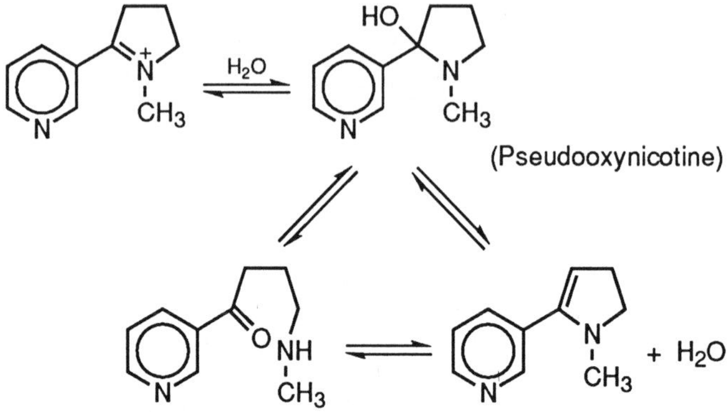

Figure 4.4 Nicotine-$\Delta^{1',2'}$ iminium ion and tautomers.

trials to isolate this possible metabolite from biological fluids. On synthetic samples at physiological (neutral) pH, these authors, using NMR spectra and chemical reactions, identified about equal amounts of N-methyl-N-4-(3′-pyridyl)-4-oxo-butylamine as the open chain form and nicotine-$\Delta^{1',2'}$ iminium ion. In preparative scale amounts pseudooxynicotine is readily dehydrated to $\Delta^{2',3'}$ dehydronicotine (Haines and Eisner, 1950), which easily undergoes polymerization and also disproportionation to nicotine and nicotyrine.

Very recently, the occurrence of the nicotine-$\Delta^{1',2'}$ iminium ion together with its $\Delta^{1',5'}$ isomer in smokers' urine was proved by formation of the cyano derivative (Neurath *et al.*, 1992). Hitherto, the hydroxylation in the 2′-position of nicotine in animal species was only documented by secondary metabolites. Allo-hydroxycotinine and allo-hydroxynorco-tinine were already identified in the urine of rats as will be mentioned later.

There is so far no direct evidence of the existence or quantitation of the methylene iminium ion or 1′-N-hydroxymethyl-nicotine as metabolic intermediate after the administration of nicotine to animal species or man (McCoy *et al.*, 1986).

The final oxidation of the tautomeric first intermediate 5′-hydroxynicotine to cotinine is catalysed by cytosolic aldehyde oxidase. Significant rates of cotinine formation from nicotine can be demonstrated only with liver post-mitochondrial supernatant fractions or with purified microsomes in the presence of purified aldehyde oxidase (Gorrod and Hibberd, 1982). The enzyme is known to provide its highest affinity for unsaturated N-heterocycles including iminium ions (Brandänge and Lindblom, 1979b). The oxidation by this pathway is blocked by strong nucleophiles like potassium cyanide by formation of the 5′-cyanonicotine (Murphy, 1973).

4.3 METABOLISM OF COTININE

4.3.1 C-OXIDATION OF COTININE

(a) (*S*)-*trans*-3′-Hydroxycotinine

As mentioned in the introduction, it was long since known that cotinine is extensively further metabolized. After intravenous administration of cotinine to male dogs, McKennis *et al.* (1959) isolated norcotinine and another metabolite from the urine. They designated this other metabolite, hydroxycotinine. Both metabolites had already been observed in the urine of dogs and humans after administration of nicotine, but were not identified in those preceding experiments (McKennis *et al.*, 1957, 1958). In attempts to synthesize this metabolite (McKennis *et al.*, 1963a), the authors isolated a compound which was clearly hydroxylated in one of the two possible positions in the pyrrolidine ring of cotinine. Without being able to authenticate, they assigned the 3′-hydroxycotinine structure to the isomer as being the most likely. Diazotization of the corresponding 3′-aminocotinine produced two isomeric pairs, a dextrorotatory enantiomer, differing from the metabolite in the high melting point (m.p. 135°C), and another dextrorotatory low melting isomer (m.p. 110–111°C), which agreed with the metabolite in melting point, chromatographic retention time and infrared spectrum, thus providing evidence for their identity. Hydroxycotinine was also identified among the nicotine metabolites produced by mouse liver tissue slices (Hansson and Schmiterlöw, 1965).

It was only in 1972, that the excretion product of cotinine in monkeys was identified in a laborious study as the *trans* isomer (Dagne, 1972, Dagne and Castagnoli, 1972). Dagne and Castagnoli administered *S*-(−)-cotinine to a male rhesus monkey and obtained a hydroxycotinine (m.p. 111°C) from the urine to which they assigned the absolute stereochemistry as *trans*-1-methyl-3-(*R*)-hydroxy-5-(*S*)-3-pyridyl-2-pyrrolidinone, corresponding to *trans*-3′-hydroxycotinine. It possesses two optically active centres, at the 3- and at the 5-position of the pyrrolidine ring. 3′-Hydroxycotinine is an extremely polar compound compared with nicotine and cotinine.

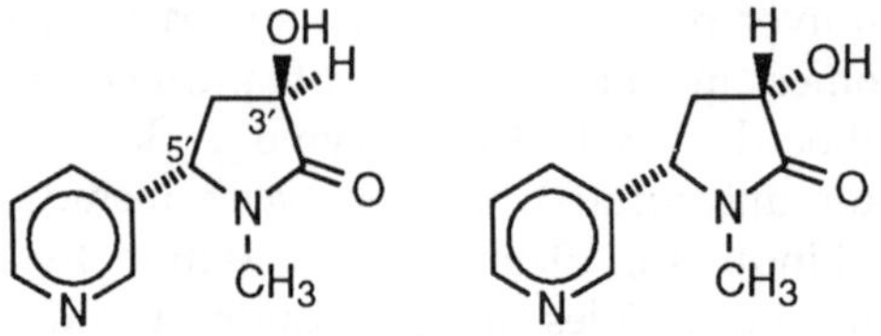

Figure 4.5 Structures of (a) (*S*)-*trans*-3′-hydroxycotinine and (b) (*S*)-*cis*-3′-hydroxycotinine.

Again 15 years later, in an expanded study, nine male volunteers smoked three different brands with varying nicotine yields in three subsequent cycles of alternating seven days of smoking 19 cigarettes per day, and wash-out phases of five days. In an attempt to set up a balance of the total excreted nicotine and four of its main metabolites, the sum of these accounted for only 18.7%, 18.0%, and 13.3% of the estimated nicotine intake in the repective runs. This marked shortfall prompted a search for further, so far not quantified, metabolites in the urine. *trans*-3′-Hydroxycotinine was found to account for a large part of the deficiency. Its amount was indeed greater than that of cotinine, formerly generally accepted as the main metabolite of nicotine, and brought the estimated balances to 39.8%, 38.2% and 31.5% in the respective studies (Neurath *et al.*, 1987, 1988; Neurath and Pein, 1987). Thus, this long since known compound surprisingly revealed its role as the most abundant metabolite in smokers' urine. Caused by the shorter half-life of 5.9 h (Neurath *et al.*, 1988; Scherer *et al.*, 1988), its mean plasma concentration, 69 ng/ml determined on the sixth day of smoking, was second only to that of cotinine. After oral administration of *trans*-3′-hydroxycotinine to two male subjects, 81% and 93% were excreted unchanged. The extreme polarity of *trans*-3′-hydroxycotinine necessitates very careful extraction of urine samples to avoid misleading results from enzyme treatments in conjugation studies.

From an infusion experiment the pharmacokinetic parameters of *trans*-3′-hydroxycotinine in smokers were established as follows: half-life ($t_{1/2\beta}$) = 5.9 (4.2–9.5) h; apparent volume of distribution (V_d) = 0.87 (0.51–1.14) l/kg; total clearance (Cl_{total}) = 1.79 (1.08–2.59) ml/min kg; renal clearance (Cl_r) = 1.31 (0.85–1.78) ml/min kg; percentage of renal clearance (Cl_r/Cl_{total}) = 75.4 $\pm$ 12.8 (60.3–98.2) (Scherer *et al.*, 1988). The high percentage of renal clearance, like the high percentage of the urinary recovery after oral administration, suggests only limited further metabolism.

The importance of *trans*-3′-hydroxycotinine in nicotine disposition in man was confirmed in subsequent studies (O'Doherty *et al.*, 1988; Jacob *et al.*, 1988; Curvall *et al.*, 1991; Richie *et al.*, 1991). The metabolite was

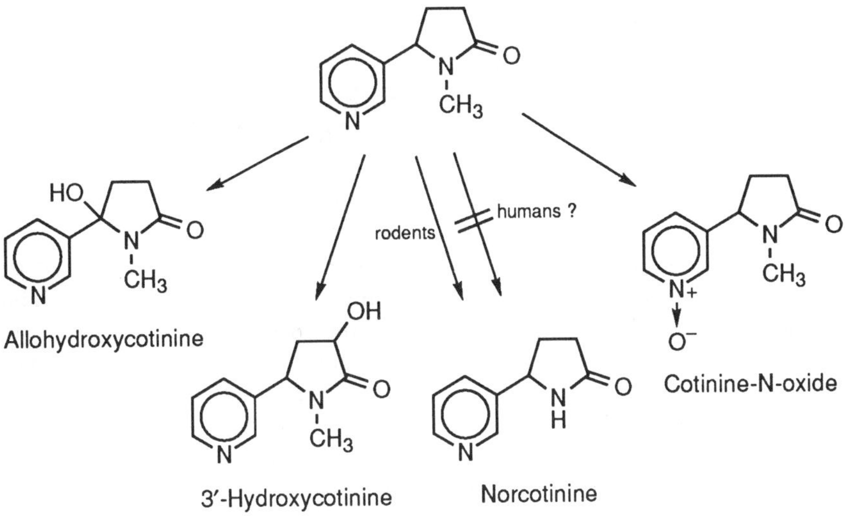

Figure 4.6 Metabolites of cotinine.

found to cause cross-reactivity with cotinine in radioimmunoassays, which are widely used as a sensitive index of tobacco smoke exposure. Radioimmunoassay resulted in 30–40% higher values in comparison to gas chromatography (Richie *et al.*, 1991).

(b) (*S*)-*cis*-3′-Hydroxycotinine

O'Doherty *et al.* (1988) tried to detect the isomeric *cis*-3′-hydroxycotinine in smokers' urine. Using high-performance liquid chromatography of a derivative, it appeared to them that the *trans* isomer was exclusively excreted. However, Voncken *et al.* (1990) were able to separate both stereoisomers of 3′-hydroxycotinine by capillary gas chromatography and detected the *cis* isomer in urine of smokers at an average level of 0.18–0.23% of the total excreted 3′-hydroxycotinine. A species dependence was observed by the same authors after intravenous administration of *S*-(−)-nicotine to rats. Of the excreted 3′-hydroxycotinine 11.2–15.8% were present as the *cis* isomer. In hamster urine the range was found to be 9.7–13.0%.

(c) *trans*-3′-Hydroxycotinine in species other than humans

Quite soon 3′-hydroxycotinine was revealed as the major urinary metabolite of both *S*-(−)- and *R*-(+)-nicotine in guinea pig, hamster and rabbit (Nwosu and Crooks, 1988). The authors found between 31.1% and 36.7% of the tritium labelled *S*-(−)-nicotine enantiomer transformed

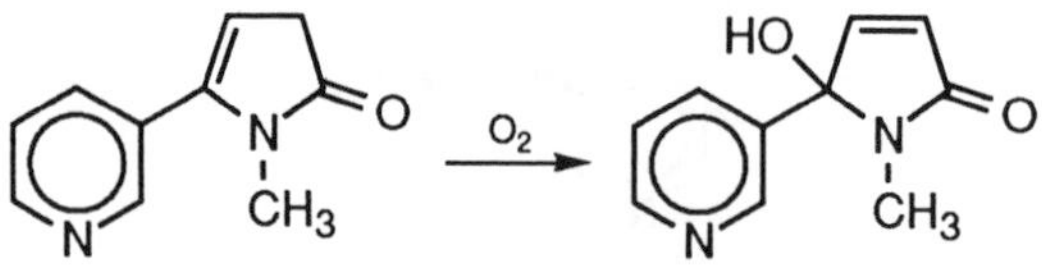

Figure 4.7 Formation of 5′-hydroxy-$\Delta^{3',4'}$ dehydrocotinine.

to 3′-hydroxycotinine in these three species. In the rat 3′-hydroxycotinine was only second in importance to nicotine-1′-N-oxide. The authors reported 9.3% of the 1′-N-oxides (not differentiated between the stereoisomers) and 6.1% of 3′-hydroxycotinine respectively of the two enantiomers. Species dependence was also observed by Kyerematen *et al.* (1988). In rabbit and hamster, 3′-hydroxycotinine was found in high amounts (44% and 36% respectively) and represented the major metabolite of nicotine in these animal species; it was only of minor importance (8%) in the rat. 3′-Hydroxycotinine was also observed to be a significant urinary metabolite of nicotine (17.8%) in the guinea pig.

(d) 5′-Hydroxycotinine (Allohydroxycotinine)

Dagne (1972) treated rats with cotinine and isolated a metabolite to which, on the basis of the mass spectral data and the evidence gathered from synthetic work, the cyclic structure of 5′-hydroxycotinine (allohydroxycotinine) was assigned. Allohydroxycotinine was also observed by McKennis *et al.* (1978) together with 3′-hydroxycotinine and norcotinine in monkey urine after administration of *S*-(−)-cotinine and *S*-(−)-cotinine-N-oxide as single or divided doses. Synthetic work (Nguyen *et al.*, 1981) provided experimental support for the statement that metabolic C-5 hydroxylation of (*S*)-cotinine leads to a tautomeric product in which the hydroxylactam structure is energetically favoured compared to the ring open form, 4-(3′-pyridyl)-4-oxo-N-methyl-butyramide, and represents 80% in water. The product isolated from the urine of rhesus monkeys treated with (*S*)-cotinine was identical to the synthetic material. For both C-oxidation products of cotinine, 3′-hydroxycotinine and 5′-hydroxycotinine, so far there are no reports on the mechanism for their metabolic formation.

(e) 5′-Hydroxy-$\Delta^{3',4'}$ dehydrocotinine (5′-(3-pyridyl)-5′-hydroxy-pyrroline-$\Delta^{3',4'}$-on-2)

This compound was recently identified as a possible product of the nicotine iminium species in the course of enzymatic studies. Monoaminoxidase B catalyses the *in vitro* oxidation of nicotine-$\Delta^{1',5'}$

iminium species to β-nicotyrine, which is a substrate for liver oxidases, and in particular lung oxidases, which are more active on a nmole cytochrome P450 basis. Analysis of the β-nicotyrine containing microsomal incubation mixtures led to the characterization of $\Delta^{4',5'}$ dehydrocotinine which is relatively stable at pH 7.6 but undergoes rapid autoxidation at acidic or basic conditions with the formation of 5′-hydroxy-$\Delta^{3',4'}$ dehydrocotinine (Castagnoli, 1991).

4.3.2 DEMETHYLATION OF COTININE

(a) Norcotinine (demethylcotinine)

Demethylation of cotinine has been reported to occur after administration of cotinine to dogs, mice and rats (McKennis *et al.* 1959, 1962; Dagne, 1972). The authors also established the absolute configuration of this metabolite as *S*-(−)-norcotinine. But norcotinine could not be detected after administration of cotinine to humans (Bowman *et al.*, 1959; Bowman and McKennis 1962), neither in smokers nor in subjects receiving cotinine, although norcotinine was present in the urine of smokers and after administration of nicotine (Bowman *et al.*, 1959). This would support its formation by α-oxidation of nornicotine as an intermediate. Norcotinine was indeed reported as a metabolite in the urine of dogs after intravenous administration of (−)-nornicotine (Wada *et al.*, 1961). Nornicotine occurs as a minor companion of nicotine in most tobacco species and is definitely formed metabolically from nicotine in humans (Jacob and Benowitz, 1991; Neurath *et al.*, 1991) but, possibly caused by further metabolism, only minor amounts are detected in the urine. Nevertheless, the metabolic route of norcotinine formation in man still remains to be definitely confirmed, i.e. whether it is derived from nornicotine by α-oxidation or by demethylation of cotinine.

The further metabolism of (−)-norcotinine was studied after oral intubation to rats by Schwartz and McKennis (1964). An intermediate, 4-(3′-pyridyl)-4-oxobutyramide, analogous to the 4-(3′-pyridyl)-4-oxomethylbutyramide arising from cotinine, undergoes subsequent metabolism to 4-(3′-pyridyl)-4-oxobutyric and 4-(3′-pyridyl)-4-hydroxybutyric acids.

(b) 5′-Hydroxynorcotinine

Meacham *et al.* (1972) proposed 5′-hydroxynorcotinine as a potential intermediate in nicotine metabolism. Kyerematen *et al.* (1988) identified this compound in rat urine and found it to have the longest half-life (9.9–11.0 h) for urinary excretion among 12 metabolites measured after intra-arterial administration of ^{14}C-labelled nicotine in rats. It remains to clarify whether the compound is derived from hydroxylation of norcoti-

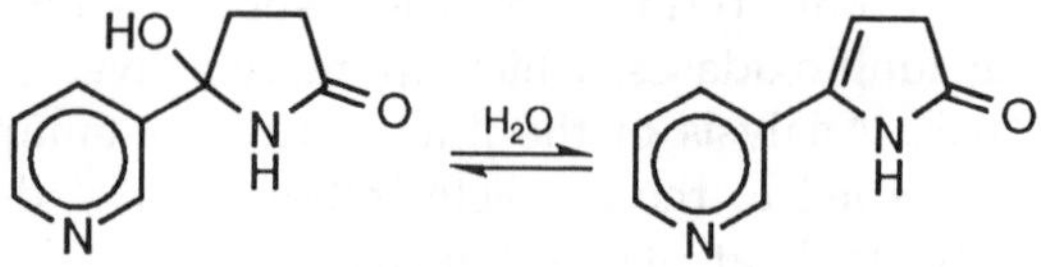

Figure 4.8 Structures of (a) 5′-hydroxynorcotinine and (b) norcotinine-$\Delta^{2',3'}$ enamine.

nine or demethylation of allohydroxynicotine. As a possible product of 5′-hydroxynicotine dehydration, on the basis of its mass spectral fragmentation, norcotinine-$\Delta^{2',3'}$ enamine appeared to be a long-lived metabolite isolated from the urine of humans and stumptailed macaques after intravenous administration of *R,S*-(±)-nicotine ([2′-^{14}C]pyrrolidine) (Kyerematen *et al.* 1990; Seaton *et al.*, 1991). A half-life for this compound of 15.1 h in smokers, and 11.2 h in nonsmokers was determined by the authors.

4.3.3 N-OXIDATION OF COTININE

(a) Cotinine-1-N-oxide

N-Oxidation of cotinine at the pyridine ring was first detected in the monkey (Dagne and Castagnoli, 1972) then later in dogs and rats (Kyerematen and Vesell, 1991). Unlike the formation of nicotine-1′-N-oxide, which is catalysed by the flavine oxidase system, cotinine is a substrate of a cytochrome P450 isozyme (Hibberd and Gorrod, 1985). This metabolic pathway is only a minor contribution to the disposition of cotinine. Shulgin *et al.* (1987) reported on the occurrence of cotinine-1-N-oxide in the 24 h urine of 25 human subjects smoking *ad libitum*. The average ratio of excreted cotinine-1-N-oxide to cotinine (on a μg/24 h basis) was 0.25, ranging from 0.15–0.25. Only 3% of the absorbed nicotine appeared in urine as cotinine-1-N-oxide (Jacob and Benowitz, 1991). Cotinine-1-N-oxide was determined by Kyerematen *et al.* (1988) to possess a longer half-life of urinary excretion in rats (7.9–8.2 h) than that of cotinine (4.8–5.3 h).

4.3.4 N-METHYLATION OF COTININE

N-Methylated quaternary pyridinium metabolites have been reported by McKennis *et al.* (1963b) to be formed from nicotine and cotinine in the dog. It was also shown that cotinine can be methylated to the N-methylcotininium (N-methyl-5′-oxonicotinium) ion in humans. In a study attempting to identify methylated metabolites in the guinea pig, an animal known to be a good N-methylater, the N-methylcotininium ion was not identified (Cundy and Crooks, 1984). Only N-methylnicotinium ion at 4% of the total ^{14}C label was detected in 24 h urine.

McKennis and his co-workers (1963a) isolated cotinine methonium ions, besides 4-(3′-pyridyl)-4-oxo-N-methylbutyramide and 3′-hydroxycotinine, after oral administration of *S*-(−)-cotinine to dogs and humans. The compound was isolated as the iodide and compared with an authentic sample, but no quantitative analytical method for the metabolite was available at that time.

4.3.5 CONJUGATION OF COTININE

The formation of conjugates of nicotine and its metabolites, in particular of cotinine and *trans*-3′-hydroxycotinine has recently been discussed. A considerable increase of nicotine and cotinine concentrations was observed in urine samples after 20 h incubation with β-glucuronidase at 37°C at pH 4.5. The average rates of free alkaloids and those amounts determined after the hydrolysis were 1.0 for nicotine and 0.5 for cotinine (Curvall *et al.*, 1991). So far, no hypothesis has been offered for the nature of possible conjugates, neither for cotinine nor for a corresponding adduct of nicotine. Recently, minor amounts of a cotinine conjugate were identified in human urine by mass spectroscopy as sulphate (Schepers *et al.*, 1992).

However, there have been several reports of conjugation of tertiary amines with glucuronic acid forming quaternary ammonium conjugates, presumably existing as an internal salt with the carboxylic acid function. Among those reported was also a pyridinium N-glucuronide formed from an analgesic, which represented 55% excretion of the dose in human urine (Hawkins, 1981).

4.3.6 CONJUGATION OF *TRANS*-3′-HYDROXYCOTININE

Conjugates of *trans*-3′-hydroxycotinine amounting to 30% were found in urine after 20 h incubation at 37°C at pH 4.5 with β-glucuronidase by Curvall *et al.* (1991). After intravenous administration of *R,S*-(±)-nicotine ([2′-^{14}C]pyrrolidine), large amounts of a long-lasting 3-hydroxycotinine glucuronide were determined in the urine of humans and stumptailed macaques by radiometric HPLC assay (Kyerematen *et al.*, 1990; Seaton *et al.*, 1991). The compound revealed an extraordinary urinary half-life of more than 20 h. This high rate of glucuronidation after administration of racemic ^{14}C-nicotine may not apply for the metabolic rate of the natural *S*-(−)-cotinine. Recently, in purified conjugates from human urine the major component was identified as the O-β-D-glucuronide of *trans*-3′-hydroxycotinine accompanied by minor amounts of sulphate (Schepers *et al.*, 1992).

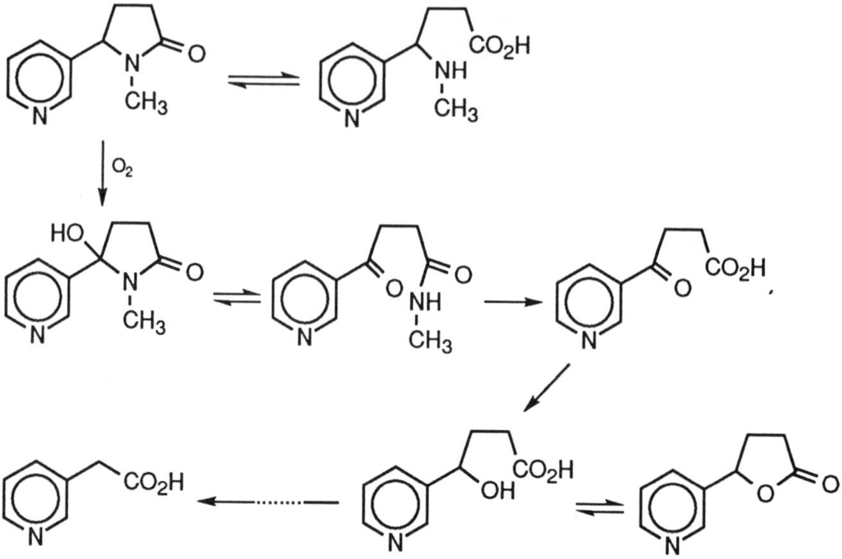

Figure 4.9 Ring cleavage of cotinine.

4.3.7 RING CLEAVAGE OF COTININE

McKennis *et al.* (1961) found evidence for the metabolism of nicotine to 3′-pyridylacetic acid in the dog. Similar results have been reported for the cat (Turner, 1969) and in rabbit liver preparations *in vitro* (Papadopoulos and Kintzios, 1963). As mentioned, different metabolic pathways have been discussed, a route via ring opening of norcotinine on the one hand, and oxidation of cotinine to allohydroxycotinine, ring opening to 4-(3′-pyridyl)-4-oxo-N-methylbutyramide on the other hand, both with 4-(3′-pyridyl)-4-oxo-butyric acid, 4-(3′-pyridyl)-4-hydroxy-butyric acid, as itself or in the form of its lactone, serving as common intermediates and final oxidative degradation of the chain to 3-pyridylacetic acid which appears to be the end product of nicotine metabolism (McKennis *et al.*, 1962, 1964a, 1964b).

Very little has been contributed recently to the knowledge and quantitation of these earlier described open chain metabolites deriving from cleavage of the pyrrolidone ring of cotinine and further degradation. Attention has to be paid to artefact formation mainly by ring-open chain tautomerism during analytical procedures. For instance the synthetic 4-(3′-pyridyl)-4-methylamino-butyric acid (McKennis *et al.*, 1957) is known to cyclize spontaneously at physiological pH (McKennis *et al.*, 1958), and the ring-open isomer may be formed as an artefact during analytical procedures. The possible product of cotinine by oxidative ring opening, 4-(3′-pyridyl)-4-oxo-N-methyl-butyramide, exists in equilibrium with 5′-hydroxycotinine which is, as mentioned, the main isomer (80%) in neutral aqueous medium. The compound has been ascribed a key role in the mammalian degradation of the pyrrolidine ring by McKennis *et al.* (1962). Both isomers led *in vivo* to the urinary excretion of 4-(3′-pyridyl)-4-oxo-butyric acid by the rat, dog or rabbit, and subsequently to the excretion of 4-(3′-pyridyl)-4-hydroxy-butyric and 3′-pyridylbutyric or –acetic acids (McKennis *et al.*, 1964a, 1964b; Schwartz and McKennis, 1963). None of these metabolites have been recently quantified in smokers' urine or blood samples. As a typical further impediment on solvent extraction of alkalinized biological fluids, these pyridyl carboxylic acids remain persistently in the aqueous phase. They are neither easily extractable from, nor easily derivatized in, the aqueous medium. A recent publication recommends freeze drying and derivatization for the determination of nicotinic acid after administration as a vasodilator (Becker and Hummel, 1990). As a further disadvantage, the acidic metabolites in their original state without derivatization, the quaternary pyridinium bases, and the nicotine and cotinine N-oxides are not capable of being analysed by gas chromatography or in its combination with mass spectroscopy (Pilotti *et al.*, 1976).

4.4 CONCLUSIONS

The balance of the excreted sum of nicotine and its known metabolites from confirmed results accounts for about 40% to 60% of the estimated or known intake of the alkaloid, if the calculation is based on the parts of well defined and reliably determined species. A further 40% is recently ascribed to those parts of metabolites which are additionally found after the treatment of urine samples with enzymes generally known to act on phase II metabolites. Conjugates of nicotine, cotinine and *trans*-3′-hydroxycotinine, are held responsible for this. Recently, some of these products have been definitely characterized and determined by direct measurements. The development of reliable extraction methods for the extreme water soluble and persistent unconjugated higher oxidized species is very problematic, but is necessary to securely preclude residues of the original metabolites.

It is felt particularly that there is a gap concerning knowledge of the further metabolic fate of nornicotine, the first metabolite of the intermediary methylene iminium species. Nornicotine is only excreted in a very small amount. No quantitative data is available on the $\Delta^{1',2'}$ iminium ion, the very recently identified tautomer of 2-hydroxynicotine in urine. The knowledge was restricted to its production by microorganisms for a long time. But the corresponding possible secondary metabolites allohydroxycotinine and 5′-hydroxynorcotinine had already been observed in the urine of animal species and man.

Knowledge of the actual structures of nicotine metabolites under different analytical conditions as in aqueous solutions at different pH values or in organic solvents is fundamental for further studies. Many of the metabolites exist as ring-chain tautomers. The respective preponderances of the actual tautomeric structures in the biological media and especially at the active sites of enzymes determine the metabolic pathways of nicotine and its subsequent metabolites and should be the aim of intense investigations.

REFERENCES

Becker, H. and Hummel, K. (1990) Ein neues Verfahren zur Bestimmung von Nikotinsäure. *Arzneim.-Forsch./Drug Res.*, **40**, 573–5.

Benowitz, N.L. (1991) Importance of nicotine metabolism in understanding the human biology of nicotine, in *Effects of Nicotine on Biological Systems* (Eds F. Adlkofer and K. Thurau), Advances in Pharmacological Sciences, Birkhäuser Verlag, Basel, pp. 19–24.

Bowman, E.R., Hanson, E., Turnbull, L.B. *et al.* (1964) Disposition and fate of (−)-cotinine-H^3 in the mouse. *J. Pharmacol. Exp. Ther.*, **143**, 301–8.

Bowman, E.R. and McKennis, H. (1962) Metabolism of (−)-cotinine in the human. *J. Pharmacol. Exp. Ther.*, **135**, 306–311.

Bowman, E.R., Turnbull, L.B. and McKennis, H. (1959) Metabolism of nicotine in the human and excretion of pyridine compounds in smokers. *J. Pharmacol. Exp. Ther.*, **127**, 91–102.

Brandänge, S. and Lindblom, L. (1979a) Synthesis, structure and stability of nicotine Δ1′(5′) iminium ion, an intermediary metabolite of nicotine. *Acta Chem. Scand.*, **B33**, 187–91.

Brandänge, S. and Lindblom, L. (1979b) The enzyme 'aldehyde oxidase' is an iminium oxidase. Reaction with nicotine-$\Delta^{(1',5')}$-iminium ion. *Biochem. Biophys. Res. Commun.*, **91**, 991–6.

Brandänge, S., Lindblom, L., Pilotti, A. *et al.* (1983) Ring-chain tautomerism of pseudooxynicotine and some other iminium compounds. *Acta Chem. Scand.*, **B 37**, 617–22.

Castagnoli, N., jun., Shigenaga, M., Carlson, T., *et al.* (1991) The *in vitro* metabolic fate of (*S*)-nicotine, in *Effects of Nicotine on Biological Systems* (Eds F. Adlkofer and K. Thurau), Advances in Pharmacological Sciences, Birkhäuser Verlag, Basel, pp. 25–33.

Cundy, K.C. and Crooks, P.A. (1984) High-performance liquid chromatographic method for the determination of N-methylated metabolites of nicotine. *J. Chromatogr.*, **306**, 291–301.

Curvall, M., Kazemi-Vala, E. and Englund, G. (1991) Conjugation pathways in nicotine metabolism, in *Effects of Nicotine on Biological Systems* (Eds F. Adlkofer and K. Thurau), Advances in Pharmacological Sciences, Birkhäuser Verlag, Basel, pp. 69–75.

Curvall, M., Kazemi-Vala, E. and Enzell, C.R. (1982) Simultaneous determination of nicotine and cotinine in plasma using capillary column gas chromatography with nitrogen-sensitive detection. *J. Chromatogr.*, **232**, 283–93.

Dagne, E. (1972) Biotransformation studies on *S*-(−)-cotinine. Dissertation,

University of California, *U.M.I Dissertation Information Service*, Ann Arbor, MI, 1987.

Dagne, E. and Castagnoli, N. (1972) Cotinine N-oxide, a new metabolite of nicotine. *J. Med. Chem.*, **15**, 840–1.

Dagne, E. and Castagnoli, N. (1972) Structure of hydroxycotinine, a nicotine metabolite. *J. Med. Chem.*, **15**, 356–60

Feyerabend, C., Levitt, T. and Russell, M.A.H. (1975) A rapid gas-liquid chromatographic estimation of nicotine in biological fluids. *J. Pharm. Pharmacol.*, **27**, 434–6.

Gorrod, J.W. and Hibberd, A.R. (1982) The metabolism of nicotine-Δ1′(5′)-iminium ion, *in vivo* and *in vitro*. *Europ. J. Drug Metab. Pharmacokin.*, **7**, 293–8.

Gorrod, J.W. and Jenner, P. (1975) The metabolism of tobacco alkaloids. *Essays in Toxicology*, **6**, 35–78.

Haines, P.G. and Eisner, A. (1950) Identification of pseudo-oxynicotine and its conversion to N-methylmyosmine. *J. Am. Chem. Soc.*, **72**, 1719–21.

Hansson, E. and Schmiterlöw, C.G. (1965) Metabolism of nicotine in various tissues, in *Tobacco Alkaloids and Related Compounds*, 4th Wenner–Gren Centre International Symposium, **4**, Ed. U.S. von Euler, Pergamon Press, Oxford, pp. 87–97.

Hawkins, D.R. (1981) Novel biotransformation pathways, in *Progress in Drug Metabolism*, Eds J.W. Bridges and L.F. Chasseaud, John Wiley & Sons, New York, **6**, 111–96.

Hibberd, A.R. and Gorrod, J.W. (1985) Comparative N-oxidation of nicotine and cotinine by hepatic microsomes, in *Biological Oxidation of Nitrogen in Organic Molecules*, Ellis Horwood, Chichester, Eds J.W. Gorrod and L.A. Damani, pp. 246–50.

Hucker, H.B., Gillette, J.R. and Brodie, B.B. (1960) Enzymatic pathway for the formation of cotinine, a major metabolite of nicotine in rabbit liver. *J. Pharm. Exptl. Ther.*, **129**, 94–100.

Jacob, P. and Benowitz, N.L. (1991) Oxidative metabolism of nicotine *in vivo*, in *Effects of Nicotine on Biological Systems* (Eds F. Adlkofer and K. Thurau), Advances in Pharmacological Sciences, Birkhäuser Verlag, Basel, pp. 35–44.

Jacob, P.III., Benowitz, N.L. and Shulgin, A.T. (1988) Recent studies of nicotine metabolism in humans. *Pharmacol. Biochem. Behav.*, **30**, 249–53.

Jacob, P.III., Wilson, M. and Benowitz, N.L. (1981) Improved gas chromatographic method for the determination of nicotine and cotinine in biologic fluids. *J. Chromatogr.*, **222**, 61–70.

Klein, A.E. and Gorrod, J.W. (1978) Age as a factor in the metabolism of nicotine. *Europ. J. Drug Metabol. Pharmacokin.*, **1**, 51–8.

Kovacic, P. (1984) Does charge transfer by diiminium play a widespread role in living systems?. *Kem. Ind.*, **33**, 473–92.

Kyerematen, G.A., Morgan, M.L., Chattopadhyay, B. *et al.* (1990) Disposition of nicotine and eight metabolites in smokers and nonsmokers: Identification in smokers of two metabolites that are longer lived than cotinine. *Clin. Pharmacol. Ther.*, **48**, 641–51.

Kyerematen, G.A., Taylor, L.H., deBethizy, J.D. *et al.* (1987) Radiometric-high-performance liquid chromatographic assay for nicotine and twelve of its metabolites. *J. Chromatogr.*, **419**, 191–203.

Kyerematen, G.A., Taylor, L.H., deBethizy, J.D. *et al.* (1988) Pharmacokinetics of nicotine and 12 metabolites in the rat, application of a new radiometric high performance liquid chromatography assay. *Drug Metab. Dispos.*, **16**, 125–9.

Kyerematen, G.A. and Vesell, E.S. (1991) Metabolism of nicotine. *Drug Metabolism Reviews*, **23**, 3–41.

Langone, J.J., Gjika, H.B. and Van Vunakis, H. (1973) Nicotine and its metabolites. Radioimmunoassays for nicotine and cotinine. *Biochemistry*, **12**, 5025–30.

McCoy, G.D., Howard, P.C. and DeMarco, G.J. (1986) Characterization of hamster liver nicotine metabolism, I. Relative rates of microsomal C and N oxidation. *Biochemical Pharmacology*, **35**, 2767–73.

McKennis, H. (1965) Disposition and fate of nicotine in animals, in *Tobacco Alkaloids and Related Compounds*, 4th Wenner–Gren Centre International Symposium, **4**, Ed. U.S. von Euler, Pergamon Press, Oxford, pp. 53–74.

McKennis, H., Bowman, E.R. and Turnbull, L.B. (1961) Mammalian degradation of (−)-nicotine to 3-pyridineacetic acid and other compounds. *Proc. Soc. Exp. Biol. Med.*, **107**, 145–8.

McKennis, H., Schwartz, S.L. and Bowman, E.R. (1964) Alternate routes in the metabolic degradation of the pyrrolidine ring of nicotine. *J. Biol. Chem.*, **239**, 3990–6.

McKennis, H., Schwartz, S.L., Turnbull, L.B. *et al.* (1964) The metabolic formation of γ-(3-pyridyl)-γ-hydroxybutyric acid and its possible intermediary role in the mammalian metabolism of nicotine. *J. Biol. Chem.*, **239**, 3981–9.

McKennis, H., Turnbull, L.B. and Bowman E.R. (1957) γ-(3-Pyridyl)-γ-methylaminobutyric acid as a urinary metabolite of nicotine. *J. Amer. Chem. Soc.*, **79**, 6342–3.

McKennis, H., Turnbull, L.B. and Bowman E.R. (1958) Metabolism of nicotine to (+)-γ-(3-pyridyl)-γ-methylamino-butyric acid. *J. Amer. Chem. Soc.*, **80**, 6597–600.

McKennis, H., Turnbull, L.B. and Bowman, E.R. (1963b) N-Methylation of nicotine and cotinine *in vivo*. *J. Biol. Chem.*, **238**, 719–23.

McKennis, H., Turnbull, L.B., Bowman, E.R. *et al.* (1963a) The synthesis of hydroxycotinine and studies on its structure. *J. Org. Chem.*, **28**, 383–7.

McKennis, H., Turnbull, L.B., Bowman, E.R., *et al.* (1959) Demethylation of cotinine *in vivo*. *J. Amer. Chem. Soc.*, **81**, 3951–4.

McKennis, H., Turnbull, L.B., Schwartz, S.L. *et al.* (1962) Demethylation in the metabolism of (−)-nicotine. *J. Biol. Chem.*, **237**, 541–5.

McKennis, H, Bowman, E.R., Yi, J.M. *et al.* (1978) Participation of pyridino-N-oxides in the metabolism of nicotine *in vivo* – A preliminary study, in *Biological Oxidation of Nitrogen* (Ed. J.W. Gorrod), Elsevier/North-Holland, Amsterdam, pp. 163–9.

Meacham, R., Bowman, E.R. and McKennis, H. (1972) Additional routes in the metabolism of nicotine to 3-pyridylacetate. *J. Biol. Chem.*, **247**, 902–8.

Morselli, P.L., Ong, H.H., Bowman, E.R. *et al.* (1967) Metabolism of (±)cotinine-$2^{14}C$ in the rat. *J. Med. Chem.*, **10**, 1033–6.

Murphy, P.J. (1973) Enzymatic oxidation of nicotine to nicotine-$\Delta^{1',(5')}$ iminium ion. *J. Biol. Chem.*, **248**, 2796–800.

Neurath, G.B., Dünger, M., Krenz, O. *et al.* (1988) *trans*-3′-Hydroxycotinine – A main metabolite in smokers. *Klin. Wochenschrift*, **66**, 2–4.

Neurath, G.B., Dünger, M., Orth, D. *et al.* (1987) *trans*-3′-Hydroxycotinine as a main metabolite of nicotine in urine of smokers. *Internat. Arch. Occup. Environ. Health*, **59**, 199–201.

Neurath, G.B., Orth, D. and Pein, F.G. (1991) Detection of nornicotine in human urine after infusion of nicotine, in *Effects of Nicotine on Biological Systems* (Eds F. Adlkofer and K. Thurau), Advances in Pharmacological Sciences, Birkhäuser Verlag, Basel, pp. 45–9.

Neurath, G.B. and Pein, F.G. (1987) Gas chromatographic determination of *trans*-3′-hydroxycotinine, a major metabolite in smokers. *J. Chromatogr.*, **415**, 400–6.

Neurath, G.B., Dünger, M. and Orth, D. (1992) Detection and determination of tautomers of 5′-hydroxynicotine and 2′-hydroxynicotine in smokers' urine. *Med. Sci. Res.*, **20**, 853–8.

Nguyen, T.-L., Gruenke, L. and Castagnoli, N. (1979) Metabolic oxidation of nicotine to chemically reactive intermediates. *J. Med. Chem.*, **22**, 259–63.

Nguyen, T.-L., Dagne, E., Gruenke, L. *et al.* (1981) The tautomeric structure of 5-hydroxycotinine, a secondary mammalian metabolite of nicotine. *J. Org. Chem.*, **46**, 758–60.

Nwosu, C.G. and Crooks, P.A. (1988) Species variation and stereoselectivity in the metabolism of nicotine enantiomers. *Xenobiotica*, **18**, 1361–72.

O'Doherty, S., Revans, A., Smith, C.L. *et al.* (1988) Determination of *cis*- and *trans*-3-hydroxycotinine by high-performance liquid chromatography. *J. High Resol. Chromatogr. and Chromatogr. Commun.*, **11**, 723–5.

Papadopoulos, N.M. and Kintzios, J.A. (1963) Formation of metabolites from nicotine by a rabbit liver preparation. *J. Pharmacol. Exp. Ther.*, **140**, 269–77.

Peterson, L.A., Trevor, A. and Castagnoli, N. (1987) Stereochemical studies on the cytochrome P450 catalysed oxidation of (*S*)-nicotine and (*S*)-nicotine Δ1′,(5′)-iminium species. *J. Med. Chem.*, **30**, 249–54.

Pilotti, A., Enzell, C.R., McKennis, H. *et al.* (1976) Studies on the identification of tobacco alkaloids, their mammalian metabolites and related compounds by gas chromatography–mass spectrometry. *Beitr. Tabakforsch.*, **8**, 339–49.

Richie, J.P., Leutzinger, Y., Axelrad, C.M. *et al.* (1991) Contribution of 3′-hydroxycotinine and glucuronide conjugates to the measurement of cotinine by RIA, in *Effects of Nicotine on Biological Systems* (Eds F. Adlkofer and K. Thurau), Advances in Pharmacological Sciences, Birkhäuser Verlag, Basel, pp. 77–81.

Schepers, G., Demetriou, D., Rustemeier, K. (1992) Nicotine phase 2 metabolites – Structure of metabolically formed *trans*-3′-hydroxycotinine glucuronide. *Med. Sci. Res.*, **20**, 863–5.

Scherer, G., Jarczyk, L., Heller, W.D. et al. (1988) Pharmacokinetics of nicotine, cotinine and 3′-hydroxycotinine in cigarette smokers. *Klin. Wochenschrift*, **66**, 5–11.

Schwartz, S.L. and McKennis, H. (1963) Studies on the degradation of the pyrrolidine ring of (−)-nicotine *in vivo*. J. Biol. Chem., **238**, 1807–12.

Schwartz, S.L. and McKennis, H. (1964) Mammalian degradation of (−)-demethylcotinine. *Nature*, **202**, 594–5.

Seaton, M., Kyerematen, G.A., Jeszenka, E.V. et al. (1991) Nicotine metabolism in stumptailed macaques, *Macaca arctoides*. *Drug Metab. Dispos.*, **19**, 946–54.

Shulgin, A.T., Jacob, P. III, Benowitz, N.L. et al. (1987) Identification and quantitative analysis of cotinine-N-oxide in human urine. *J. Chromatogr.*, **423**, 365–72.

Turner, D.M. (1969) The metabolism of ^{14}C-nicotine in the cat. *Biochem. J.*, **115**, 889–96.

Voncken, P., Rustemeier, K. and Schepers, G. (1990) Identification of *cis*-3′-hydroxycotinine as a urinary nicotine metabolite. *Xenobiotica*, **20**, 1353–56.

Wada, E., Bowman, E.R., Turnbull, L.B. et al. (1961) Norcotinine (demethylcotinine) as a urinary metabolite of nornicotine. *J. Med. Pharm. Med.*, **4**, 21–30.

Wada, E. and Yamasaki, K.J. (1954) Degradation of nicotine by soil bacteria. *J. Am. Chem. Soc.*, **76**, 155–7.

N-Oxidation, N-methylation and N-conjugation reactions of nicotine

5

P.A. Crooks

5.1 INTRODUCTION

N-Oxidation, N-methylation and N-conjugation reactions of nicotine are metabolic transformations that result in the formation of the corresponding quaternary ammonium product, which is usually more polar in nature and more water soluble than the parent base. In these reactions, the generation of a positively charged tetrahedral nitrogen centre represents a higher oxidation state, hence such metabolic transformations can be considered collectively as N-oxidation reactions (Damani, 1985). In the case of nicotine, two tertiary amino centres are present in the molecule, thus allowing two possible sites of oxidation. The relative basicities of these nitrogen centres are somewhat different, i.e., pKa (pyridino N) = 3.04, pKa (pyrrolidino N) = 7.84. The pyridino moiety contains an aromatic azaheterocyclic nitrogen which is considerably less basic than the pyrrolidino nitrogen, due to the effect of sp^2 hybridization on base strength, since as the 's' character of an orbital increases, the electrons in that orbital are bound more tightly to the nucleus. In this respect, the regiospecificity of N-oxidation is an important consideration, especially as it relates to the structural and chemical features of the nicotine molecule, and the active site chemistry of the metabolizing enzymes involved. In addition, the presence of a chiral centre at C-2′ of the nicotine molecule introduces the interesting possibility of diastereo-

Nicotine and Related Alkaloids: Absorption, distribution, metabolism and excretion. Edited by J.W. Gorrod and J. Wahren. Published in 1993 by Chapman & Hall, London. ISBN 0 412 55740 1

meric metabolic products resulting from N-1′-oxidation reactions. This chapter will review current knowledge of the different types of metabolic N-oxidation reactions that nicotine undergoes, and will also address what is presently known about the enzymic basis for these biotransformation reactions, especially as it relates to the structure and chemical properties of the nicotine molecule.

5.2 N-OXIDATION REACTIONS OF NICOTINE

The N-oxidation pathway constitutes an important route of nicotine biotransformation leading to the formation of N-oxide derivatives of nicotine and its metabolites. Five possible N-oxide structures of each nicotine enantiomer can exist (Figure 5.1). N-1′-Oxidation of the pyrrolidine N-atom of either *S*-(−)- or *R*-(+)-nicotine affords, in each case, two diastereomeric products, due to the prochiral nature of the tertiary amino group (Figure 5.2). All of the five N-oxides of *S*-(−)-nicotine have been synthesized and characterized by nuclear magnetic resonance spectroscopy (Phillipson and Handa, 1975).

The N-1′-oxo compounds have been identified as *in vivo* metabolites of both nicotine enantiomers in rabbit (Papadopoulos, 1964), cat (Turner, 1969), guinea pig (Booth and Boyland, 1970; Cundy *et al.* 1984), mouse (Thompson *et al.*, 1985), rat (Kyerematen *et al.*, 1987), stumptailed macaques (Seaton *et al.*, 1991) and humans (Beckett *et al.*, 1971a; Neurath and Pein, 1987). N-1′-Oxonicotine has also been identified in the urine of smokers (Kyerematen and Vesell, 1991). Excretion of 2′*S*,N-1′-oxonicotine in the urine of smokers over 24 h has previously been shown to be about half that of cotinine (Beckett *et al.*, 1971a). However, recent studies (Byrd *et al.*, 1992a) indicate that urinary N-1′-oxide levels in both smokers and nonsmokers are quite low.

N-1′-Oxidation of *S*-(−)-nicotine was initially investigated using incubates of 9000 × *g* supernatant preparations from rabbit liver in the presence of NADPH and O_2 (Papadopoulos and Kintzios, 1963; Papadopoulos, 1964). The N-1′-oxide metabolite was isolated and structurally characterized as a mixture of two optically active stereoisomers. Booth and Boyland (1970, 1971) demonstrated that *S*-(−)-nicotine was converted into two diastereomeric N-1′-oxides by a 2650 × *g* supernatant of guinea pig liver, and that the ratio of the two isomers produced by the reaction varied with tissue and species. *In vitro* formation of N-1′-oxo metabolites of nicotine have subsequently been demonstrated in hamster (McCoy *et al.*, 1986), rabbit, mouse and rat (Jenner *et al.*, 1973; Jenner and Gorrod, 1973b).

As is the case with most xenobiotics and drugs that possess an aliphatic tertiary amino group, nicotine-N-1′-oxide formation occurs via a direct two-electron oxidation of the pyrrolidine N-atom, mediated via

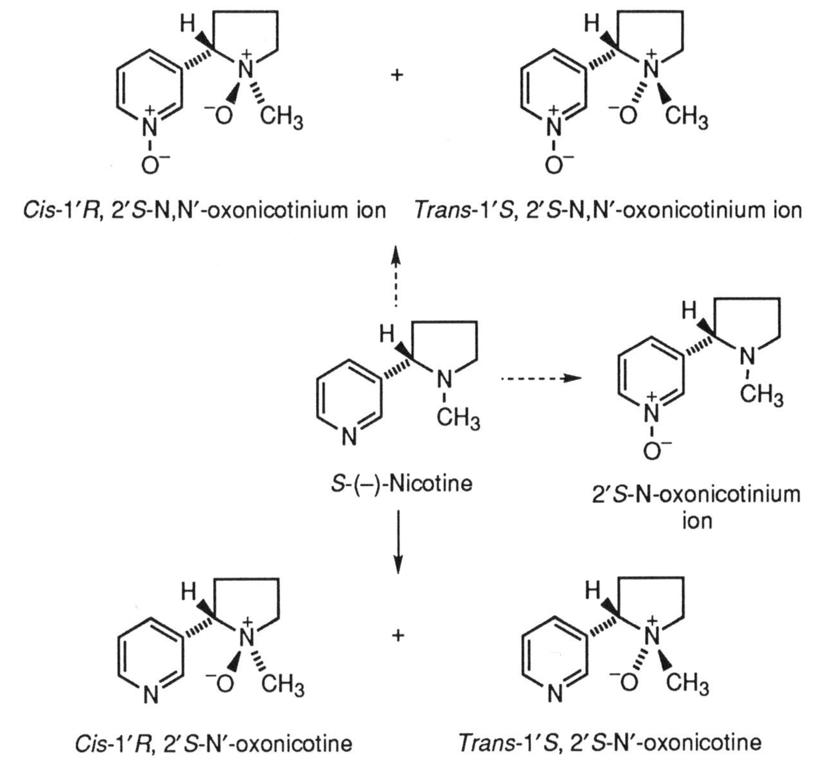

Figure 5.1 Possible N-oxide metabolites of nicotine. Solid arrows indicate established pathways, broken arrows indicate postulated pathways.

an ionic, rather than a radical mechanism. Oxygen transfer involves the liver enzyme system flavin-containing monooxygenase (FMO), mediated via enzyme-bound hydroperoxyflavin (Figure 5.2) (Ziegler, 1985). The involvement of cytochrome P450 in the N-1'-oxidation of nicotine would appear to be precluded on mechanistic grounds (Ziegler, 1985), and this has been substantiated experimentally (Hibberd and Gorrod, 1985; Damani *et al.*, 1988). However, recent investigations (Williams *et al.*, 1990) have reported that the formation of nicotine-N-1'-oxide in rabbit lung vesicles is mediated in part by rabbit lung P450 II, an isozyme of cytochrome P450.

A marked stereoselectivity has been observed in the formation of N-1'-oxide metabolites from nicotine enantiomers. Hepatic preparations from rats, mice, hamsters and guinea pigs generally produce more of the N-1'*R*,2'*S*-*cis* diastereomer than the N-1'*S*,2'*S*-*trans* diastereomer after incubation with *S*-(−)-nicotine; however, incubations with the *R*-(+)-enantiomer of nicotine afforded more N-1'*R*,2'*R*-*trans* product than the

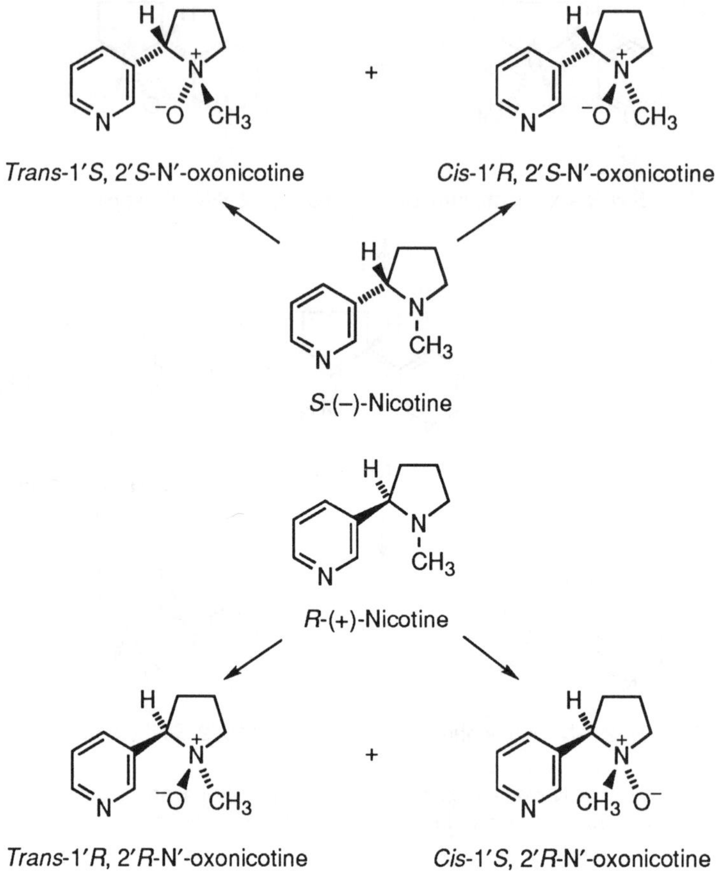

Figure 5.2 The diastereomers of N-1′-oxonicotine arising from metabolic N-oxidation of the pyrrolidino nitrogen atom of *S*-(−)-nicotine and *R*-(+)-nicotine.

Cis-1′*S*, 2′*R*-N-Methyl-N′-oxonicotinium diacetate

Trans-1′*R*, 2′*R*-N-Methyl-N′-oxonicotinium diacetate

Figure 5.3 N-1-Methyl-N-1′-oxonicotinium diacetate diastereomers.

N-1′*S*,2′*R*-*cis*-N-oxide (Figure 5.2) (Jenner *et al.*, 1973; Jenner and Gorrod, 1973). Thus, product formation from either enantiomer leads predominantly to the metabolite with the N-1′*R* configuration. This has led to the postulation that the FMO active site is able to distinguish between the two faces of the pyrrolidine ring, leading to preferential binding of one of them, followed by addition of oxygen in a defined direction (Testa and Jenner, 1976). Recently, Damani *et al.* (1988) have shown that guinea pig liver microsomes incubated with *S*-(−)-nicotine resulted in a *trans:cis* N-1′-oxide ratio of 0.27, while the use of *R*-(+)-nicotine as substrate afforded a *trans:cis* ratio of 2.32. These data parallel those of Beckett *et al.* (1971a) who used fortified (i.e. containing added co-factors) 10,000 × *g* hepatic preparations from male guinea pigs, affording *trans:cis* ratios of 0.4 and 2.4 with *S*-(−)- and *R*-(+)-nicotine, respectively, as substrates.

Several studies indicate that N-1′-oxidation of nicotine is mediated via flavin-containing monooxygenase, and not cytochrome P450. Cysteamine and methimazole, substrates for the former enzyme, strongly inhibit the microsomal conversion of nicotine to nicotine-N-1′-oxide, and antibodies against fP_2, SKF 525A, and metyrapone had little or no effect on nicotine-N-1′-oxide formation (McCoy *et al.*, 1986; Gorrod *et al.*, 1971; Nakayama *et al.*, 1987; Jenner *et al.*, 1973a).

N-1′-Oxidation of nicotine isomers by purified pig liver FAD-containing monooxygenase has recently been reported (Damani *et al.*, 1988). These studies demonstrated a clear stereoselectivity in the formation of the diastereomeric N-1′-oxides. The K_m values for *S*-(−)- and *R*-(+)-nicotine were found to be 181 and 70 μM, respectively, whereas the V_{max} value for both enantiomers was 22 nmol/min nmol of FAD (Table 5.1). The data in Table 5.2, in contrast to the results with microsomal and hepatic preparations, show that *S*-(−)-nicotine does not exhibit stereoselectivity in the formation of *cis*-N-1′*R*,2′*S* and *trans*-N-1′*S*,2′*S* N-1′-oxide diastereomers with the purified porcine enzyme. In addition,

Table 5.1 Kinetic constants for N-1′-oxygenation of nicotine enantiomers catalysed by purified porcine liver flavin-containing monooxygenase

Substrate	K_m (μM)[a,b]	V_{max} (nmol/min.nmol FAD)[b]	
		N-octylamine	
		Minus	Plus
S-(−)-nicotine	181	22	37
R-(+)-nicotine	70	22	41

[a]Values were calculated from double reciprocal plots of initial rates vs. substrate concentration at pH 7.5 and 37°C.
[b]N = 3, values did not vary by more than 2%.

Table 5.2 *In vitro* N-1′-oxidation of *R*-(+)- and *S*-(−)-nicotine and *R*-(+)-N-1-methylnicotinium ion by various oxygenase preparations

	Substrates						
	S-(−)-nicotine products[a,b]			*R*-(+)-nicotine products[a,b]			*R*-(+)-NMN[c] products
Enzyme System	Trans-NNO[d]	*Cis*-NNO	*Trans/Cis*	*Trans*-NNO	*Cis*-NNO	*Trans/Cis*	NMNO[e] (*Trans/Cis*)
Guinea pig liver microsomal preparation	7.5	27.7	0.27	22.8	9.8	2.32	n.d.[f]
Porcine liver flavin monooxygenase	9.6	9.96	1.0	39.6	–	–	n.d.
Sheep seminal cylooxygenase	n.d.	n.d.	n.d.	n.d.	n.d.	–	n.d.

[a]nmoles product (total) recovered.
[b]values are the mean of duplicate assays that did not vary by more than ± 10%.
[c]NMN = N-1-methylnicotinium ion.
[d]NNO = nicotine-N-1′-oxide.
[e]NMNO = N-1-methyl-N′-oxonicotinium ion.
[f]n.d. = not detected.

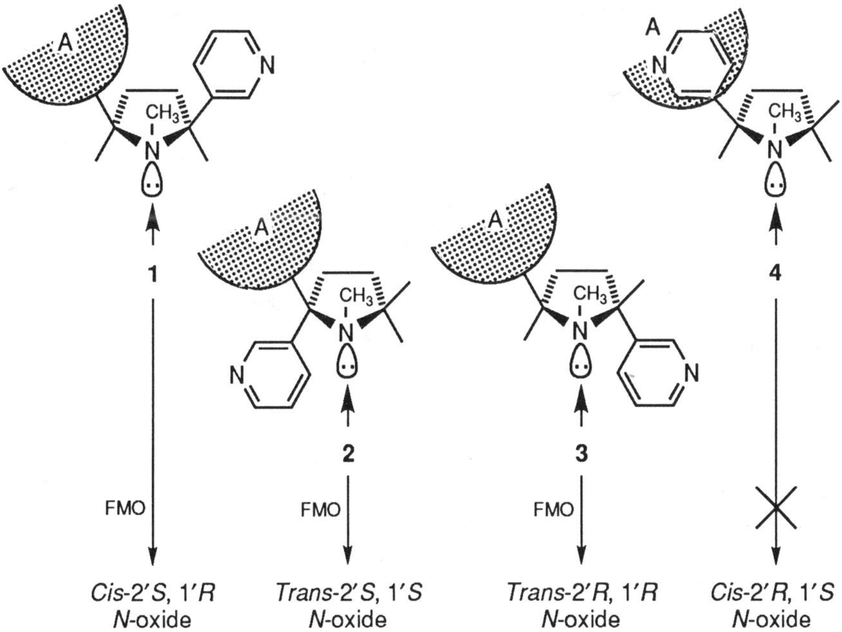

Figure 5.4 Stereospecific N-1′-oxidation of nicotine isomers by porcine liver flavin-containing monooxygenase, showing steric hindrance at the active site. **1** = *cis* conformer of *S*-(−)-nicotine, **2** = *trans* conformer of *S*-(−)-nicotine, **3** = *trans* conformer of *R*-(+)-nicotine, **4** = *cis* conformer of *R*-(+)-nicotine. (Reproduced by permission of the American Society for Pharmacology and Experimental Therapeutics.)

there is a clear stereospecificity when *R*-(+)-nicotine is utilized as substrate; with this enantiomer only the *trans*-N-1′*R*,2′*R* N-1′-oxide is formed. These results would appear to indicate, if one assumes that the N-1′-methyl group and pyrrolidine ring bind in a constant manner at the active site of porcine FMO, that attack by the reactive hydroperoxy flavin species occurs as indicated in Figure 5.4. Thus, the generation of specific stereochemistry at the N-1′ position is dependent upon the relative stereochemistry of the N-1′-methyl group and the adjacent 2′-(3-pyridyl) moiety. A *cis* orientation of the 3-pyridyl and N-1′-methyl groups will result in the formation of an N-1′-oxide product with *cis* stereochemistry, and the converse will be true for the *transoid* forms of the nicotine enantiomers. In the above model, exclusive formation of the *trans*-2′*R*,N-1′*R* N-oxide may indicate that the *transoid* conformer of *R*-(+)-nicotine is accommodated by the FMO active site, whereas the *cisoid* conformer of *R*-(+)-nicotine is not. Thus a region near the active site may sterically hinder the binding of *R*-(+)-nicotine when the N-1′-methyl and 3-pyridyl groups are in a *cis*-orientation. In the model illustrated in Figure 5.4, the binding of the other three structures is allowed because the 3-pyridyl group is positioned outside this region of steric hinderance.

The metabolism of nicotine enantiomers in isolated guinea pig hepatocytes has recently been investigated (Crooks *et al.*, 1992c). Table 5.3 shows the amount of *cis*- and *trans*-N-1′-oxides formed from either *R*-(+)- or *S*-(−)-nicotine in these studies. The data clearly indicate that the *trans* isomer is formed predominantly when *R*-(+)-nicotine is used as substrate, while the *cis* isomer is preferentially formed from *S*-(−)-nicotine. The *trans:cis* ratio within each series of diastereomers remains fairly constant throughout the time course examined. These results are consistent with the data obtained from the guinea pig microsomal preparations (Damani *et al.*, 1988). In addition, this study also shows that the formation of N-1′-oxide metabolites is somewhat faster with the *R*-(+)-enantiomer compared with the *S*-(−)-enantiomer over a 60 min incubation period (Table 5.4), even though the $t_{1/2}$ value for *R*-(+)-nicotine is lower than that for the *S*-(−)-enantiomer, in the hepatocyte culture. This appears to be due to the stereoselectivity exhibited by other metabolic pathways that compete with N-1′-oxidation.

Recently, two new N-1′-oxide metabolites have been isolated from the urine of guinea pigs which had been administered *R*-(+)-nicotine (Pool *et al.*, 1986). The structures of these metabolites (Figure 5.3) were determined by thermospray and fast atom bombardment mass spectrometry, and NMR spectroscopy, and were shown to be the diastereomers of N-1-methyl-N-1′-oxonicotinium ion. These two metabolites have also been isolated from the urine of guinea pigs which had been administered *R*-(+)-N-1-methylnicotinium salt. In this case, the diastereomeric ratio of the N-1′-oxides was 1.6 to 1.0 (*trans:cis*, respectively). N-1-Methyl-N-1′-oxonicotinium ion has not been detected in guinea pig urine following administration of *S*-(−)-nicotine, due to the stereospecificity exhibited in the N-1-methylation of nicotine by guinea pig, which involves exclusively the *R*-(+)-enantiomer (see later). These results suggest that N-1-methylation of *R*-(+)-nicotine in the guinea pig precedes N-1′-oxidation. However, it is interesting, but not unexpected, that *R*-(+)-N-1-methylnicotinium ion is not a substrate for either guinea pig liver microsomal preparation, porcine liver flavin monooxygenase or sheep seminal cycloxygenase (Damani *et al.*, 1988). It is likely that N-1-methylnicotinium ion exists as a di-cation under the conditions of the incubation, and it has previously been shown (Ziegler, 1985) that diamines in which both nitrogens are protonated at physiological pH are not substrates for FMO. Cytochrome P450, by virtue of its reaction mechanism, would not be expected to catalyse the formation of N-1-methyl-N-1′-oxo-nicotinium ion from N-1-methylnicotinium ion, because the adjacent α-carbons around the N-1′-atom are not quaternary in nature.

Table 5.3 Effect of time on the metabolisms of *S*-(−)- and *R*-(+)-nicotine enantiomers to their diastereomeric N-1′-oxides by guinea pig hepatocytes

Time (min)	Picomoles of product[a,b] formed from *R*-(+)-nicotine			Picomoles of product[a,b] formed from *S*-(−)-nicotine		
	Trans-1′*R*,2′*R*	*Cis*-1′*S*,2′*R*	*Trans*/*cis* ratio	*Trans*-1′*S*,2′*S*	*Cis*-1′*R*,2′*S*	*Trans*/*cis* ratio
5	1.44 (±0.21)	0.50 (±0.10)	2.88	0.51 (±0.05)	1.65 (±0.02)	0.31
15	2.02 (±0.80)	0.99 (±0.15)	2.04	0.60 (±0.11)	2.21 (±0.15)	0.27
30	2.51 (±0.10)	0.88 (±0.02)	2.85	0.68 (±0.03)	2.56 (±0.14)	0.27
45	2.85 (±0.63)	0.94 (±0.02)	3.03	0.44 (±0.06)	1.93 (±0.33)	0.23
60	2.70 (±0.10)	0.82 (±0.02)	3.30	0.55 (±0.05)	2.51 (±0.11)	0.22

[a]Data represent the mean of triplicate incubations ± S.E.M., and are the sum of cell lysate and supernatant values.
[b]Incubates contained 1×10^6 cells/ml, and 1 μCi [^{3}H]-nicotine enantiomer (sp. activity 76 Ci/mmol/ml).

Table 5.4 Effect of time on the formation of N-1′-oxides of *R*-(+)- and *S*-(−)-nicotine by guinea pig hepatocytes

	Picomoles N-1′-oxide[a,b]	
Time (min)	*R*-(+)-	*S*-(−)-
0	0	0
1	1.05 (±0.21)	0.19 (±0.12)
5	2.21 (±0.12)	2.09 (±0.17)
15	3.27 (±0.02)	2.65 (±0.10)
30	3.23 (±0.10)	2.79 (±0.03)
45	3.31 (±0.06)	2.81 (±0.06)
60	3.37 (±0.09)	2.82 (±0.10)

[a]Data represent the mean of triplicate incubations ± S.E.M., and are the sum of cell lysate and supernatant values.
[b]Incubates contained 1 × 10^6 cells/ml, and 1 μCi [^{3}H]-nicotine enantiomer (sp. activity 76 Ci/mmol/ml).

Other N-oxides postulated to be metabolically formed from nicotine have been reported (Schievelbein, 1982). These include N-1-oxonicotine, and N-1,N-1′-dioxonicotine (Figure 5.1). The latter biotransformation product accounts for approximately 5% of the total *in vivo* N-oxidation pathway. It is likely that N-oxidation at the pyridyl N atom in nicotine is a cytochrome P450 dependent reaction, based upon model studies with pyridino compounds (Gorrod and Damani, 1979a, 1979b).

Cotinine-N-1-oxide has been shown to be a urinary metabolite of *S*-(−)-nicotine in several animal species (Kyerematen *et al.*, 1988; Seaton *et al.*, 1991; Kyerematen *et al.*, 1987) and in smokers (Kyerematen and Vesell, 1991; Shulgin *et al.*, 1987), and its *in vivo* formation from cotinine has been reported (Cundy and Crooks, 1987). Evidence from studies with hamster and guinea pig hepatic microsomes indicates that N-1-oxidation of cotinine occurs via the cytochrome P450 system (Hibberd and Gorrod, 1985).

Nicotine-N-1′-oxides are known to undergo reduction to nicotine both *in vitro* and *in vivo* (Booth and Boyland, 1971; Dajani *et al.*, 1972) and this process is also under some form of stereochemical control. The N-1′*R*,2′*S*-*cis* diastereomer is reduced significantly more rapidly than the N-1′*S*,2′*S*-*trans* isomer *in vitro* by hepatic microsomal and by intestinal reductase enzymes (Sepkovic *et al.*, 1986; Jenner *et al.* 1973; Dajani *et al.*, 1972). Thus, it would appear that differences observed in the stereochemistry of nicotine N-1′-oxidation in *in vivo* experiments, might result at least in part from the involvement of stereoselective reduction processes that utilize the N-1′-oxide diastereomers as substrates. It has been demonstrated that oral administration of N-1′-oxonicotine to man results in the reductive formation of nicotine, which can be detected in the urine, with a decreased excretion rate than that seen after oral administration of nicotine (Beckett *et al.*, 1971a). However, when N-1′-oxonicotine is administered intravenously, the same N-1′-oxide is excreted unchanged. This leads to the conclusion that the reduction of N-1′-oxonicotine observed on oral administration results from biotransformation by gut microflora, and not by mammalian metabolizing enzymes. Nevertheless, it has been postulated that this back-conversion of N-1′-oxonicotine to nicotine in smokers may create a nicotine 'reservoir' that may serve to reinforce the tobacco habit (Sepkovic *et al.*, 1986).

There is evidence that the N-1′-oxides of nicotine may have the capacity to be directly nitrosated to form several potent tobacco-specific carcinogens, such as N-1′-nitrosonornicotine (NNN), 4-(methylnitrosamino)-1-(3-pyridyl)-1-butanone and 4-(methylnitrosamino)-4-(3-pyridyl)-butanol (Sepkovic *et al.*, 1986; Hecht and Hoffman, 1988). It has been postulated that the above nitrosamines could be formed from nitrite present in saliva or nitrogen oxides in inhaled mainstream tobacco smoke reacting with nicotine and other tobacco alkaloids. However, no evidence of the *in vivo* formation of nitrosated products from nicotine metabolism is available.

5.3 N-METHYLATION REACTIONS OF NICOTINE

The biological N-methylation of nicotine is a pathway of nicotine metabolism that, until recently, has received relatively little attention. Theoretically, the N-methylation of nicotine can result in the formation of three products, as a result of either N-1-methylation, N-1′-methylation or N-1,N-1′-dimethylation (Figure 5.5). McKennis *et al.* (1963) were the first to report that the *in vivo* N-1-methylation of nicotine and cotinine to their respective N-1-methyl quaternary ammonium metabolites, was a major route of biotransformation in the dog and in humans, respectively. More recent studies (Cundy *et al.*, 1984, 1985a, 1985b, 1985c) have shown that N-1-methylation of nicotine in the guinea pig is both a regiospecific and stereospecific metabolic pathway involving only the *R*-(+)-enantiomer. In these experiments, *R*-(+)-N-1-methylnicotinium salt metabolically formed *in vivo* from *R*-(+)-nicotine, was fully characterized and its absolute stereochemistry determined from chiral lanthanide shift NMR analysis. *S*-(−)-Nicotine was not converted to the corresponding N-1-methylated metabolite in this animal species and no N-1′-methylated urinary metabolites could be detected after exposure to either of the nicotine enantiomers.

This observed regiospecificity is interesting since, chemically, one would expect the more basic pyrrolidino nitrogen to be the site of methylation. However, there are examples of biological N-methylation reactions which also exhibit similar regiospecificity to nicotine N-methylation, e.g. the imidazole-N-methylation of histamine by histamine N-

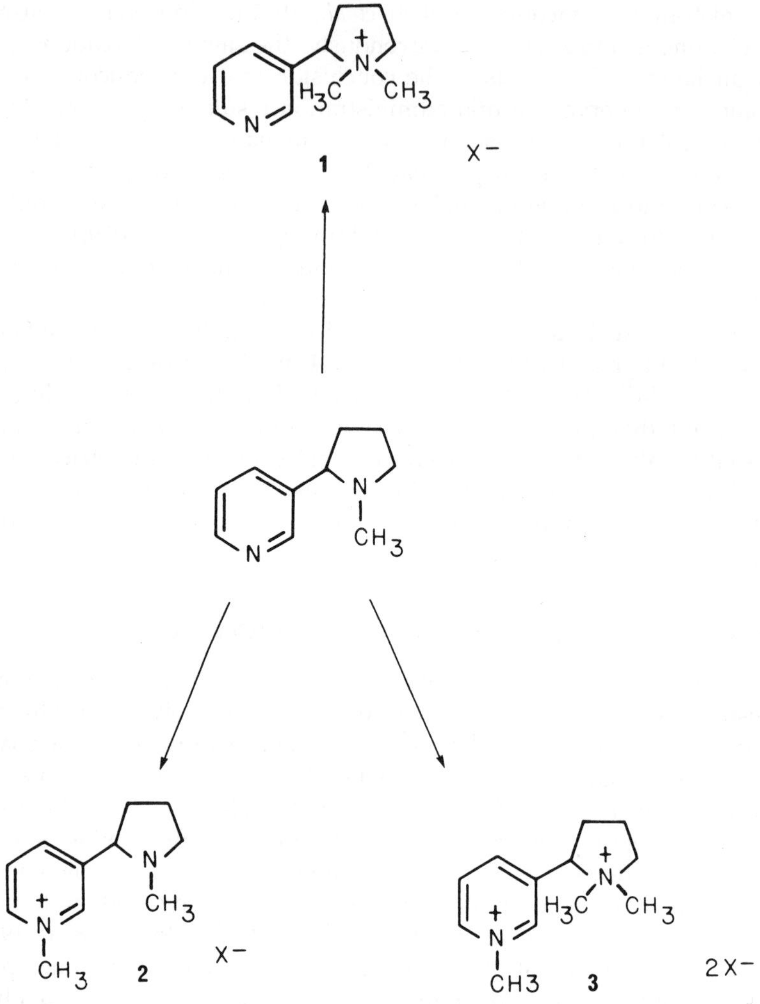

Figure 5.5 Potential N-methylated metabolites of nicotine. **1** N-1′-methylnicotinium ion; **2** N-1-methylnicotinium ion; **3** N-1,N-1′-dimethylnicotinium ion.

methyltransferase (HNMT). HNMT does not catalyse the transfer of a methyl group from *S*-adenosylmethionine (SAM) cofactor to the side chain amino group of histamine. N-Methylation of nicotine exclusively at the pyridyl nitrogen may be due to a number of factors, which include steric considerations, ionic factors (at physiological pH, nicotine exists 73% in the mono-cationic form, protonated on the pyrrolidine nitrogen), and the nature of the active site of the nicotine N-methyltransferase enzyme and its mode of interaction with nicotine.

Although only the *S*-(−)-enantiomer of nicotine is present in tobacco products, nevertheless, the above N-methylation pathway may be of relevance in the metabolism of nicotine in smokers, since it is known that *S*-(−)-nicotine undergoes significant racemization to the *R*-(+)-enantiomer during the smoking of a cigarette (Klus and Kuhn, 1977; Crooks *et al.*, 1992a). Thus smokers may be exposed to varying amounts of *R*-(+)-nicotine. This stereospecificity in the N-1-methylation of nicotine enantiomers appears to be species dependent (Nwosu and Crooks, 1988).

Stereospecificity in the formation of N-1-methylnicotinium ion from nicotine has also been demonstrated *in vitro* in the guinea pig (Cundy *et al.*, 1985b), and the enzyme responsible has been identified as a cytosolic S-adenosylmethionine dependent azaheterocycle N-methyltransferase (Cundy and Crooks, 1985). The enzyme has apparent Km values of 14.2 and 13.2 μM for *R*-(+)-nicotine and S-adenosylmethionine, respectively, and like most SAM dependent methyltransferases (Crooks, 1989), it exhibits product inhibition by the feed-back inhibitor S-adenosylhomocysteine (Ki = 32.2 μM). S-(−)-Nicotine is not a substrate for this methyltransferase, although surprisingly it acts as a competitive inhibitor (Ki = 62.5 μM) of the N-1-methylation of its optical antipode (Cundy *et al.*, 1985a). Azaheterocycle N-methyltransferase activity is widely distributed in guinea pig tissues, being found in liver, lung, spleen and brain (Cundy *et al.*, 1985b); specific activity is increased significantly in tissue homogenates that have undergone dialysis, presumably due to the loss of a heat-stable, small molecular weight endogenous inhibitor.

The N-methylation of nicotine *in vitro* in the rabbit has also been demonstrated (Damani *et al.*, 1986). Purification of the N-methyltransferase activity from rabbit cytosol has resulted in the isolation of two SAM-dependent amine N-methyltransferases (A and B), which exhibit broad and overlapping specificities for a large number of amines (Ansher and Jacoby, 1986; Crooks *et al.*, 1988a), including a significant number of aromatic azaheterocycles where N-methylation results in the formation of quaternary ammonium metabolites. Both A and B forms of the enzyme methylate *R*-(+)-nicotine at the pyridyl N atom, whereas S-(−)-nicotine does not undergo this biotransformation with either of the purified enzymes. Surprisingly, *R*-(+)-nicotine is methylated at both the pyrrolidine nitrogen and the pyridine nitrogen when an impure enzymic preparation containing both A and B methyltransferase activities is utilized to afford both the N-1-monomethylated and N-1′-monomethylated metabolites; under similar conditions S-(−)-nicotine was methylated only at the pyridyl N atom (Damani *et al.*, 1986).

The broad substrate specificity observed with rabbit N-methyltransferases suggests that these enzymes may play a significant role in the biotransformation of nitrogen-containing xenobiotics *in vivo*, especially if the methylated product is a polar, charged, water-soluble quaternary ammonium salt.

Incubation of human liver cytosol with either *R*-(+)-nicotine or S-(−)-nicotine results in the formation of the corresponding N-1-methylquaternary ammonium metabolite (Crooks and Godin, 1987). In this case, a substrate stereoselectivity was observed in that the turnover number for the methylation of the S-(−)-isomer was 0.25 pmole/mg protein/h, whereas that for the *R*-(+)-isomer was 2.11 pmole/mg protein/h. The latter substrate exhibited an apparent Km value of 20.1 μM. The apparent high activity of this enzyme in human liver and its ability to N-1-methylate both nicotine enantiomers is puzzling in view of the reported very low levels of N-1-methylated metabolites of nicotine present in the urine of smokers (Kyerematen *et al.*, 1990; Byrd *et al.*, 1992c).

Isolated guinea pig hepatocytes are not able to N-methylate either of the enantiomers of nicotine (Crooks *et al.*, 1992b) presumably because this cell type is devoid of nicotine N-methyltransferase. However, cultured guinea pig pulmonary alveolar macrophages, which are among the first lung cells to interact with the inhaled particles of cigarette smoke, and which have been shown previously to possess a number of methyltransferase activities (Zuckerman *et al.*, 1982; Pacheco *et al.*, 1985; Gairola *et al.*, 1987), avidly N-methylate *R*-(+)-nicotine (Gairola *et al.*, 1988) to afford N-1-methylnicotinium ion. Interestingly, *S*-(−)-nicotine is not N-1-methylated by guinea pig pulmonary alveolar macrophages, but this enantiomer does appear to inhibit the N-1-methylation of *R*-(+)-nicotine in this cell system.

The primary oxidative metabolites of nicotine all retain the basic pyridine N-atom in their structure, and as such can be considered as candidates for subsequent N-1-methylation reactions *in vivo*. N-1-methylcotinium ion has been detected as a urinary metabolite of *S*-(−)-nicotine in humans (McKennis *et al.*, 1963), and of *R*-(+)-nicotine (but not *S*-(−)-nicotine) in guinea pig (Cundy *et al.*, 1984). The isolation of N-1-methylnornicotinium ion from the urine of guinea pigs that have been administered *R*-(+)-nicotine provides a unique example of both N-demethylation and N-methylation occurring within the same molecule (Sato and Crooks, 1985). The isolation and characterization of N-1-methyl-N-1′-oxonicotinium ion, a urinary metabolite of *R*-(+)-nicotine in the guinea pig, has also recently been described (Pool *et al.*, 1986; Crooks *et al.*, 1988b). This metabolite was shown to be a mixture of the *cis*-N-1′*S*,2′*R*- and *trans*-N-1′*R*,2′*R*- diastereomers, formed in the ratio 1.6:1.0, respectively; *S*-(−)-nicotine under similar conditions did not afford this urinary metabolite. The formation of these metabolites is summarized in Figure 5.6.

It is interesting to note that N-1-methylated metabolites of three of

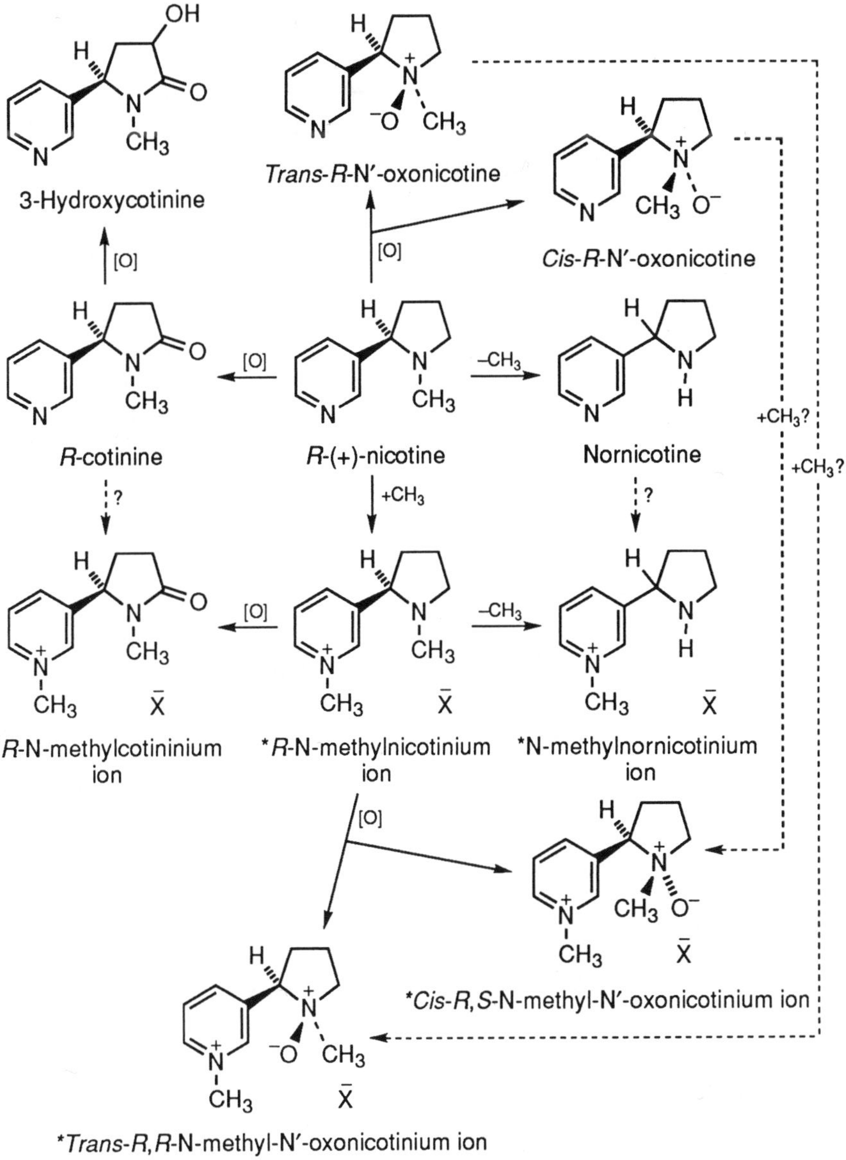

Figure 5.6 Metabolic pathways to N-1-methylated urinary metabolites of nicotine in the guinea pig. Asterisks indicate that optical antipodes of these compounds are not formed as urinary metabolites of *S*-(−)-nicotine. (Reproduced by permission of the American Society for Pharmacology and Experimental Therapeutics.)

the primary oxidative biotransformation products of *R*-(+)-nicotine have been identified in guinea pig urine after administration of *R*-(+)-nicotine. However, *in vitro* studies in the guinea pig have failed to demonstrate the ability of these primary oxidation products to act as substrates for guinea pig 'nicotine N-methyltransferase' (Crooks, 1992). Administration of *R*-(+)-N-methylnicotinium ion to guinea pigs affords the following urinary metabolites: N-1-methylcotinium ion (17.5%), N-1-methylnornicotinium ion (8.0%), N-1-methyl-N-1′-oxonicotinium ion (9.0%) and N-1-methylnicotinium ion (60.9% unmetabolized) (Pool and Crooks, 1985). These results indicate that subsequent metabolism of *R*-(+)-N-1-methylnicotinium ion, after *in vivo* formation from *R*-(+)-nicotine, involves its oxidative biotransformation to several quaternary ammonium metabolites. It is important to note in this respect, that guinea pig liver microsomal preparations were unable to oxidize *R*-(+)-N-1-methylnicotinium ion to the corresponding N-1′-oxide (Damani *et al.*, 1988), this compound is also inactive as a substrate for purified porcine flavin-containing monooxygenase and sheep seminal flavin-peroxidase (cyclooxygenase).

The use of double isotope studies to determine the *in vivo* stability of the N-methyl groups of *R*-(+)-N-1-methylnicotinium ion in the guinea pig has led to some intriguing results (Pool and Crooks, 1987). Analysis of urinary metabolites of *R*-(+)-[^{3}H-N-1′-CH_3; ^{14}C-N-1-CH_3]-N-1-methylnicotinium acetate clearly indicated that *in vivo* N-1-demethylation, to form nicotine, is an unlikely metabolic step. However, unexpected loss of the ^{3}H label in urinary *R*-(+)-N-1-methylnicotinium ion and N-1-methyl-N-1′-oxonicotinium ion suggested that an additional pathway to the direct N-1′-oxidation pathway may be operating in the formation of N-1-methyl-N-1′-oxonicotinium ion. This pathway may involve an initial N-1′-demethylation to N-1-methylnornicotinium ion followed by N-1′-methylation back to N-1-methylnicotinium ion, and then N-1′-oxidation to N-1-methyl-N-1′-oxonicotinium ion (Figure 5.7). An alternative path-way may involve the formation of an iminium species which is reductively converted back to N-1-methylnicotinium ion followed by N-1′-oxidation (Figure 5.8), however, there appears to be no precedent for the reductive step in this latter hypothetical pathway.

5.4 GLUCURONIDATION REACTIONS OF NICOTINE

Attempts to account for all the nicotine absorbed in *in vivo* studies, by summation of urinary nicotine and its metabolites have generally not been successful (Neurath et al., 1988; Zhang et al., 1990). The recently utilized technique of HPLC analysis with radiometric detection (Crooks and Cundy, 1988) has resulted in the identification of nicotine and eight of its metabolites excreted in the urine after an intravenous dose of

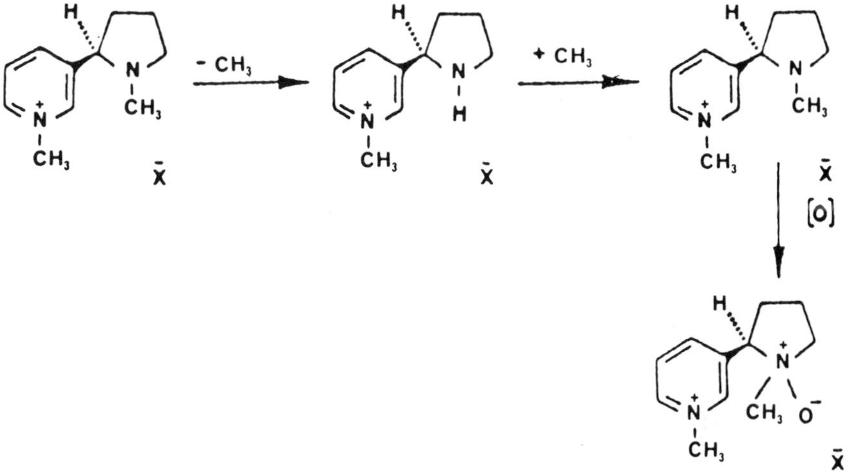

Figure 5.7 Demethylation pathway proposed for the formation of N-1-methyl-N-1′-oxonicotinium ion in the guinea pig.

racemic nicotine, which constitutes 80% of the administered dose and about 100% of the radiolabel in the urine (Kyerematen *et al.*, 1990). More importantly, this study indicated that one of the urinary metabolites might be a glucuronide conjugate of 3-hydroxycotinine. Recently, there have been several reports of quantitation of nicotine metabolites in smokers' urine using β-glucuronidase coupled with analysis of the liberated free metabolite (Curvall *et al.*, 1989; Haley *et al.*, 1990; Byrd *et al.*, 1990). Byrd *et al.* (1992a, 1992b) have shown that the glucuronide metabolites of nicotine include the quaternary ammonium conjugates nicotine glucuronide and cotinine glucuronide as well as 3-hydroxycotinine glucuronide (Figure 5.9), and together constitute almost 30% of total nicotine metabolites in smokers' urine (Figure 5.10). Thus, these new biotransformation products represent a major class of nicotine metabolite.

Recent studies by Caldwell *et al.* (1992) have established that a major metabolite of nicotine in smokers' urine is S-(−)-cotinine N-1-glucuronide. The identity of this quaternary ammonium conjugate was established by chemical synthesis and characterization of the metabolite, and application of a cation exchange HPLC method to enable the collection of a fraction containing S-(−)-cotinine glucuronide from smokers' urine. The electrospray mass spectrum of this fraction contained ions consistent with presence of the glucuronide metabolite, and the concentrated fraction, when subjected to enzymatic hydrolysis by β-glucuronidase, afforded S-(−)-cotinine. In addition, a thermospray ion-exchange liquid chromatographic mass spectral method was used for

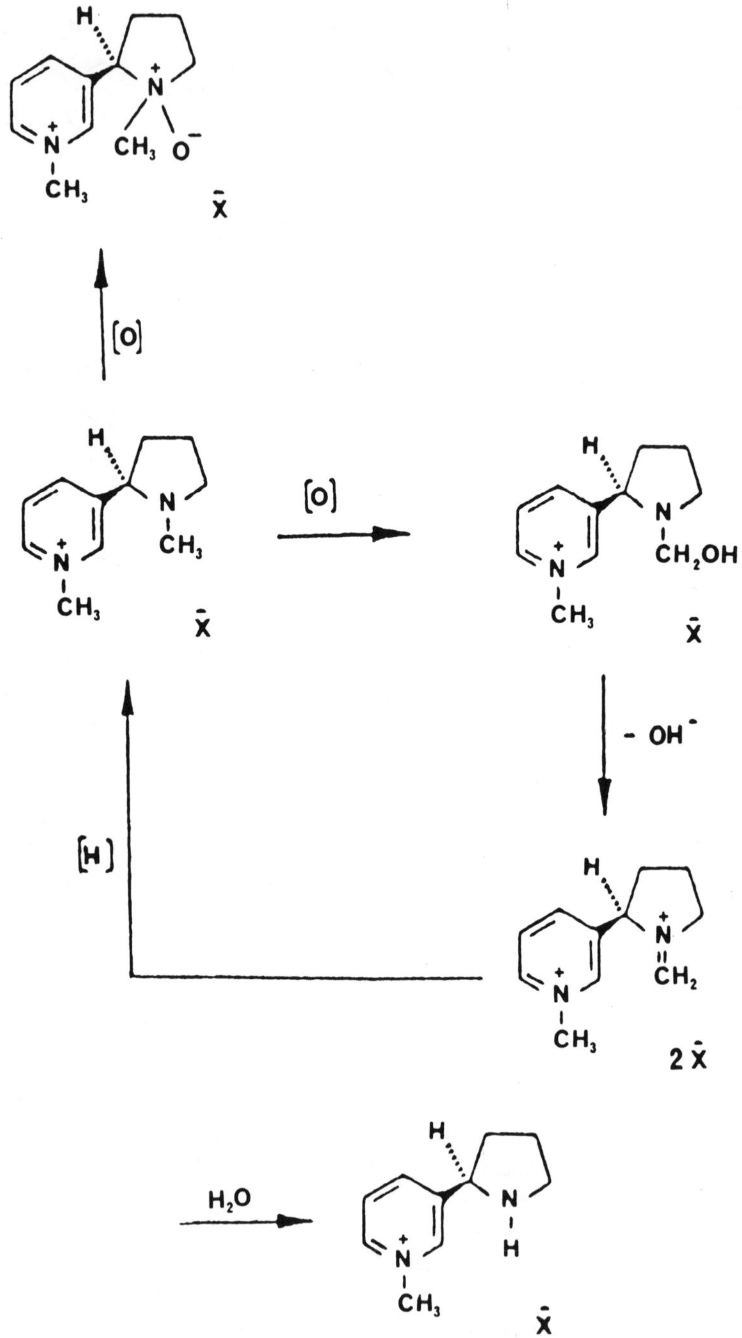

Figure 5.8 Reductive pathway proposed for the formation of N-1-methyl-N-1′-oxo-nicotinium ion in the guinea pig.

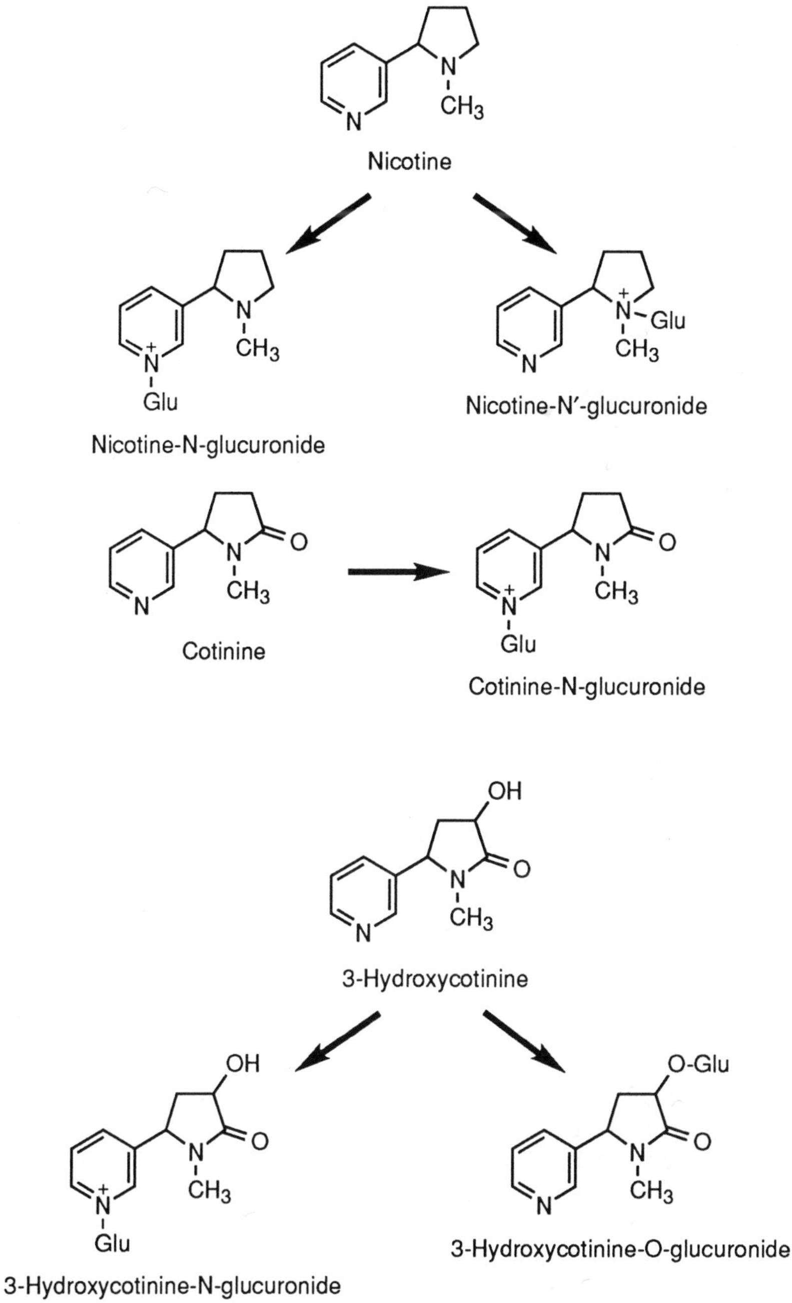

Figure 5.9 Potential glucuronide metabolites of nicotine, cotinine and 3-hydroxycotinine.

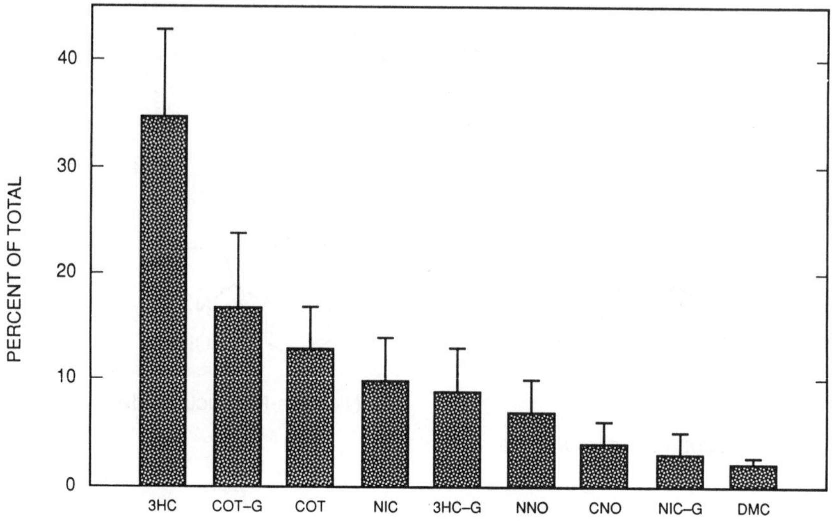

Figure 5.10 Urinary metabolite distribution observed in seven smokers. 3HC = 3-hydroxycotinine; COT-G = cotinine glucuronide; COT = cotinine; NIC = nicotine; 3HC-G = 3-hydroxycotinine glucuronide; NNO = nicotine-N-1′-oxide; CNO = cotinine-N-1-oxide; NIC-G = nicotine glucuronide; DMC = demethylcotinine.

the direct determination of S-(−)-cotinine glucuronide in smokers' urine (Figure 5.11). A further refinement of this method has utilized electrospray LC–MS and a d_3-labelled analogue of S-(−)-cotinine glucuronide as an internal standard (Byrd *et al.*, 1992d). Electrospray MS proved to be a gentler analytical technique than thermospray MS (cotinine N-1-glucuronide is thermally unstable under thermospray conditions, and decomposes almost completely to give the aglycon, cotinine), and afforded a significant molecular ion (Figure 5.12). The limit of detection using this method of analysis was 0.70 nmol/ml.

The synthesis and characterization of S-(−)-nicotine N-1-glucuronide, and the *cis*- and *trans*-isomers of 3-hydroxycotinine-N-1-glucuronide, as well as their d_3-labelled forms, have all been completed (Crooks *et al.*, 1992b). The availability of these metabolic standards will be invaluable in characterizing the remaining glucuronide conjugates of nicotine metabolism. At the moment, it is conceivable that '3-hydroxycotinine glucuronide' could be the *trans*- and/or *cis*-isomer of either an O- or an N-glucuronide or a mixture of both. Similarly, 'nicotine glucuronide' may be an N-1-glucuronide or an N-1′-glucuronide or a mixture of both (Figure 5.9). Recent studies indicate that the '3-hydroxycotinine glucuronide' identified in the urine of smokers, is neither a *cis* nor a *trans* N-1-glucuronide, based upon electrospray LC–MS analysis (Byrd *et al.*, 1992c), whereas 'nicotine glucuronide' in smokers' urine co-elutes with

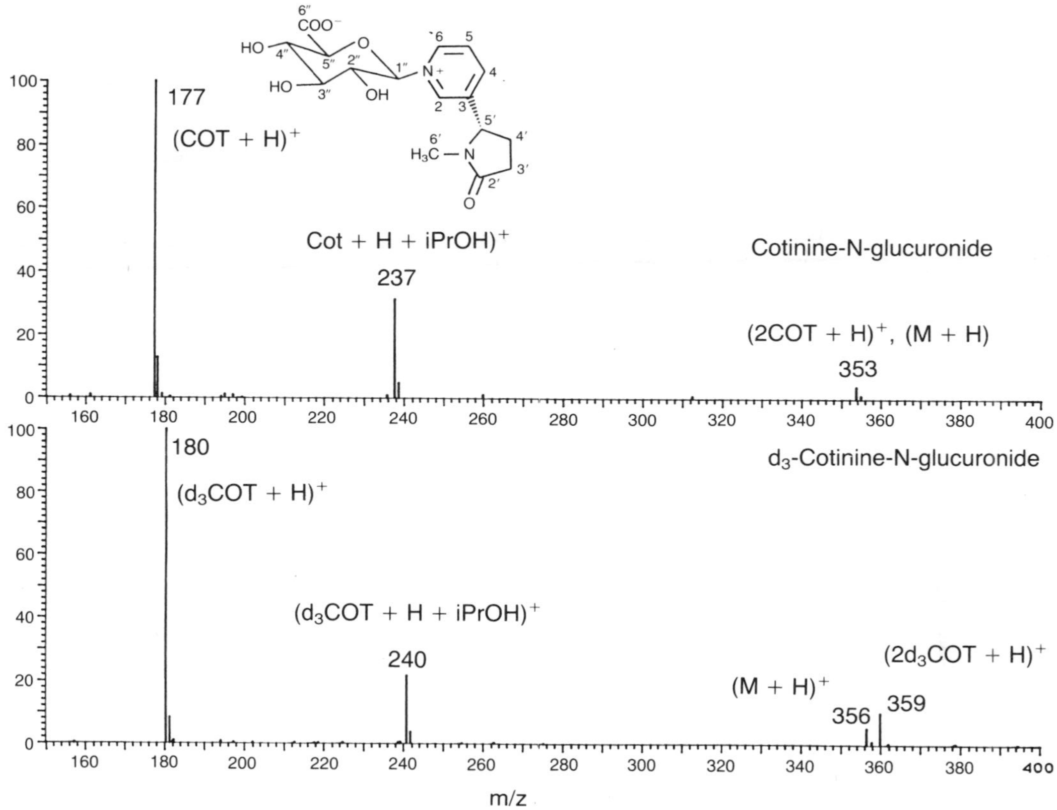

Figure 5.11 Thermospray mass spectrum of urinary cotinine glucuronide and [methyl-d_3]-cotinine glucuronide.

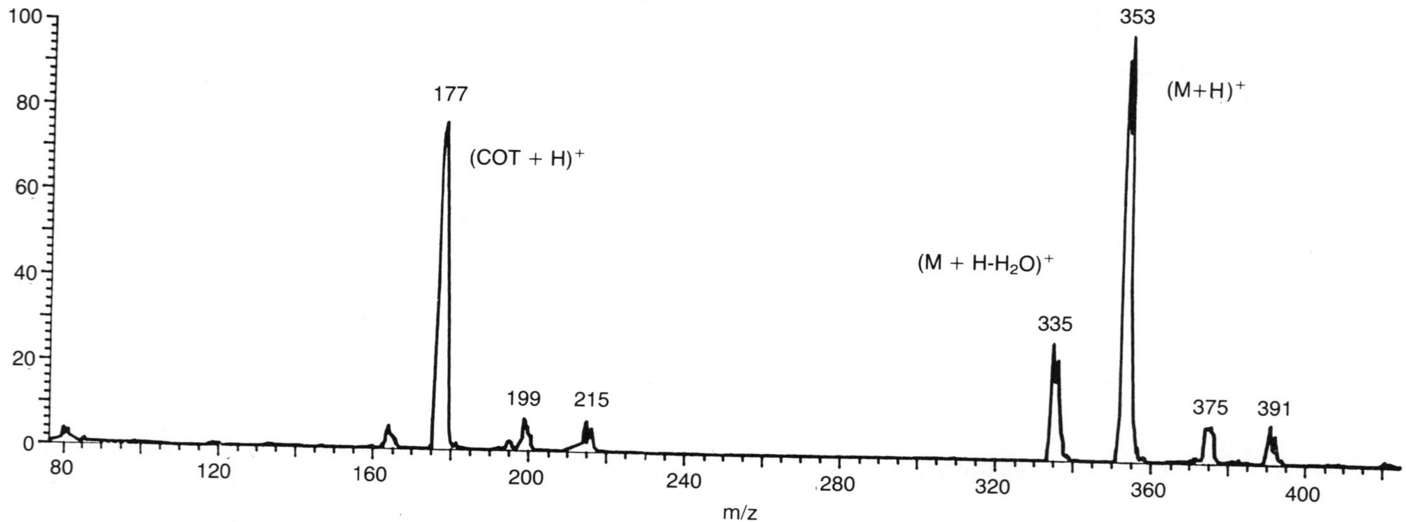

Figure 5.12 Electrospray mass spectrum of urinary cotinine glucuronide.

a synthetic standard of nicotine-N-1-glucuronide on cation-exchange high-performance liquid chromatography, demonstrating an interesting regiospecificity in the N-glucuronidation of nicotine. Thus, glucuronidation at the pyridino-N atom, rather than at the more basic pyrrolidino-N atom occurs, which follows the same pattern as nicotine N-methylation, and may result from stereochemical factors at the active site of the glucuronyl transferase enzyme, i.e. the less sterically hindered but least basic pyridino N-atom may be more accessible to the UDP-glucuronic acid co-factor.

Over the last two decades, nicotine and cotinine have been used to assess exposure to mainstream and environmental tobacco smoke (Neurath *et al.*, 1987; Davis, 1986; Jarvis, 1987), even though in smokers, urinary nicotine and cotinine account for only 10–40% of the nicotine absorbed (Neurath *et al.*, 1988). Since the amount of nicotine excreted is dependent upon the smoking status of the individual, i.e. smokers excrete twice the amount of metabolized nicotine than nonsmokers (Kyerematen *et al.*, 1990), accurate assessment of nicotine exposure requires the quantification of all urinary nicotine and nicotine metabolites. Thus, the elucidation of the final pieces in the nicotine metabolism story, viz. the glucuronide conjugates, is of significance in this regard. In fact, urinary '3-hydroxycotinine glucuronide' appears to be a relatively long-lived metabolite, the kinetic disposition of which is unaltered by cigarette smoking (Kyerematen and Vesell, 1991; Kyerematen *et al.*, 1990). This had led to the proposal that '3-hydroxycotinine glucuronide' may be an even more sensitive and reliable index than cotinine for determining possible exposure to tobacco smoke (Kyerematen and Vesell, 1991).

5.5 NICOTINAMIDE NUCLEOTIDE ANALOGUES OF NICOTINE AND COTININE

Nicotinamide nucleotide analogues of both nicotine and cotinine (i.e. nicotine-TPN and cotinine-TPN) have been isolated and purified from incubation mixtures containing nicotine or cotinine, TPN and a rabbit liver microsomal fraction (Shen and Vunakis, 1974) (Figure 5.13). Presumably these conjugates are formed via the action of DPNase, an enzyme that exists in most animal tissues, and is able to catalyse the exchange of the nicotinamide moiety of DPN and TPN for other pyridine containing compounds (Kaplan, 1960). The nicotine–TPN conjugate could also be formed in incubation mixtures containing pig brain or beef spleen DPNase instead of the liver microsomal preparation. However, nicotine–TPN cannot substitute for TPN in TPN-dependent reactions, but it does exhibit competitive inhibition against two such

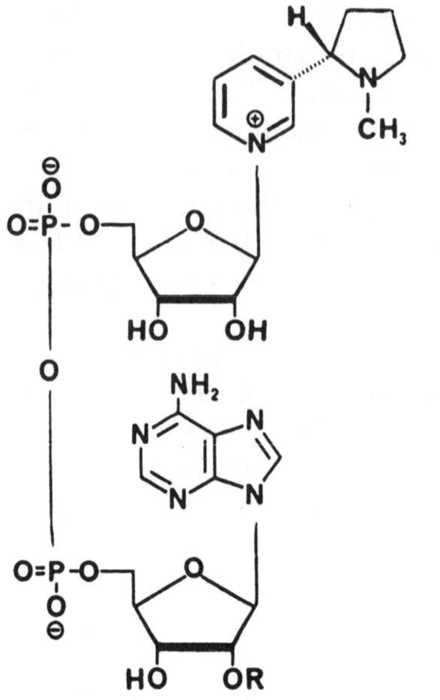

1

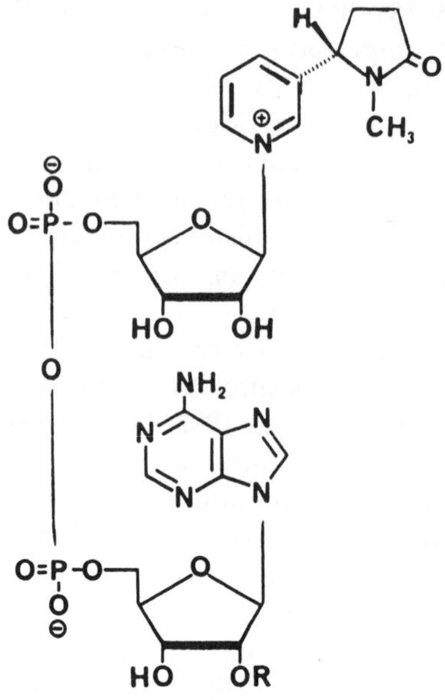

2

Figure 5.13 TPN and DPN analogues of nicotine and cotinine. **1** (R═H), nicotine-DPN; **1** (R═PO_3^{2-}), nicotine-TPN; **2** (R═H)cotinine-DPN; **2** (R═PO_3^{2-}), cotinine-TPN.

TPN-dependent enzymes, namely glucose-6-phosphate dehydrogenase and glutamate dehydrogenase. Both the TPN and DPN conjugates of nicotine and cotinine are extensively metabolized by liver microsomes to the corresponding mononucleotide (Shen *et al.*, 1977a). It has been suggested that nicotine and cotinine mononucleotide resulting from the enzymic hydrolysis of the corresponding TPN conjugates may accumulate inside the cell in some rabbit tissues, such as liver and lung, and that although these analogues may account for a minor proportion of compounds which result during the metabolism of nicotine, they are potentially among the most important (Shen *et al.*, 1977a). A radioimmunoassay has been developed for the detection of TPN and DPN conjugates of nicotine and cotinine (Shen *et al.*, 1977b). Using this analytical methodology, rabbits injected with cotinine were found to have pmole levels per gram of wet tissue of cotinine nucleotide analogues in extracts from liver, kidney and lung. Cotinine nucleotides were also found in tissues of rabbits injected with nicotine. There is, as yet, no data available on the presence of these interesting nucleotide conjugates in the tissue of humans that have been exposed to nicotine.

REFERENCES

Ansher, S.S. and Jakoby, W.B. (1986) Amine N-methyltransferase from rabbit liver. *J. Biol. Chem.*, **267**, 3996–4001.

Beckett, A.H., Gorrod, J.W. and Jenner, P. (1971a) The analysis of nicotine-1′-N-oxide in urine, in the presence of nicotine and cotinine, and its application to the study of *in vivo* nicotine metabolism in man. *J. Pharm. Pharmacol.*, **23**, 55S-61S.

Beckett, A.H., Gorrod, J.W. and Jenner, P. (1971b) The effect of smoking on nicotine metabolism *in vivo* in man. *J. Pharm. Pharmacol.*, **23**, 55–61.

Booth, J. and Boyland, E. (1970) The metabolism of nicotine into two optically active stereoisomers of nicotine-1′-oxide by animal tissue *in vitro* and by cigarette smokers. *Biochem. Pharmacol.*, **19**, 733–42.

Booth, J. and Boyland, E. (1971) Enzymic oxidation of (−)-nicotine by guinea pig tissues *in vitro. Biochem. Pharmacol.*, **20**, 407–15.

Byrd, G.D., Chang, K.M., Greene, J.M. *et al.* (1990) Determination of nicotine and its metabolites in urine by thermospray LC/MS. 44th Tobacco Chemists' Research Conference, Winston-Salem, NC, October 1990.

Byrd, G.D., Chang, K.M., Greene, J.M. *et al.* (1992a) Determination of nicotine and its metabolites in urine by thermospray LC/MS, in *The Biology of Nicotine* (Eds P. Lippiello, A.C. Collins, J.A. Gray and J.H. Robinson), Raven Press, New York, pp. 71–83.

Byrd, G.D., Chang, K.M., Greene, J.M. *et al.* (1992b) Evidence for urinary excretion of glucuronide conjugates of nicotine, cotinine and 3′-hydroxycotinine in smokers. *Drug Metab. Disposit.*, in press.

Byrd, G.D., Caldwell, W.S., deBethizy, J.D. *et al.* (1992c) unpublished results.

Byrd, G.D., Caldwell, W.S., Uhrig, M.S. *et al.* (1992d) Cotinine glucuronide, a major urinary metabolite in smokers. 40th ASMS Conference on Mass Spectrometry and Allied Topics, Washington DC, June 1992.

Caldwell, W.S., Greene, J.M., Byrd, G.D., *et al.* (1992) Characterization of the glucuronide conjugate of cotinine: a previously unidentified major metabolite of nicotine in smokers' urine. *Chem. Res. Toxicol.*, **5**, 280–5.

Crooks, P.A. (1989) Interaction of sulphur xenobiotics with *S*-adenosyl-L-methionine-dependent methyltransferase reactions, in *Sulphur Drugs and Related Organic Chemicals – Chemistry, Biochemistry and Toxicology*, Vol. 2, Part B, Chapter 6, (Ed. L.A. Damani), Ellis Horwood Ltd, Chichester, UK, pp. 47–85.

Crooks, P.A. (1992) unpublished observations.

Crooks, P.A. and Cundy, K.C. (1988) High performance liquid chromatography of radiolabelled nicotine enantiomers and their metabolism in the guinea pig to N-methylated products, in *Flow Through Radioactive Detectors in HPLC: Progress in HPLC Vol. 4* (Eds H. Parvez, M. Kessler and S. Parvez), VNU International Science Press, Utrecht, Holland, pp. 129–147.

Crooks, P.A. and Godin, C.S. (1987) N-Methylation of nicotine enantiomers by human liver cytosol. *J. Pharm. Pharmacol.*, **40**, 153–4.

Crooks, P.A., Godin, C.S., Damani, L.A. *et al.* (1988a) Formation of quaternary amines by N-methylation of azaheterocycles with homogeneous amine N-methyltransferases. *Biochem. Pharmacol.*, **37**, 1673–7.

Crooks, P.A., Pool, W.F. and Damani, L.A. (1988b) N-Methyl-N′-oxonicotinium ion: synthesis and stereochemistry. *Chemistry and Industry (London)*, 95–6.

Crooks, P.A., Godin, C.S. and Pool, W.F. (1992a) Enantiomeric purity of nicotine in tobacco smoke condensate, *Med. Sci. Res.* **20**, 879–80.

Crooks, P.A., Ravard, A., Bhatti, B.S. *et al.* (1992b) Synthesis of N-glucuronic acid conjugates of cotinine, *cis*-3-hydroxycotinine and *trans*-3-hydroxycotinine, **20**, 881–3.

Crooks, P.A., Sherratt, A.J. and Godin, C.S. (1992c) N′-Oxidative metabolism of nicotine enantiomers in isolated guinea pig hepatocytes. *Med. Sci. Res.*, **20**, 909–11.

Cundy, K.C. and Crooks, P.A. (1985) *In vitro* characteristics of guinea pig lung aromatic azaheterocycle N-methyltransferase. *Drug Metab. Disposit.* **13**, 658–63.

Cundy, K.C. and Crooks, P.A. (1987) Biotransformation of primary nicotine metabolites II. Metabolism of [^{3}H]-*S*-(−)-cotinine in the guinea pig: determination of *in vivo* urinary metabolites by high-performance liquid radiochromatography. *Xenobiotica*, **17**, 785–92.

Cundy, K.C., Crooks, P.A. and Godin, C.S. (1985a) Remarkable substrate-inhibitor properties of nicotine enantiomers towards a guinea pig lung aromatic azaheterocycle N-methyltransferase. *Biochem. Biophys. Res. Commun.*, **128**, 312–6.

Cundy, K.C., Godin, C.S. and Crooks, P.A. (1984) Evidence of stereospecificity in the *in vivo* methylation of [^{14}C] (±)-nicotine in the guinea pig. *Drug Metab. Disposit.*, **12**, 755–9.

Cundy, K.C., Godin, C.S. and Crooks, P.A. (1985b) Stereospecific *in vitro* N-methylation of nicotine in guinea pig tissues by an *S*-adenosylmethionine-dependent N-methyltransferase. *Biochem. Pharmacol.*, **34**, 281–4.

Cundy, K.C., Sato, M. and Crooks, P.A. (1985c) Stereospecific *in vivo* N-methylation of nicotine in the guinea pig. *Drug Metab. Disposit.*, **13**, 175–85.

Curvall, M., Vala, E.K., England, G. *et al.* (1989) Urinary excretion of nicotine and its major metabolites. 43rd Tobacoo Chemists' Research Conference, Richmond, VA, October 1989.

Dajani, R.M. Gorrod, J.W. and Beckett, A.H. (1972) Hepatic and extrahepatic reduction of nicotine-1′-N-oxide in rats. *Biochem. J.*, **130**, 88P.

Damani, L.A. (1985) Oxidation of tertiary heteroaromatic amines, in *Biological Oxidation of Nitrogen in Organic Molecules – Chemistry, Toxicology and Pharmacology* (Eds J.W. Gorrod and L.A. Damani), Ellis Horwood, Chichester, England, pp. 205–18.

Damani, L.A., Pool, W.F., Crooks, P.A. *et al.* (1988) Stereoselectivity in the N′-oxidation of nicotine isomers by flavin-containing monooxygenase. *Mol. Pharmacol.*, **33**, 702–5.

Damani, L.A., Shaker, M.S., Godin, C.S. *et al.* (1986) The ability of amine N-methyltransferases from rabbit liver to N-methylate azaheterocycles. *J. Pharm. Pharmacol.*, **38**, 547–50.

Davis, R.A. (1986) The determination of nicotine and cotinine in plasma. *J. Chromatogr. Sci.*, **24**, 134–41.

Gairola, C., Godin, C.S., Houdi, A.A. *et al.* (1988) Inhibition of histamine N-methyltransferase activity in guinea pig pulmonary alveolar macrophages by nicotine. *J. Pharm. Pharmacol.*, **40**, 724–6.

Gairola, C., Houdi, A.A., Godin, C.S. *et al.* (1987) Stereospecific N-methylation of nicotine by intact guinea pig pulmonary alveolar macrophages, in *Tobacco Smoking and Nicotine. A Neurobiological Approach* (Eds W.R. Martin, G.R. VanLoon, E.T. Iwamoto and L. Davis), Plenum Press, New York, p. 497.

Gorrod, J.W. and Damani, L.A. (1979a) Some factors involved in the N-oxidation of 3-substituted pyridines by microsomal preparations *in vitro*. *Xenobiotica*, **9**, 209–18.

Gorrod, J.W. and Damani, L.A. (1979b) The effect of various potential inhibitors, activators and inducers on the N-oxidation of 3-substituted pyridines *in vitro*. *Xenobiotica*, **9**, 219–26.

Gorrod, J.W., Jenner, P., Keysell, G. *et al.* (1971) Selective inhibition of alternative pathways of nicotine metabolism *in vitro*. *Chem. Biol. Interact.*, **3**, 269–70.

Haley, N.J., Leutzinger, Y., Axelrod, C.M. *et al.* (1990) Contribution of *trans*-3′-hydroxycotinine and glucuronide conjugates of nicotine metabolites to the measurement of cotinine by RIA. Symposium of XIth International Conference of Pharmacology (International Symposium on Nicotine), Hamburg, June, 1990.

Hecht, S.S. and Hoffman, D. (1988) Tobacco specific nitrosamines, an important group of carcinogens in tobacco and tobacco smoke. *Carcinogenesis*, **9**, 875–84.

Hibberd, A.R. and Gorrod, J.W. (1985) Comparative N-oxidation of nicotine and cotinine by hepatic microsomes, in *Biological Oxidation of Nitrogen in Organic Molecules – Chemistry, Toxicology and Pharmacology* (Eds J.W. Gorrod and L.A. Damani), Ellis Horwood, Chichester, England, pp. 246–50.

Jarvis, J.M. (1987) Uptake of environmental tobacco smoke, in *Environmental Carcinogens – Methods of Analysis and Exposure Measurement – Passive Smoking* (Eds I.K. O'Neil, K.D. Brunneman, B. Dodet and D. Hoffmann), International Agency for Research on Cancer, Lyon, France, pp. 43–58.

Jenner, P., Gorrod, J.W. and Beckett, A.H. (1973) Species variation in the metabolism of *R*-(+)- and *S*-(−)-nicotine by α-C and N-oxidation *in vitro*. *Xenobiotica*, **3**, 573–80.

Jenner, P. and Gorrod, J.W. (1973) Comparative *in vitro* hepatic metabolism of some tertiary N-methyl tobacco alkaloids in various species. *Res. Commun. Chem. Pathol. Pharmacol.*, **6**, 829–43.

Kaplan, N.O. (1960) in *Enzymes* (Eds P.D. Boyer, H. Lardy and K. Myrbäck), 2nd edn, vol. 3, Academic Press, New York, p. 105.

Klus, H. and Kuhn, H. (1977) A study of the optical activity of smoke nicotines. *Fachliche Oesterr. Tabakregie*, **17**, 331–6.

Kyerematen, G.A., Morgan, M.L., Chattopadhyay, B. *et al.* (1990) Disposition of nicotine and eight metabolites in smokers and nonsmokers: identification in smokers of two metabolites that are longer lived than cotinine. *Clin. Pharmacol. Ther.*, **48**, 641–51.

Kyerematen, G.A., Owens, G.F., Chattopadhyay, B. *et al.* (1988) Sexual dimorphism of nicotine metabolism and distribution in the rat. *Drug Metab. Disposit.*, **16**, 823–8.

Kyerematen, G.A., Taylor, L.H., deBethizy, J.D. *et al.* (1987) Radiometric-high-performance liquid chromatographic assay for nicotine and twelve of its metabolites. *J. Chromatogr.*, **419**, 191–203.

Kyerematen, G.A. and Vesell, E.S. (1991) Metabolism of nicotine. *Drug Metab. Rev.*, **23**, 3–41.

McKennis jun., H., Turnbull, L.B. and Bowman, E.R. (1963) N-Methylation of nicotine and cotinine *in vivo*. *J. Biol. Chem.*, *238*, 719–23.

McCoy, G.D., Howard, P.C. and DeMarco, G.J. (1986) Characteristics of hamster liver nicotine metabolism 1. *Biochem. Pharmacol.*, **35**, 2767–73.

Nakayama, H., Nakashima, T. and Kurogochi, Y. (1987) Formation of two major nicotine metabolites in livers of guinea pigs. *Biochem. Pharmacol.*, **36**, 4313–7.

Neurath, G.B., Dunger, M., Orth, D. *et al.* (1987) *Trans*-3′-hydroxycotinine as a main metabolite in urine of smokers. *Int. Arch. Occup. Environ. Health*, **59**, 199–201.

Neurath, G.B., Dunger, M., Krenz, O. *et al.* (1988) *Trans*-3′-hydroxycotinine, a main metabolite in smokers. *Klin. Wochenscht.*, **666**, 2–4.

Neurath, G.B. and Pein, F.G. (1987) Gas chromatographic determination of *trans*-3′-hydroxycotinine, a major metabolite of nicotine in smokers, *J. Chromatog.*, **415**, 400–6.

Nwosu, C.G. and Crooks, P.A. (1988) Species variation and stereoselectivity in the metabolism of nicotine enantiomers. *Xenobiotica*, **18**, 1361–72.

Pacheco, Y., Fonlupt, P., Bensoussan, P. *et al.* (1985) Membrane phosphatidylethanolamine methylase in blood leukocytes and alveolar macrophages of asthmatic patients. *Rev. Pneumol. Clin.*, **41**, 47–56.

Papadopoulos, N.M. (1964) Nicotine-1′-Oxide. *Arch. Biochem. Biophys.*, **106**, 182–5.

Papadopoulos, N.M. and Kintzios, J.A. (1963) Formation of metabolites from nicotine by a rabbit liver preparation. *J. Pharmacol. Exp. Ther.*, **140**, 269–77.

Phillipson, J.D. and Handa, S.S. (1975) Nicotine N-Oxides. *Phytochemistry*, **14**, 2683–90.

Pool, W.F. and Crooks, P.A. (1985) Biotransformation of primary nicotine metabolites. 1. *In vivo* metabolism of *R*-(+)-[^{14}C-NCH$_3$]-N-methylnicotinium ion in the guinea pig. *Drug Metab. Disposit.*, **13**, 578–81.

Pool, W.F. and Crooks, P.A. (1987) Biotransformation of primary nicotine metabolites; metabolism of *R*-(+)-[^{3}H-N′-CH$_3$; ^{14}C-N-CH$_3$]-N-methylnicotinium acetate – the use of double isotope studies to determine the *in vivo* stability of the N-methyl groups of N-methylnicotinium ion. *J. Pharm. Pharmacol.*, **40**, 758–62.

Pool, W.F., Houdi, A.A., Damani, L.A. *et al.* (1986) Isolation and characterization of N-methyl-N′-oxonicotinium ion, a new urinary metabolite of *R*-(+)-nicotine in the guinea pig. *Drug Metab. Disposit.*, **14**, 574–9.

Sato, M. and Crooks, P.A. (1985) N-Methylnornicotinium ion, a new *in vivo* metabolite of *R*-(+)-nicotine. *Drug Metab. Disposit.*, **13**, 348–53.

Schievelbein, H. (1982) Nicotine, resorption and fate. *Pharmacol. Ther.*, **18**, 233–47.

Shulgin, A.T., Jacob III, P., Benowitz, N.L. *et al.* (1987) Identification and quantitative analysis of cotinine-N-oxide in human urine, *J. Chromatogr.*, **423**, 365–72.

Seaton, M., Kyerematin, G.A., Morgan, M. *et al.* (1991) Nicotine metabolism in stumptailed macaques, *Macaca arctoides*. *Drug Metab. Disposit.*, **19**, 946–54.

Sepkovic, D.W., Haley, N.J., Axelrod, C.M. *et al.* (1986) Short term studies on the *in vivo* metabolism of N-oxides of nicotine in rats, *J. Toxicol. Environ. Health*, **18**, 205–14.

Shen, W.-C., Franke, J. and Vunakis, H.V. (1977a) Nicotinamide nucleotide analogues of nicotine and cotinine – enzymic studies. *Biochem. Pharmacol.*, **26**, 1835–40.

Shen, W.-C., Greene, K.M. and Vunakis, H.V. (1977b) Detection by radioimmunoassay of nicotinamide nucleotide analogues in tissues of rabbits injected with nicotine and cotinine. *Biochem. Pharmacol.*, **26**, 1841–6.

Shen, W.-C. and Vunakis, H.V. (1974) The formation and characterization of the nicotine analog of triphosphopyridine nucleotide. *Biochemistry*, **13**, 5362–7.

Testa, B. and Jenner, P. (1976) *Drug Metabolism. Chemical and Biochemcial Aspects*, Marcel Dekker Inc., New York, pp. 262–3.

Thompson, J.A., Norris, K.J. and Peterson, D.R. (1985) Isolation and analysis of N-oxide metabolites of tertiary amines. *J. Chromatogr.*, **341**, 349–59.

Turner, D.M. (1969) The metabolism of [^{14}C]-nicotine in the cat. *Biochem. J.*, **115**, 889–96.

Williams, D.E., Shigenaga, M.K. and Castagnoli, N. (1990) The role of cytochrome P450 and flavin-containing monooxygenase in the metabolism of (*S*)-nicotine by the rabbit lung. *Drug Metab. Disposit.* **18**, 418–428.

Zhang, Y., Jacob, P. and Benowitz, N.L. (1990) Determination of nornicotine in smokers' urine by gas chromatography following reductive alkylation to N-propylnornicotine. *J. Chromatogr.*, **525**, 349–57.

Ziegler, D.M. (1985) Molecular basis for N-oxygenation of *sec*- and *tert*-amines, in *Biological Oxidation of Nitrogen in Organic Molecules – Chemistry, Toxicology and Pharmacology* (Eds J.W. Gorrod and L.A. Damani), Ellis Horwood, Chichester, England, pp. 43–52.

Zuckerman, S.H., O'Dean, R.F., Olson, J.M. *et al.* (1982) Protein carboxymethylation during *in vitro* culture of human peripheral blood monocytes and pulmonary alveolar macrophages. *Mol. Immunol.*, **19**, 281–6.

Extrahepatic metabolism of nicotine and related compounds by cytochromes P450* 6

K. Vähäkangas and O. Pelkonen

6.1 INTRODUCTION

Although the liver is the main site of metabolism of foreign chemicals – generally regarded as the most important organ in terms of kinetics and kinetics-dependent effects of xenobiotics – practically every tissue in the body contains at least some enzymes capable of metabolizing at least some exogenous substances (Gorrod, 1978). The gastrointestinal tract may contribute significantly to the first pass phenomenon of compounds absorbed in the gut. Respiratory tract, including nasal mucosa, and lungs form the first portal of entry for inhaled compounds. Skin is the barrier which also harbours xenobiotic-metabolizing enzymes, and kidney, through which the excreted chemicals and metabolites pass, is a metabolizing organ in its own right. Consequently, appropriate enzymes in any tissue may be of significance in the metabolic activation of drugs, carcinogens and other toxic substances.

In this review the authors try to summarize what is known about the extrahepatic metabolism of nicotine and related compounds, and its significance. Although relatively few direct studies on the metabolism of nicotine by extrahepatic tissues are available, another approach is to

The latest P450 nomenclature (Nebert *et al.*, 1991) will be used in this chapter.

Nicotine and Related Alkaloids: Absorption, distribution, metabolism and excretion. Edited by J.W. Gorrod and J. Wahren. Published in 1993 by Chapman & Hall, London. ISBN 0 412 55740 1

identify the enzymes participating in nicotine metabolism, search their tissue localization and try to envisage the possibility of nicotine metabolism in those tissues harbouring the respective enzymes. Thus, the authors present available data on the extrahepatic distribution of P450 isozymes in an attempt to predict the role of different tissues in the metabolism and possible effects of nicotine.

Since the main carcinogen(s) of tobacco smoke is (are) still unidentified, it is of interest in this context that both nicotine and so-called tobacco-specific nitrosamines (such as 4-(methylnitrosamino)-1-(3-pyridyl)-1-butanone, NNK; and N′-nitrosonornicotine, NNN) being derivatives of nicotine, are metabolically activated. A brief summary of what is known of these activation pathways in extrahepatic tissues is also given.

6.2 PATHWAYS AND ENZYMES OF NICOTINE METABOLISM

Both *in vitro* studies at different levels of cellular and tissue organization and *in vivo* studies, mainly urinary metabolite profiles, indicate that the primary metabolism of nicotine is catalysed predominantly by two microsomal monooxygenase systems, the cytochrome P450-dependent monooxygenase and the flavin-containing monooxygenase (Gorrod and Jenner, 1975). The oxidation at the 5′-carbon by P450 monooxygenase yields initially the $\Delta^{1',5'}$ iminium ion, which is further converted by cytosolic aldehyde oxidase to cotinine. The oxidation at the pyrrolidine nitrogen by the flavin monooxygenase results in isomeric N′-oxides. Cotinine is further hydroxylated at 3′-position, resulting in a major human metabolite, 3′-hydroxycotinine, and is also N-oxidized. The hydroxylation and N-oxidation of cotinine are also catalysed by cytochrome P450.

Other primary metabolites are nornicotine, N-methyl-nicotinium ion, nicotine glucuronide and nicotine-$\Delta^{4',5'}$ enamine (Kyerematen and Vesell, 1991). The contribution of each pathway to nicotine biotransformation differs markedly, not only between species, but also within a given species according to numerous host factors.

Recent studies indicate that nicotine can be metabolized to products which bind covalently to tissue macromolecules both in liver and lung tissues (Shigenaga *et al.*, 1988; Kyerematen and Vesell, 1991). Since there is a close correlation between the covalent binding and the formation of nicotine-$\Delta^{1',5'}$ iminium ion, this metabolite is likely to be the activated form. That $\Delta^{1',5'}$ iminium ion is an electrophilic metabolite and as such capable of covalent binding to tissue macromolecules is also confirmed by the studies of Williams *et al.* (1990b). In addition, production of free radicals may also be involved (Kyerematen and Vesell, 1991).

In this review most attention is paid to the role of P450 isozymes in the extrahepatic metabolism of nicotine. The reason for this restriction is simply that P450 isozymes catalyse the first, rate-limiting reaction in the metabolism of nicotine, and thus are at the crucial position in the further metabolism and ensuing biological consequences. Other enzymes are, however, mentioned where their inclusion seems warranted.

6.3 CYTOCHROME–P450 ISOZYME SPECIFICITY OF NICOTINE METABOLISM

Cytochrome–P450 enzymes are important in oxidative metabolism of both exogenous and endogenous compounds (Guengerich, 1987; Nebert *et al.* 1991). The P450 superfamily contains more than 150 genes which each produce a unique protein. Currently this superfamily can be divided into 27 families (indicated by numbers, e.g. P4501), of which ten are found in mammals, and each of the families divides further into subfamilies (indicated by letters, e.g. P4501A; single genes are again indicated by numbers, such as P4501A1) (Nebert *et al.*, 1991). Within a family the protein sequence is over 40% identical and within a subfamily over 55% identical. An orthologous P450 gene in two species refers to a gene that is known to correspond to an ancestral gene. Prediction of the catalytic activities on the basis of orthologous genes in other species has been a routine method in the P450 field. There are several complications in such predictions, however. First, Nebert *et al.* (1991) point to the fact that in several subfamilies gene duplications and conversions make the orthologous assignments impossible (subfamilies 2A, 2B, 2C, 2D, 3A and 4A). Secondly, Lindberg and Negishi (1989) have shown that a change of only one amino acid can alter the catalytic activity. Furthermore, genes within the same subfamilies may be regulated differently. With these reservations in mind the current knowledge on P450 isozymes in nicotine metabolism is considered.

The first clues as to the P450 isozyme specificity of nicotine metabolism came from *in vivo* animal studies, which demonstrated that phenobarbital induces nicotine elimination (Stålhandske, 1970). This finding has been amply confirmed (Rudell *et al.*, 1987; Foth *et al.*, 1990). It is of interest to note that a short phenobarbital treatment of patients undergoing surgery resulted in an enhanced nicotine oxidation in hepatocytes isolated from biopsy samples, suggesting that humans are also responsive to phenobarbital (Kyerematen *et al.*, 1990). In earlier animal studies nicotine metabolism was found to be enhanced also in 2-acetylaminofluorene and benzo(a)pyrene induced liver microsomes (Yamamoto *et al.*, 1966). In humans, cigarette smoking increases nicotine elimination, but the specific chemicals causing this induction are not known

(Nakayama, 1988). Induction data, especially that with phenobarbital, however, does not give unequivocal indication of participating isozymes, because inducers are known to affect more than one isozyme usually (Waxman and Azaroff, 1992).

Early indirect studies indicated that the formation of the $\Delta^{1',5'}$ iminium ion from nicotine is catalysed by P450 isozyme(s) belonging to the 2B subfamily (Nakayama *et al*, 1982, 1985; Rudell *et al.* 1987). Later studies on purified enzymes from rabbit and rat liver have confirmed that the PB-inducible isozyme(s) is (are) relatively active in catalysing the primary C-oxidation of nicotine (McCoy *et al.*, 1989; see Table 6.1, Hammond *et al.* 1991). A study of Williams *et al.* (1990a) indicated that a rabbit nasal P450 isozyme, termed Nasal P450 NMa, is several times more active than the PB-induced isozyme. However, the exact identity of the nasal isozyme has not yet been elucidated.

Table 6.1 Nicotine C-oxidation by P450 isozymes

Isozyme	Inducer	Rabbit			Rat	Human	
		Isozyme	Activity				
			1[a]	2[b]	3[c]	4[d]	5[e]
1A1	PAH	Rabbit 6	7.4		0		
1A2	PAH	Rabbit 4	1.4			0	
2A6						21	0
2B1	PB				0.17		
2B4	PB	Rabbit 2	28.5	2.14			
2B6						166	4.8
2C3	Constit.	Rabbit 3B	29.4				
2C6					0		
2C8						16	0
2C9						46	0
2D6						0	5.3
2E1	Ethanol	Rabbit 3a	4.1			50	0
2F1						15	
3A3						0	
3A4						0	0
3A5						0	
4B1		Rabbit 5		0			

[a]nmol product/min/mol cytochrome P450.
[b]nmol $\Delta^{1',5'}$ iminium ion/min/nmol P450.
[c]nmol continine/nmol P450.
[d]nmoles product/min/mg cell lysate.
[e]pmol nicotine/min/mg microsomal protein.
1 McCoy *et al.* 1989, purified and reconstituted.
2 Williams *et al.* 1990b, purified and reconstituted.
3 Hammond *et al.* 1991, purified and reconstituted.
4 Flammang *et al.* 1992, cDNA expressed, HepG2 cells.
5 McCracken *et al.* 1992, cDNA expressed, B-lymphoblastoid cells.

It is interesting to observe that rabbit purified isozymes inducible by polycyclic aromatic hydrocarbons (1A) and ethanol (2E) catalyse nicotine oxidation, albeit at a much reduced level than 2B or 2C isozymes (Table 6.1). Purified rat liver 1A1, on the other hand, did not catalyse nicotine oxidation (Hammond *et al.*, 1991) implicating species differences in nicotine oxidation. It was also of interest that none of the studied rabbit liver P450 isozymes was able to catalyse nicotine-1′-N-oxidation (McCoy *et al.*, 1989).

Additionally, human P450 isozymes seem to be incapable of 1′-N-oxidation (Flammang *et al.*, 1992). However, two recent studies using cDNA directed expression systems, one with HepG2 cells (Flammang *et al.*, 1992) and the other with B-lymphoblastoid cell lines (McCracken *et al.*, 1992) show that cDNA expressed human isozymes are capable of C-oxidation of nicotine. According to Flammang *et al.* (1992) seven of 12 cDNA expressed human isozymes studied catalysed C-oxidation, 2B6 being the most active (Table 6.2). Of the studied isozymes 4B1, which had low but measurable activity, is expressed in human lung. P4501A1 was not studied by Flammang *et al.* (1992), and there are no other studies of human 1A1 isoenzyme with respect to nicotine metabolism, either. However, some animal studies indicate that it may take part in nicotine metabolism in lung. Consequently, the role CYP1A1 is of considerable interest. McCracken *et al.* (1992) found a somewhat different spectrum of activity; 2D6 was the most active isozyme in cotinine formation, 2B6 was almost as active and the other isozymes studied (2A6, 2C8, 2C9, 2E1, 3A4) were inactive.

In a perfused rat lung model inhibition of benzo(a)pyrene metabolism by nicotine has been investigated. In one study on isolated perfused lung from 3-methylcholanthrene-pretreated rats a minimal inhibition by nicotine of benzo(a)pyrene elimination was observed (Vähäkangas *et al.*, 1979). Using a similar type of inducer, 5,6-benzoflavone (β-naphthoflavone), but with different timing and substrate concentrations, Foth *et al.* (1988) found a clear inhibition of benzo(a)pyrene elimination by nicotine. Both of these inducers are supposed to induce especially CYP1A1 in extrahepatic tissues (Gonzalez, 1989; Nebert, 1989). These observations may have some interesting consequences in terms of the environmental regulation of nicotine metabolism.

6.4 EXTRAHEPATIC METABOLISM OF NICOTINE IN RELATION TO THE LOCALIZATION OF ENZYMES POTENTIALLY PARTICIPATING IN NICOTINE METABOLISM

When discussing this topic it should be borne in mind that even within the same species the enzyme pattern participating in a specific pathway may differ from one organ to another (e.g. benzo(a)pyrene activation,

Table 6.2 Extrahepatic expression of P450 isozymes

Organ	Human	Rabbit	Rat	Mouse
Lung	1A1[a], 4B1[a], 2F1[a]	2B4[b], 2B1[d], 4B1[f] 2?(5)[d], 1A1[d]	1A1[d],2A1[d], 2B1[d,e], 2E[e]	1A[d]
Kidney		1A1[d], 1A2[d], 2B1[d], 2C3[d]	1A[d], 2B[d], 3?[d], 2E[e]	
Skin	(1A1)		1A[d], 2B[d], 3A1[d]	
Extrahepatic vascular endothelium	(1A1)	1A1[d], 1A2[d] 2B1[d], 2?(5)[d]	1A1[d]	1A[d]
Placenta	1A1	(1A1)	(1A1)	(1A1)
Nasal mucosa		NMa[c], NMb[c]	1A?[d], 2B[d]	
Pancreas			1A[d], 2B[d]	
Intestine	2C[e], 2E1[e], 2D6[e], 3A4[e]		1A1[e], 2B[e]	
Prostate			1A[d]	

The level of certainty varies in different cases; depiction without parentheses refers to immunochemical and immunohistochemical data, sometimes to Northern data. Depiction with parentheses refers to enzymatic data. Useful general reviews of P450 isozymes are: for rat, Soucek and Gut (1992); for human, Guengerich (1992). Thorough reviews on earlier literature are contained in the book edited by Guengerich (1987).

[a]Wheeler and Guenthner, 1991; [b]Williams *et al.*, 1990b; [c]Williams *et al.*, 1990a; [d]Anderson *et al.*, 1989; [e]De Waziers *et al.*, 1990; [f]Flammang *et al.*, 1992.

Vähäkangas *et al.*, 1989) and that the corresponding isozymes may not catalyse the same activities in different species (Gonzalez, 1989; Waxman *et al.* 1991). It is striking in this context that even one amino acid difference may change the catalytic activity of a P450 isozyme (Lindberg and Negishi, 1989). Accordingly, several authors have called for caution in extrapolation of animal data to humans (Nebert *et al.*, 1991; Shimada and Guengerich, 1990; Vähäkangas *et al.*, 1989).

Early *in vitro* studies indicated that lung and kidney microsomes from several species including man were capable of nicotine metabolism (Hansson and Schmiterlöw, 1964; Booth and Boyland, 1971; Hill *et al.*, 1972, reviewed by Scheline, 1978). In rat, enzyme-catalysed anaerobic reduction of nicotine-1′-N-oxide was described in small intestine, kidney, heart and lung tissue fractions in addition to liver (Dajani *et al.*, 1975). However, in the 1970s the knowledge of extrahepatic metabolism of tobacco alkaloids was still minimal (Scheline, 1978). After these early studies several interesting observations have been made on, for example, nasal mucosa. Extrahepatic expression and localization of total P450 and specific isozymes have been dealt with in a number of reviews and congress proceedings (e.g. Gram, 1980; Rydström *et al.*, 1983; Anderson *et al.*, 1989) and a general overview is presented in Table 6.2.

6.4.1 NASAL MUCOSA

The levels of P450s are relatively high in nasal preparations of several mammalian species (Hadley and Dahl, 1982; Larsson *et al.*, 1989; Dahl and Hadley, 1991) with specific contents ranging from 8–41% of that found in the liver. The activity of nasal microsomes toward a number of xenobiotics, when expressed on the basis of P450 content, are actually significantly higher than activities measured in liver microsomes. For example, known or suspected nasal carcinogens hexamethylphosphoramide and N-nitrosodiethylamine are metabolized at high rates (Ding and Coon, 1988).

As described above, rabbit nasal mucosa contains tissue-specific P450 isozymes, NMa and NMb (CYP2G1, Ding *et al.*, 1991), of which the first is extremely active in nicotine primary metabolism, catalysing both C- and N-oxidation at rates which are several times higher than those of the PB-inducible isozyme. Neither one of these isozymes has been detected in other tissues, so one would anticipate a specific role for them in terms of olfaction or other physiological function of nasal tissues. However, nothing certain is known at present.

6.4.2 LUNGS

It has been demonstrated that the major P450 isozyme in the lungs of rat and rabbit is identical to, or a close relative of, the major PB-inducible isozyme in the liver, CYP2B1 (rat) or CYP2B4 (rabbit) (Philpot and Smith, 1984; Guengerich, 1990; Wheeler and Guenthner, 1991). The corresponding isozyme is apparently lacking in human lung tissue. Other isozymes which have been detected in the lungs include those identified in human lung, 1A1, 4B1 and 2F1 (Wheeler *et al.*, 1990; Wheeler and Guenthner, 1991).

In vitro studies have demonstrated that mouse lungs are capable of metabolizing nicotine. The isolated perfused rat lung system is also active in the clearance of nicotine, and this occurs at practically the same level as in perfused rat liver (Foth *et al.*, 1991). The clearance is almost doubled after the pretreatment of rats with phenobarbital, whereas pretreatment with β-naphthoflavone had no significant effect (Foth *et al.*, 1988). Isolated perfused canine lung is also able to oxidize nicotine (Turner *et al.*, 1975).

On the basis of the catalytic selectivity and activity of the studied P450 isozymes and of their presence in the lungs, it seems clear that subfamily 2B isozymes are responsible for nicotine oxidation in the lungs of rats and rabbits. The contribution of other isozymes present is probably negligible. With respect to another potentially important enzyme metabolizing nicotine, FAD-monooxygenase, rabbit lung enzyme exhibit subtrate specificities, which distinguish it from the liver enzyme (Poulsen *et al.*, 1986). Although rabbit lungs contain quite a high level of FAD-monooxygenase, the enzyme seems not to be significant in nicotine metabolism, because it has very low activity of N-oxidation toward (*S*)-nicotine (Williams *et al.*, 1990b).

6.4.3 GUT MUCOSA

P4503A enzymes have been shown to be the predominant isozymes in enterocytes of different species (Watkins *et al.*, 1987; Watkins, 1990). For example, in an uninduced state, 3A enzymes appear to account for >70% of the total P450 present in human jejunal mucosa (Watkins *et al.*, 1987). However, studies on the purified rabbit isozyme have demonstrated that this isozyme is not very active in oxidizing nicotine (see above). Actually, it seems probable that the C-oxidation activity of nicotine by 3A isoforms can be demonstrated only in artificially high substrate concentrations and so the *in vitro* activity has no bearing in terms of *in vivo* metabolism. On this basis one would conclude that gut mucosa does not contribute in a major way to the primary metabolism of nicotine.

6.4.4 GUT CONTENTS

There is some evidence that intestinal microflora is capable of catalysing some transformations of nicotine and cotinine. In particular, the reduction of some metabolites back to nicotine and cotinine have been observed (Dajani *et al.*, 1975; Scheline, 1978).

6.4.5 KIDNEY

In vitro studies have shown that the ability of mouse kidney to oxidize nicotine is rather limited (Scheline, 1978).

6.5 EXTRAHEPATIC ACTIVATION OF TOBACCO-SPECIFIC NITROSAMINES

Metabolic activation is the prerequisite for the carcinogenic activity of tobacco-specific nitrosamines (Hecht *et al.*, 1983, Hoffman and Hecht, 1985; Hecht and Hoffman, 1988). In several studies it has been shown that the target organs of tobacco-induced carcinogenesis, which are also target organs for tobacco-specific nitrosamines in animal studies, are capable of metabolism leading to tissue-bound metabolites (Table 6.3).

Cumulating evidence points to the α-hydroxylation pathway of nitrosamine metabolism as one for the metabolic activation, leading to metabolites capable of covalent binding to DNA and responsible for the carcinogenic effect (Hecht and Hoffman, 1988). The extent of α-hydroxylation in human tissue explants was only 1/10 to 1/100 of that in animal tissues and while there were no differences in the extent between different tissues studied, there was a ten-fold variation between individuals (Castonguay *et al.*, 1983).

Nicotine inhibits strongly NNN (N-nitrosonornicotine) metabolism by cultured rat oral tissue (Murphy and Heiblum, 1990). The identity of the enzyme in oral tissue is not known. Nicotine also interferes with the α-carbon hydroxylation and nitrogen oxidation, but not with the carbonyl reduction of NNK (4-(methylnitrosamino)-1-(3-pyridyl)-1-butanone) in hamster lung explants (Schuller *et al.*, 1991). Nicotine thus competes with tobacco specific nitrosamines for the same enzymatic systems in more than one extrahepatic tissue *in vitro* and may exert the same effect *in vivo*, implicating inhibition of the potent carcinogenic effect of these nitrosamines by nicotine.

Recently Crespi *et al.* (1991) have used cell lines stably expressing human P450 isozymes to study the isozyme pattern taking part in the activation of tobacco-specific nitrosamines. NNK was found to be activated by CYP2D6, 1A2, 2A6 and 2E1, while NNN, which is less carcinogenic than NNK, and NNA (1-(N-methyl-N-nitrosamino)-1-(3-pyridinyl)-4-butanal), which is non-carcinogenic, were not activated by CYP2D6. On the basis of animal experiments and less complete human studies, it is known that CYP1A2 is expressed only in the liver (Boobis *et al.*, 1990) and the expression of CYP2D6 is also largely restricted to

Table 6.3 Examples of extrahepatic tissues capable of activation of tobacco-specific nitrosamines

Species	Tissue	Preparation	Compound	Reference
Hamster	Lung, trachea, ethmoturbinal	Whole body	NNK	Tjälve and Castonguay, 1983
Rat	Nasal mucosa,	Tissue culture	NNK, NNN	Brittebo *et al.*, 1983
	oesophagus	Tissue culture	NNN	Hecht *et al.*, 1983
Human	Buccal mucosa, trachea, bronchi, lung, oesophagus, urinary bladder	Tissue culture	NNK, NNN	Castonguay *et al.*, 1983

NNK = 4-(methylnitrosamino)-1-(3-pyridyl)-1-butanone.
NNN = N-nitrosonornicotine.

this tissue. CYP2A6 and 2E1 may also be present in extrahepatic tissues, although in relatively low amounts (de Waziers *et al.*, 1990).

6.6 CONTRIBUTION OF EXTRAHEPATIC METABOLISM TO THE PHARMACOKINETICS AND TOXICITY OF NICOTINE AND RELATED COMPOUNDS

It is important to consider if extrahepatic metabolism contributes to the overall kinetics and elimination of nicotine and/or its main metabolites, and furthermore, whether there is any indication of participation of extrahepatic pathways in toxification.

The rate of cotinine elimination is accelerated in cigarette smokers compared to nonsmokers (Beckett and Triggs, 1967; Barlow *et al.*, 1987). On this basis one would anticipate that either P4501A1 or 1A2 participates in the further metabolism of cotinine, although some other drug metabolizing enzymes, such as certain forms of glucuronosyl transferases, are induced by cigarette smoking. Because P4501A1 is increased significantly only in extrahepatic tissues (Boobis *et al.*, 1990), this would mean, if cotinine is metabolized by this isozyme, that in smokers extrahepatic tissues would contribute significantly to cotinine metabolism. However, this possibility remains to be demonstrated.

Extrahepatic metabolism may be of considerable toxicological significance, because in many instances toxicity is assumed to result from target–organ activation of the substance. With nicotine, N-oxidation is potentially important because N-oxidized products are easily nitrosated to tobacco-specific carcinogens. Although this pathway exists in extrahepatic tissues of rabbit, none of the human P450 enzymes studied metabolizing nicotine are capable of N-oxidation (Flammang *et al.*, 1992). There is evidence strongly suggesting the involvement of flavin-containing monooxygenases in the N-oxidation of nicotine (Nakayama *et al.*, 1987; McCoy *et al.*, 1989; Damani *et al.*, 1988). If this is the case, the suggestion that in addition to animal extrahepatic tissue, FMO exists in human kidney (Dolphin *et al.*, 1991) is important when thinking of human tobacco exposure. C-Oxidation leads to activated nicotine-$\Delta^{1',5'}$ iminium ion capable of covalent binding, which is probably catalysed by P450s in extrahepatic tissues giving this metabolic pathway special toxicological importance.

Compounds participating in cigarette smoke carcinogenesis are still largely unknown (IARC, 1986). The mechanism is most probably an extensive and complex interaction between the thousands of chemicals occurring in cigarette smoke. Since neither polycyclic aromatic hydrocarbons (PAHs) nor nitrosamines can account for all the carcinogenic potency of cigarette smoke, others, like nicotine should be considered. Table 6.4 shows an overview of the enzymes on the one hand capable

Table 6.4 The main enzymes catalysing nicotine metabolism in lung tissue in different species

Species	Nicotine	
	C-oxidation	N-oxidation
Rabbit	IA1[a,b], IIB4[c]	IIB4[c]
Rat	II[d], IIB1[b,e]	
Human	IIF1[f,g], IVB1[6]	

[a]McCoy *et al.*, 1989; [b]Anderson *et al.*, 1989; [c]Williams *et al.*, 1990b; [d]Foth *et al.*, 1991; [e]Hammond *et al.*, 1991; [f]Flammang *et al.*, 1992; [g]Wheeler and Guenthner, 1991.

of the activation pathway of nicotine (C-oxidation) and on the other hand existing in lung tissue. Since P4501A1 exists in human lung (Wheeler and Guenthner, 1991), and is capable of nicotine C-oxidation in rabbits (McCoy *et al.*, 1989), it may take part in nicotine oxidation in human lung. Unfortunately, the only extensive study of human P450 isozymes capable of nicotine metabolism did not include this isozyme (Flammang *et al.*, 1992). The fact that smoking induces nicotine metabolism (Nakayama, 1988) gives further impact to the possible cocarcinogenic effect of nicotine in cigarette smoke induced carcinogenesis in human lung.

The formation of 3-(2,3-dihydro-1-methyl-2-pyrrolyl)pyridine from nicotine by prostaglandin H synthetase has been described (Mattamal *et al.*, 1987). This metabolite was found in the urine from a nicotine treated rabbit and also in a male smoker. Lung is especially active in prostaglandin synthesis, and thus lung may be an important site of this pathway in humans as well.

6.7 UNANSWERED IMPORTANT QUESTIONS AND SUGGESTIONS FOR FUTURE WORK

There is still only scant information regarding which enzymes and to what extent these enzymes participate in nicotine metabolism in extrahepatic tissues. Animal studies give some indications of this, although even in animals there is still much work to do. There is at least one active metabolite formed. Whether others exist and are also formed in human extrahepatic tissues needs to be studied. Since nicotine is capable of acting as a cocarcinogen in animals (Chen and Squier, 1990), it is possible that it does the same in humans. Knowledge of the mechanism involved is important, and although this may be connected with the metabolism of nicotine, it may occur as a consequence of its ability to function as a mitogen in lung (Schuller *et al.*, 1991).

6.8 CONCLUSIONS

It has been shown that nicotine is metabolized by extrahepatic tissues in experimental animals, and most probably also in humans, but the participating organs and enzymes are currently poorly identified. Although a P4502B isozyme, which is the major metabolizing enzyme leading to reduced half-life of nicotine in animals pretreated with phenobarbital, probably does not exist in human lung (Wheeler and Guenthner, 1991), other enzyme mediated pathways of nicotine metabolism do. These may have important consequences as to the carcinogenic effect of smoking rather than serve as major detoxicating routes. This conclusion is based on the following: (1) nicotine from cigarette smoke enters the body through the lung, which contains enzymes capable of nicotine metabolism (P450, aldehyde oxidase, prostaglandin synthetase); (2) P4501A1, capable of nicotine C-oxidation in rabbits (McCoy *et al.*, 1989) leading to DNA-binding via $\Delta^{1'5'}$ iminium ion, exists in human lung; (3) cDNA expressed human P4504B1 is found in human lung and is capable of C-oxidation of nicotine.

REFERENCES

Anderson, L.M., Ward, J.M., Park, S.S. *et al.* (1989) Immunohistochemical localization of cytochromes P450 polyclonal and monoclonal antibodies. *Pathol. Immunopathol. Res.*, **8**, 61–94.

Barlow, R.D., Thompson, P.A. and Stone, R.B. (1987) Simultaneous determination of nicotine, cotinine and five additional nicotine metabolites in the urine of smokers using pre-column derivatisation and high-performance liquid chromatography. *J. Chromatogr.*, **419**, 375–80.

Beckett, A.H. and Triggs, E.J. (1967) Enzyme induction in man caused by smoking. *Nature*, **216**, 587.

Boobis, A.R., Sesardic, D., Murray, B.P. *et al.* (1990) Species variation in the response of the cytochrome P450-dependent monooxygenase system to inducers and inhibitors. *Xenobiotica*, **20**, 1139–61.

Booth, J. and Boyland, E. (1971) Enzymatic oxidation of (−)nicotine by guinea-pig tissues *in vitro. Biochem. Pharmacol.*, **20**, 407–15.

Brittebo, E.B., Castonguay, A., Furuya, K. *et al.* (1983) Metabolism of tobacco-specific nitrosamines by cultured rat nasal mucosa. *Cancer Res.*, **43**, 4343–8.

Castonguay, A., Stoner, G.D., Schut, H.A.J. *et al.* (1983) Metabolism of tobacco-specific N-nitrosamines by cultured human tissues. *Proc. Natl. Acad. Sci.* USA, **80**, 6694–7.

Chen, Y-P. and Squier, C.A. (1990) Effect of nicotine on 7,12-dimethylbenz(a)anthracene carcinogenesis in hamster cheek pouch. *J. Natl. Cancer Inst.*, **82**, 861–4.

Crespi, C.L., Penman, B.W., Gelboin, H.V. *et al.* (1991) A tobacco-derived nitrosamine, 4-(methylnitrosamino)-1-(3-pyridyl)-1-butanone, is activated by multiple human cytochrome P450s including the polymorphic human cytochrome P4502D6. *Carcinogenesis*, **12**, 1197–201.

Dahl, A.R. and Hadley, W.M. (1991) Nasal cavity enzymes involved in xenobiotic metabolism: Effects on the toxicity of inhalants. *CRC Crit. Rev. Toxicol.*, **21**, 345–72.

Dajani, R.M., Gorrod, J.W. and Beckett, A.H. (1975) *In vitro* hepatic and extrahepatic reduction of (−)nicotine-1′-N-oxide in rats. *Biochem. Pharmacol.*, **24**, 109–17.

Damani, L.A., Pool, W.P., Crooks, P.A. *et al.* (1988) Stereoselectivity in the N′-oxidation of nicotine isomers by flavin-containing monooxygenase. *Mol. Pharmacol.*, **33**, 702–5.

De Waziers, I., Cugnenc, P.H., Yang, C.S., *et al.* (1990) Cytochrome P450 isoenzymes, epoxide hydrolase and glutathione transferases in rat and human hepatic and extrahepatic tissues. *J. Pharmacol. Exp. Ther.*, **253**, 387–94.

Ding, X. and Coon, M.J. (1988) Purification and characterization of two unique forms of cytochrome P450 from rabbit nasal microsomes. *Biochemistry*, **27**, 8330–7.

Ding, X., Porter, T.D., Peng, H-M. *et al.* (1991) cDNA and derived amino acid sequence of rabbit nasal cytochrome P450NMb (P450IIG1). A unique isozyme possibly involved in olfaction. *Arch. Biochem. Biophys.*, **285**, 120–5.

Dolphin, C., Shephard, E.A., Povey, S. *et al.* (1991) Cloning and chromosomal mapping of a human flavin-containing monooxygenase (FMO1) primary sequence. *J. Biol. Chem.*, **266**, 12 379–85.

Flammang, A.M., Gelboin, H.V., Aoyama, T. *et al.* (1992) Nicotine metabolism by cDNA-expressed human cytochrome P450s. *Biochem. Archives*, **8**, 1–8.

Foth, H., Looschen, H., Neurath, H. *et al.* (1991) Nicotine metabolism in isolated perfused lung and liver of phenobarbital- and benzoflavone-treated rats. *Arch. Toxicol.*, **65**, 68–72.

Foth, H., Rudell, U., Ritter, G. *et al.* (1988) Inhibitory effect of nicotine on benzo(a)pyrene elimination and marked pulmonary metabolism of nicotine in isolated perfused rat lung. *Klin. Wochenschr.*, **66**, 98–104.

Foth, H., Walther, U.I. and Kahl, G.F. (1990) Increased hepatic nicotine elimination after phenobarbital induction in the conscious rat. *Toxicol. Appl. Pharmacol.*, **105**, 382–92.

Gonzalez, F.J. (1989) The molecular biology of cytochrome P450s. *Pharmac. Rev.*, **40**, 243–88.

Gorrod, J.W. (1978) Extra-hepatic metabolism of drugs, in *Drug metabolism in man.* (Eds J.W. Gorrod and A.H. Beckett), Taylor & Francis Ltd., London, pp. 157–74.

Gorrod, J.W. and Jenner, P. (1975) The metabolism of tobacco alkaloids, in *Essays in Toxicology*, vol. 6 (Ed. W.J. Hayes, Jr), Academic Press, New York, pp. 35–78.

Gram, T.E. (Ed.) (1980) *Extrahepatic metabolism of drugs and other foreign compounds.* MTP Press Ltd., Falcon House, Lancaster, England.

Guengerich, F.P. (Ed) (1987) *Mammalian Cytochromes P450.* Vols I & II. CRC Press, Boca Raton, Florida.

Guengerich, F.P. (1990) Purification and characterization of xenobiotic-metabolizing enzymes from lung tissue. *Pharmac. Ther.*, **45**, 299–307.

Guengerich, F.P. (1992) Human cytochrome P450 enzymes. *Life Sci.* **50**, 1471–8.

Hadley, W.M. and Dahl, A.R. (1982) Cytochrome P450 dependent monooxygenase activity in rat nasal epithelial membranes. *Toxicol. Lett.*, **10**, 417–22.

Hammond, D.K., Bjercke, R.J., Langone, J.J. *et al.* (1991) Metabolism of nicotine by rat liver cytochromes P450. Assessment utilizing monoclonal antibodies to nicotine and cotinine. *Drug Metab. Disp.*, **19**, 804–8.

Hansson, E. and Schmiterlöw, C.G. (1964) Metabolism of nicotine in various tissues, in: *Tobacco alkaloids and related compounds* (Ed. U.S. von Euler), Pergamon Press, pp. 87–98.

Hecht, S.S., Castonguay, A., Rivenson, A. *et al.* (1983) Tobacco specific nitrosamines: Carcinogenicity, metabolism, and possible role in human cancer. *J. Environ. Sci. Health* CI(1), 1–54.

Hecht, S.S. and Hoffmann, D. (1988) Tobacco-specific nitrosamines, an important group of carcinogens in tobacco and tobacco smoke. *Carcinogenesis*, **9**, 875–84.

Hill, D.L., Laster, W.R. and Struck, R.F. (1972) Enzymatic metabolism of cyclophosphamide and nicotine and production of a toxic cyclophosphamide metabolite. *Cancer Res.*, **32**, 658–65.

Hoffmann, D. and Hecht, S.S. (1985) Nicotine-derived N-nitrosamines and tobacco-related cancer: Current status and future directions. *Cancer Res.*, **45**, 935–44.

IARC (1986) *Tobacco Smoking*, IARC, Lyon, France.

Kyerematen, G.A., Morgan, M., Warner, G. *et al.* (1990) Metabolism of nicotine by hepatocytes. *Biochem. Pharmacol.*, **40**, 1747–56.

Kyerematen, G.A. and Vesell, E.S. (1991) Metabolism of nicotine. *Drug Metab. Rev.*, **23**, 3–41.

Larsson, P., Pettersson, H. and Tjälve, H. (1989) Metabolism of aflatoxin B1 in the bovine olfactory mucosa. *Carcinogenesis*, **10:** 111–8.

Lindberg, R.L.P. and Negishi, M. (1989) Alteration of mouse cytochrome P450coh substrate specificity by mutation of a single amino-acid residue. *Nature*, **339**, 632–4.

Mattamal, M.B., Lakshmi, V.M., Zenser, T.V. *et al.* (1987) Lung prostaglandin H synthase and mixed-function oxidase metabolism of nicotine. *J. Pharmacol. Exp. Ther.* **242**, 827–32.

McCoy G.D., DeMarco, G.J. and Koop, D.B. (1980) Microsomal nicotine metabolism: A comparison of relative activities of six purified rabbit cytochrome P450 isoenzymes. *Biochem. Pharmacol.*, **38**, 1185–8.

McCracken, N.W., Cholerton, S. and Idle J.R. (1992) Cotinine formation by cDNA-expressed human cytochromes P450, *Medical Science Research*, **20**, 877–8.

Murphy, S.E. and Heiblum, R. (1990) Effect of nicotine and tobacco-specific nitrosamines on the metabolism of N′-nitrosonornicotine and 4-(methylnitrosamino)-1-(3-pyridyl)-1-butanone by rat oral tissue. *Carcinogenesis*, **11**, 1663–6.

Nakayama, H. (1988) Nicotine metabolism in mammals. *Drug Metab. Drug Interact.*, **6**, 95–122.

Nakayama, H., Fujihara, S., Nakashima, T. *et al.* (1987) Formation of two major nicotine metabolites in livers of guinea-pigs. *Biochem. Pharmacol.*, **36**, 4313–7.

Nakayama, H., Nakashima, T. and Kurogochi, Y. (1982) Participation of cytochrome P450 in nicotine oxidation. *Biochem. Biophys. Res. Commun.*, **108**, 200–5.

Nakayama, H., Nakashima, T. and Kurogochi, Y. (1985) Cytochrome P450-dependent nicotine oxidation by liver microsomes of guinea pigs. Immunochemical evidence with antibody against phenobarbital-inducible cytochrome P450. *Biochem. Pharmacol.*, **34**, 2281–6.

Nebert, D.W. (1989) The Ah locus: genetic differences in toxicity, cancer, mutation, and birth defects. *Crit. Rev. Toxicol.*, **20**, 153–74.

Nebert, D.W., Nelson, D.R., Coon, M.R. *et al.* (1991) The P450 superfamily: Update on new sequences, gene mapping, and recommended nomenclature. *DNA Cell. Biol.*, **10**, 1–14.

Philpot, R.M. and Smith, B.R. (1984) Role of cytochrome P450 and related enzymes in the pulmonary metabolism of xenobiotics. *Environ. Health Perspect.*, **55**, 359–67.

Poulsen, L.L., Taylor, K., Williams, D.E. *et al.* (1986) Substrate specificity of the rabbit lung flavin-containing monooxygenase for amines: oxidation products of primary alkylamines. *Mol. Pharmacol.*, **30**, 680–5.

Rudell, U., Foth, H. and Kahl, G.F. (1987) Eight-fold induction of nicotine elimination in perfused rat liver by pretreatment with phenobarbital. *Biochem. Biophys. Res. Commun.*, **148**, 192–8.

Rydström, J., Montelius, J., Bengtsson, M. (Eds) (1983) *Extrahepatic drug metabolism and chemical carcinogenesis*. Elsevier Science Publishers, Amsterdam, The Netherlands.

Scheline, R.R. (1978) *Mammalian metabolism of plant xenobiotics.*, Academic Press, London.

Schuller, H.M., Castonguay, A., Orloff, M. *et al.* (1991) Modulation of the uptake and metabolism of 4-(methylnitrosamino)-1-(3-pyridyl)-1-butanone by nicotine in hamster lung. *Cancer Res.*, **51**, 2009–14.

Shigenaga, M.K., Trevor, A.J. and Castagnoli, N. (1988) Metabolism-dependent covalent binding of (*S*)-[5-^{3}H]nicotine to liver and lung microsomal macromolecules. *Drug. Metab. Disp.*, **16**, 397–402.

Shimada, T. and Guengerich, P. (1990) Inactivation of 1,3–1,6,- and 1,8-dinitropyrene by cytochrome P450 enzymes in human and rat liver microsomes. *Cancer Res.*, **50**, 2036–43.

Soucek, P. and Gut, I. (1992) Cytochromes P450 in rats – structures, functions, properties and relevant human forms. *Xenobiotica* **22**, 83–104.

Stålhandske, T. (1970) Effects of increased liver metabolism of nicotine on its uptake, elimination and toxicity in mice. *Acta Physiol. Scand.*, **80**, 222–34.

Tjälve, H. and Castonguay, A. (1983) The *in vivo* tissue disposition and *in vitro* target-tissue metabolism of the tobacco-specific carcinogen 4-(methylnitrosamino)-1-(3-pyridyl)-1-butanone in Syrian golden hamsters. *Carcinogenesis*, **4**, 1259–65.

Turner, D.M., Armitage, A.K., Briant, R.H. *et al.* (1975) Metabolism of nicotine by the isolated perfused dog lung. *Xenobiotica*, **5**, 539–51.

Vähäkangas, K., Raunio, H., Pasanen, M. *et al.* (1989) Comparison of the formation of benzo(a)pyrene diolepoxide-DNA adducts *in vitro* by rat and human microsomes: Evidence for the involvement of P450IAI and P450IA2. *J. Biochem. Toxicol.*, **4**, 79–86.

Vähäkangas, K., Nevasaari, K., Pelkonen, O. *et al.* (1979) Effects of various *in vitro* inhibitors of benzo(a)pyrene metabolism in isolated rat lung perfusion. *Acta Pharmacol. Toxicol.*, **45**, 1–8.

Watkins, P.B. (1990) Role of cytochromes P450 in drug metabolism and hepatotoxicity. *Seminars in Liver Disease*, **10**, 235–50.

Watkins, P.B., Wrighton, S.A., Schuetz, E.G. *et al.* (1987) Identification of glucocorticoid-inducible cytochromes P450 in the intestinal mucosa of rats and man. *J. Clin. Invest.*, **80**, 1029–36.

Waxman, D.J. and Azaroff, L. (1992) Phenobarbital induction of cytochrome P450 gene expression. *Biochem. J.*, **281**, 577–92.

Waxman, D.J., Lapenson, D.P., Aoyama, T. *et al.* (1991) Steroid hormone hydroxylase specificities of eleven cDNA-expressed human cytochrome P450s. *Arch. Biochem. Biophys.*, **290**, 160–6.

Wheeler, C.W. and Guenthner, T.M. (1991) Cytochrome P450-dependent metabolism of xenobiotics in human lung. *J. Biochem. Toxicol.*, **6**, 163–9.

Wheeler, C.W., Park, S.S. and Guenthner, T.M. (1990) Immunochemical analysis of a cytochrome P450IAI homologue in human lung microsomes. *Mol. Pharmacol.*, **38**, 634–43.

Williams, D.E., Ding, X. and Coon, M.J. (1990a) Rabbit nasal cytochrome P450 NMa has high activity as a nicotine oxidase. *Biochem. Biophys. Res. Commun.*, **166**, 945–52.

Williams, D.E., Shigenaga, M.K. and Castagnoli, N. (1990b) The role of cytochromes P450 and flavin-containing monooxygenase in the metabolism of (*S*)-nicotine by rabbit lung. *Drug Metab. Disp.*, **18**, 418–28.

Yamamoto, L., Nagai, K., Kimura, H. *et al.* (1966) Nicotine and some carcinogens in special reference to the hepatic drug-metabolizing enzymes. *Japan. J. Pharmacol.*, **16**, 183–90.

The metabolic fate of the minor tobacco alkaloids

7

X. Liu, P. Jacob III and N. Castagnoli Jr

7.1 INTRODUCTION

This chapter is concerned with the metabolic fates of the so-called minor tobacco alkaloids. Nicotine* (**1**, see Figure 7.1) is the most abundant (Schmeltz and Hoffman, 1977) and pharmacologically potent (Clark *et al.*, 1965) alkaloid present in commercial tobacco products but is only one of a complex mixture of tobacco alkaloids. When considering the potential health consequences of exposure to tobacco products, it may be as well to remember that about 2200 components have been separated from tobacco plant extracts (Schmeltz and Hoffman, 1977) and almost 500 compounds in tobacco smoke have been characterized by gas chromatographic techniques (Schumacher *et al.*, 1977).

The present review will focus on literature reports describing the metabolic fates of cotinine (**2**) and β-nicotyrine (**3**), for which the most detailed information is available. Other compounds of interest include nornicotine (**4**), myosmine (**5**), anabasine (**6**), N-methylanabasine (**7**) and anatabine (**8**) about which much less is known. N-Nitrosonornicotine (**9**) and related tobacco derived nitrosamines which have been detected in tobacco are not included in this review. The subject of carcinogenic tobacco related nitrosamines has been treated extensively in the recent literature (Hoffman *et al.*, 1990; Andersen *et al.*, 1991; Batsch and Montesano 1984; Djordjevic *et al.*, 1989).

*The absolute stereochemistry of the chiral centre in naturally occurring nicotine and most of the related tobacco alkaloids and their metabolites is *S* and is represented as such in the structures, although specific designations of *R* and *S* are not included in the text.

Nicotine and Related Alkaloids: Absorption, distribution, metabolism and excretion. Edited by J.W. Gorrod and J. Wahren. Published in 1993 by Chapman & Hall, London. ISBN 0 412 55740 1

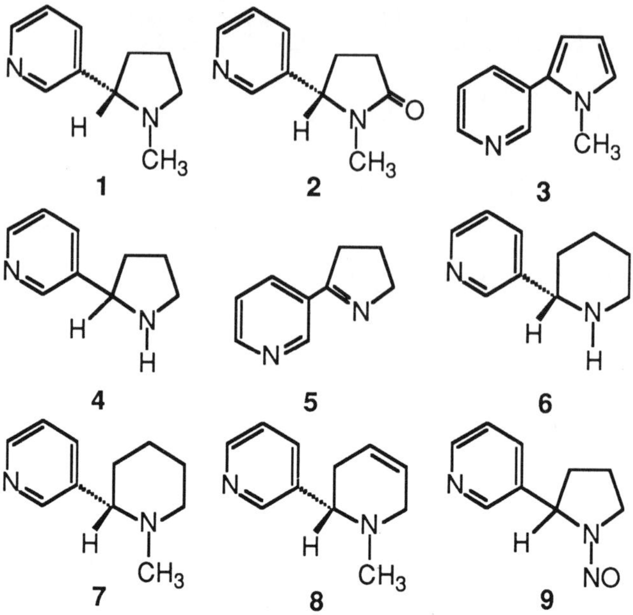

Figure 7.1 Structures of nicotine and the minor tobacco alkaloids.

7.2 THE METABOLIC FATE OF COTININE

After nicotine, the metabolic fate of the γ-lactam cotinine (**2**) has been examined more extensively than any other tobacco alkaloid (Figure 7.2). Cotinine is present in tobacco and may be formed during the fermentation process (Frankenburg and Vaitekunas, 1957). Irradiation of nicotine in the presence of TiO_2 or 9,10-dicyanoanthracene also is reported to yield cotinine in addition to β-nicotyrine and myosmine (Yamada *et al.*, 1986). Interest in the metabolic fate of cotinine is a direct consequence of its pivotal role in nicotine metabolism. Studies in humans indicate that as much as 70% of nicotine is metabolized via cotinine (Jacob *et al.*, 1988). The biotransformation proceeds by a two-step process involving the initial cytochrome P450 catalysed oxidation of nicotine to the iminium species **10** (Hill *et al.*, 1972; Murphy, 1973; Peterson and Castagnoli, 1988) followed by a second two-electron oxidation catalysed by aldehyde oxidase which generates the lactam (Brandange and Lindblom, 1979; Hibberd and Gorrod, 1981; Gorrod and Hibberd, 1982). Recent studies indicate that an NAD dependent dehydrogenase also may convert **10** to cotinine (Obach and Van Vunakis, 1990).

Cotinine undergoes extensive further metabolism *in vivo*. As with the majority of the metabolic pathways described for nicotine, McKennis and co-workers were the first to examine the metabolic fate of cotinine

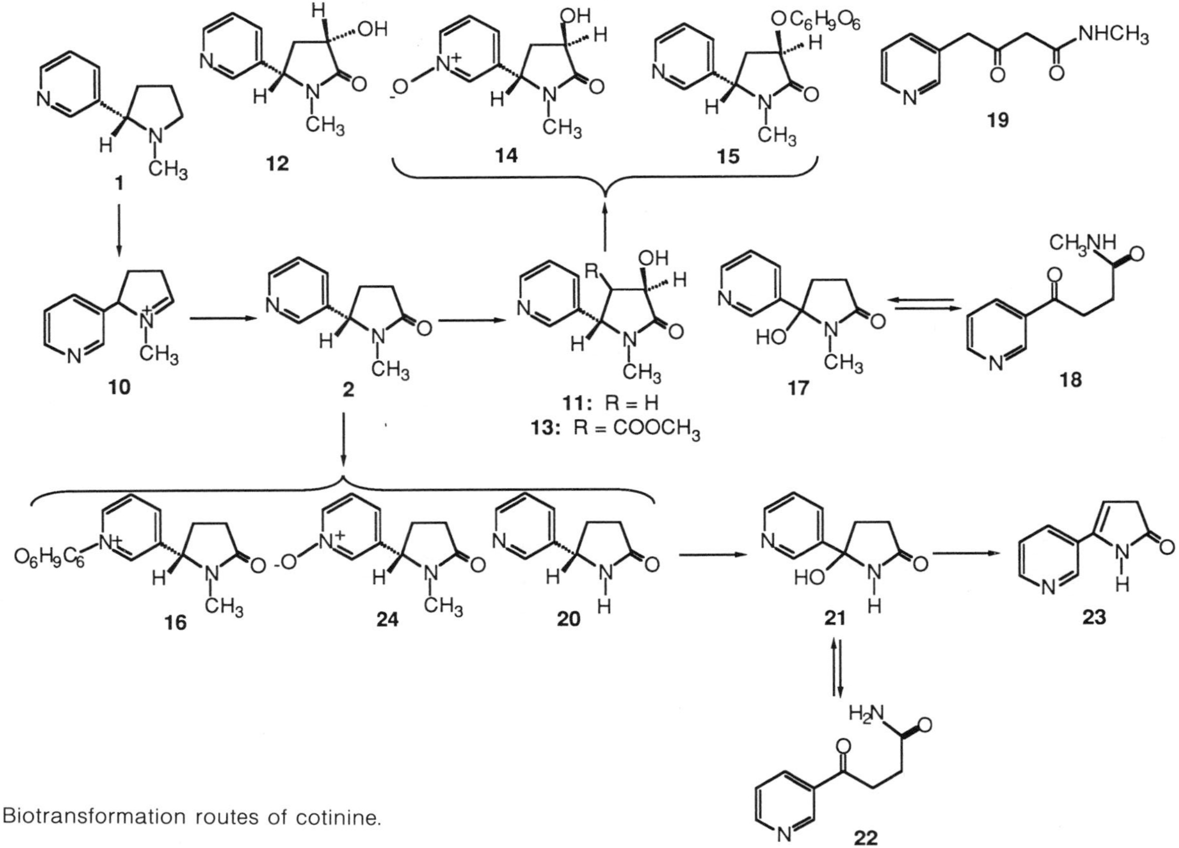

Figure 7.2 Biotransformation routes of cotinine.

(Bowman *et al.*, 1964). A major product isolated from the urine of humans administered large doses of cotinine was tentatively assigned the carbinolamide structure of 3-hydroxycotinine without specifying the configuration about the newly introduced chiral centre at C3 (Bowman and McKennis, 1962). Subsequent studies established that 3-hydroxycotinine isolated from the urine of monkeys treated with cotinine has the *trans* stereochemistry as shown in **11** (Dagne and Castagnoli, 1972a). Although some debate continues to appear in the literature (Kyerematen *et al.*, 1987; Benowitz and Jacob, 1991; Kyerematen *et al.*, 1991a), the majority of quantitative studies indicates that **11** is the principal metabolite excreted in the urine of humans and experimental animals administered nicotine (Foth *et al.*, 1992; Neurath *et al.*, 1987; Voncken *et al.*, 1989; O'Doherty *et al.*, 1988; Cundy and Crooks, 1987; Kyerematen and Vesell, 1991b). A careful analysis of human urine using synthetic standards of both the *cis*-(**12**) and *trans*-isomers failed to detect significant levels of the *cis* isomer (Jacob *et al.*, 1990; Jacob *et al.*, 1988). A recent report describes the synthesis of the 4-methoxycarbonyl analogue **13** of *trans*-3-hydroxycotinine which is to be used to generate antibodies to this metabolite for future studies (Desai and Amin, 1991).

The further metabolism of *trans*-3-hydroxycotinine by rabbit liver preparations is reported to yield the corresponding N-oxide **14** and cotinine (Rashid *et al.*, 1983). The reductive conversion of *trans*-3-hydroxycotinine to cotinine represents a novel metabolic transformation. Recent reports describe the formation of glucuronides of both cotinine and *trans*-3-hydroxycotinine (Kyerematen and Vesell, 1991b; Kyerematen *et al.*, 1990). Presumably it is the carbinol moiety of *trans*-3-hydroxycotinine that is converted to the corresponding O-glucuronide (**15**). The glucuronidation of cotinine, however, must involve quaternization of the pyridyl nitrogen atom to generate the zwitterionic species **16**. A recent publication reports the synthesis of **16** and its conversion to cotinine upon treatment with β-glucuronidase (Caldwell *et al.*, 1992). The identical material was isolated from smokers' urine and is considered by the authors to be a major urinary metabolite of nicotine.

Cotinine is also metabolized to an isomeric carbinolamide, namely 5-hydroxycotinine (**17**). This compound has been synthesized and its chemical behaviour has been examined with the aid of modern spectroscopic instrumentation (Nguyen *et al.*, 1981). The results establish that **17** is in equilibrium with the corresponding ketoamide **18**. ^{1}H NMR and IR studies indicate that the carbinolamide **17** is the principal if not exclusive form of this compound in lipophilic environments. The ketoamide **18** has been identified as a urinary metabolite of nicotine and cotinine (Gorrod and Jenner, 1975). An early report by McKennis (Bowman and McKennis, 1962) which described the open chain compound as the isomeric β-ketoamide **19** was amended later (McKennis *et al.*, 1962).

The enzymatic pathway or pathways leading to the carbinolamides **11** and **17** remains poorly defined. Since these reactions involve a net insertion of oxygen between carbon and hydrogen, it may be reasonable to speculate that they are catalysed by members of the cytochrome P450 superfamily of enzymes (Ortiz de Montellano, 1986). Consistent with a cytochrome P450 pathway, the *in vivo* conversion of cotinine to *trans*-3-hydroxycotinine is accompanied by an apparent deuterium isotope effect of 8 (Dagne *et al.*, 1974). Unfortunately, evidence for the *in vitro* formation of *trans*-3-hydroxycotinine from nicotine and cotinine is scarce (Stålhandske, 1970). A recent study on enzyme induction employing perfused organs and whole cell preparations does not report *trans*-3-hydroxycotinine as a cotinine or nicotine metabolite (Foth *et al.*, 1992). Clearly more detailed mechanistic studies are required to provide a better understanding of the metabolic pathway responsible for these important biotransformations.

Desmethylcotinine (**20**) has been identified in the urine of dogs administered cotinine (McKennis *et al.*, 1959; Harke *et al.*, 1974). Since nornicotine also is metabolized to **20** (Wada *et al.*, 1961) more than one pathway may be involved. Like cotinine, desmethylcotinine apparently is converted to 5-hydroxydesmethylcotinine (**21**) (Kyerematen *et al.*, 1988) which is in equilibrium with the corresponding open chain keto-amide **22**. An unexpected metabolite is the pyrrolinone **23** that is reported to be a long-lived metabolite of nicotine (Kyerematen *et al.*, 1990). The authors and others have found that pyrrolinones of this type tend to be quite unstable (see β-nicotyrine metabolism discussed below). Since **23** has been identified exclusively by GC–MS, it may be possible that it arises from the thermal degradation of **21**.

Finally, the pyridine moiety of cotinine undergoes N-oxidation on the pyridine ring to give the corresponding N-oxide **24**. This conversion has been observed *in vivo* in human (Shulgin *et al.*, 1987) the guinea pig (Cundy and Crooks, 1987) and monkey (Dagne and Castagnoli, 1972b) and *in vitro* in hamster and guinea pig (Hibberd and Gorrod, 1985). It is worth noting that when subjected to gas chromatographic analysis, **24** (and perhaps other pyridine-N-oxides) undergoes reduction via an unknown mechanism to form cotinine (Dagne and Castagnoli, 1972b). Consequently, care must be taken when thermal separation techniques are employed in the analysis of these types of metabolites.

7.3 β-NICOTYRINE

β-Nicotyrine is present in both tobacco products and tobacco smoke (Schmeltz and Hoffman, 1977). This pyrrolic derivative is also formed by photochemical processes (Yamada *et al.*, 1986) and is a potential metabolite of nicotine (Stålhandske and Slanina, 1982). Of particular

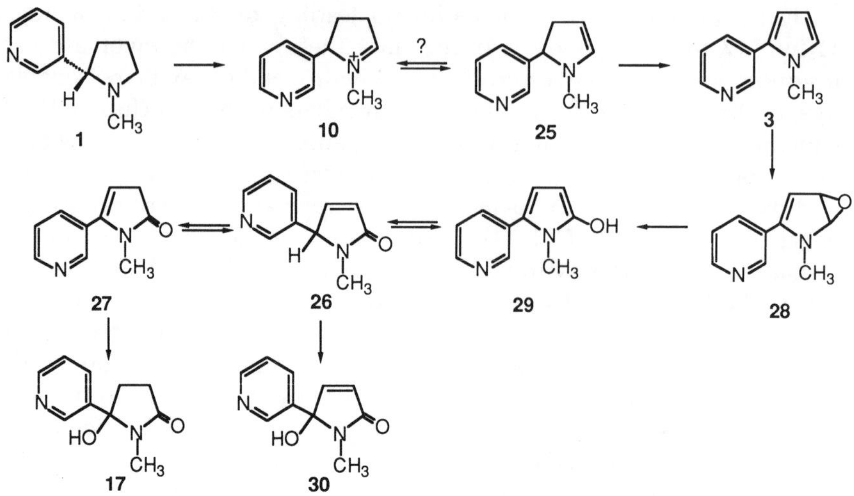

Figure 7.3 Possible bioformation and *in vitro* biotransformation routes for β-nicotyrine.

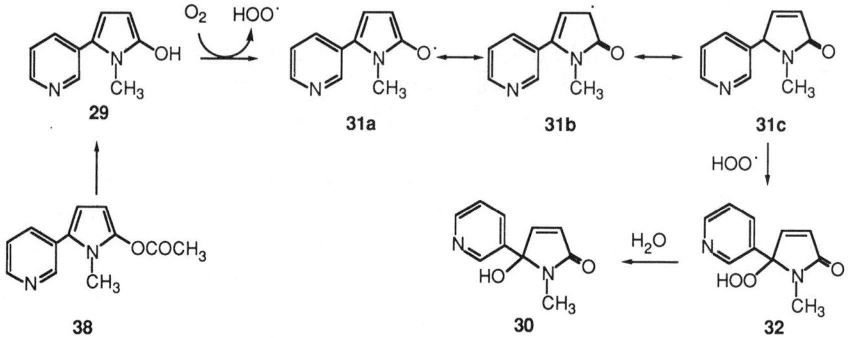

Figure 7.4 Proposed autoxidation route for pyrrolinones derived from the metabolism of β-nicotyrine.

interest is the possible metabolic formation of β-nicotyrine in tissues such as the lung and brain which are known to contain cytochrome P450 oxidase activity but which lack aldehyde oxidase activity. Under these circumstances it is possible that the iminium ion intermediate **10**, formed in the initial two-electron oxidation of nicotine, may undergo further oxidation to β-nicotyrine via the corresponding enamine **25** in a reaction catalysed by an amine oxidase such as monamine oxidase (Shigenaga, 1989a).

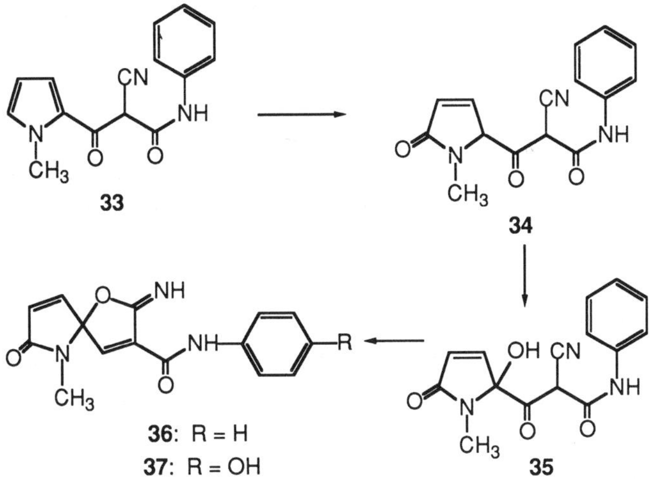

Figure 7.5 Biotransformation route for the pyrrole-containing compound prinomide.

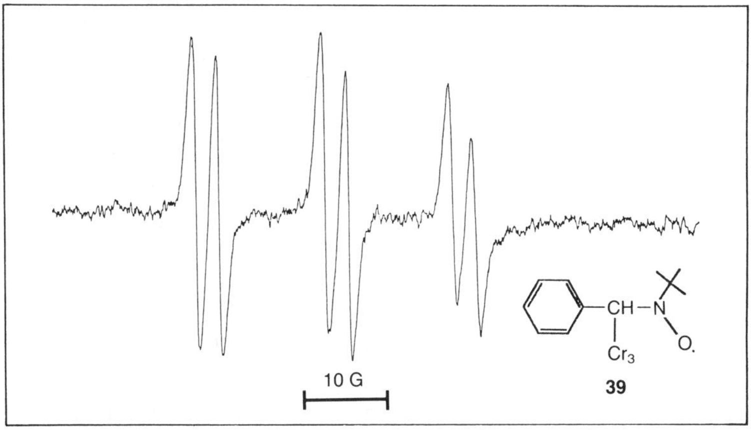

Figure 7.6 Electron spin resonance spectrum of the 5-acetoxypyrrole derived carbon centred radical as its *tert*-butyl-α-phenylnitroxide spin adduct.

The pyrrole moiety which is present in β-nicotyrine is a π-electron excessive heteroaromatic system and therefore would be expected to be susceptible to oxidative metabolic transformations. Analysis of urine samples established that β-nicotyrine is extensively metabolized in the human (Beckett *et al.*, 1972). *In vitro* results confirm the extent to which this system is biotransformed (Jenner and Gorrod, 1973). Studies pursued in our laboratory with rabbit liver and lung microsomal preparations demonstrated the NADPH supported disappearance of this compound and the simultaneous appearance of an unstable metab-

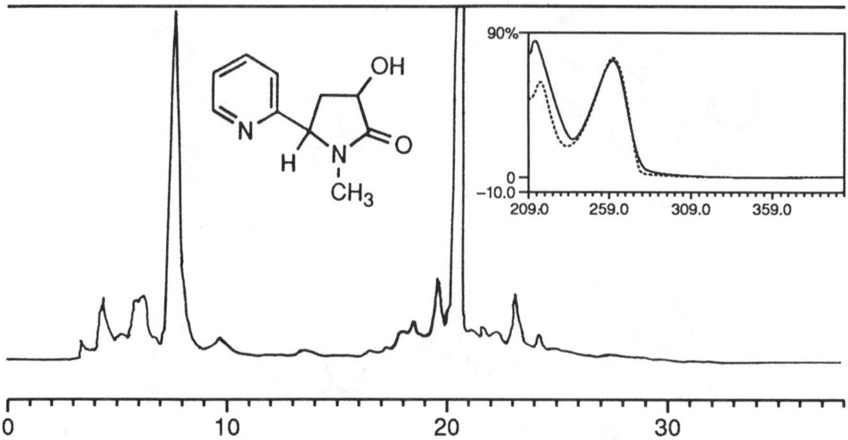

Figure 7.7 HPLC trace of urine extract obtained from a rabbit treated with 50 mg/kg i.p. β-nicotyrine. The inset shows the diode array UV spectra of the metabolite (solid line) and corresponding spectrum of synthetic *trans*-3-hydroxycotinine (broken line).

olite that eventually was characterized as the corresponding mixture of pyrrolinones **26** and **27** (Shigenaga *et al.*, 1989b). The postulated pathway leading to these products is shown in Figure 7.3. The initial step is thought to be a cytochrome P450 catalysed oxidation which generates the unstable arene oxide **28**. Subsequent rearrangement of **28** via the hydroxypyrrole intermediate **29** gives **26** and **27**. These compounds are known to be in equilibrium with each other (and presumably with the hydroxypyrrole tautomer **29**) since in the presence of D_2O deuterium is incorporated into the 3 and 5 positions. Similar conclusions have been reached with studies on the corresponding 1-methylpyrrolin-2-ones (compounds **26, 27** and **29** in which the pyridyl moiety is replaced with a proton) (Baker and Sifniades, 1979).

In addition to the pyrrolinones, the authors have identified, with the aid of synthetic standards, smaller quantities of 5-hydroxycotinine (**17**) and the 5-hydroxypyrrolinone **30** in NADPH supplemented rabbit liver microsomal preparations. Compound **17** is likely to be generated via hydration of the 4-pyrrolinone intermediate **27** during the acid work-up of the incubation mixture (see below) while the hydroxypyrrolinone **30** is postulated to result from an autoxidative reaction proceeding via the 5-hydroxypyrrole intermediate **29** as summarized in Figure 7.4. It is visualized that the highly electron rich hydroxypyrrole loses an electron to dioxygen to generate the hydroperoxy radical (HOO•) and the radical species **31** which is stabilized by the three resonance forms **31a, 31b** and **31c**. Recombination of the hydroperoxy radical with **31c** generates intermediate **32** which, following hydrolysis, yields the

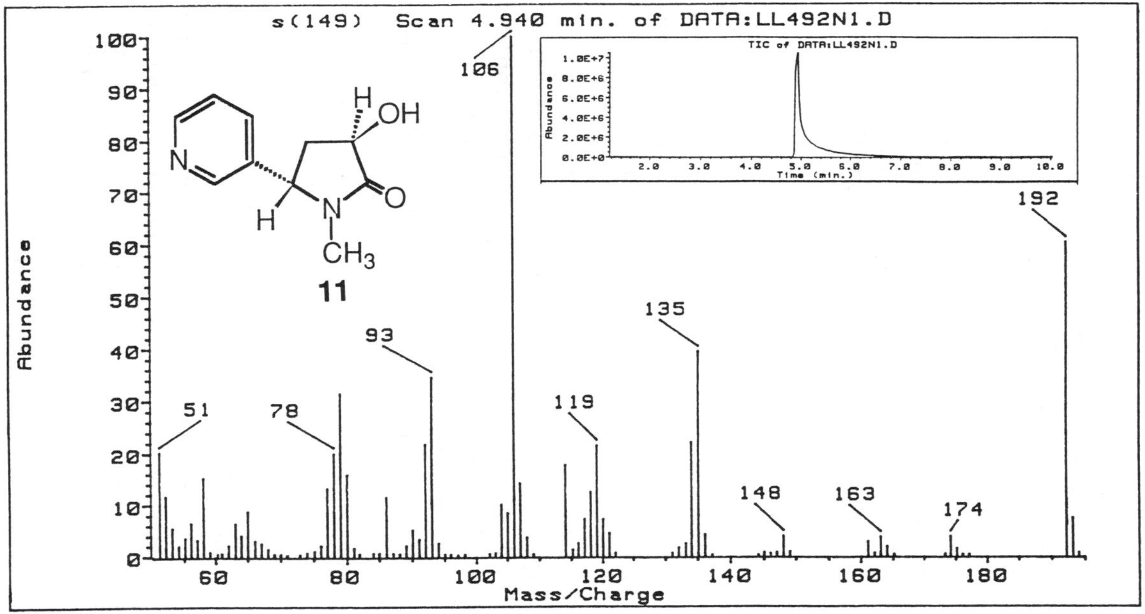

Figure 7.8 GC–EI mass spectrum of the β-nicotyrine derived metabolite which is identical to that of *trans*-3-hydroxycotinine. Inset shows the ion chromatogram monitored at m/z 192, the parent ion for the compound.

hydroxypyrrolinone. This proposal is consistent with the initial observation that, upon standing under mildly basic conditions, the initial mixture of products detected in the β-nicotyrine incubation mixture undergoes spontaneous oxidation to yield **30** (Shigenaga *et al.*, 1989b).

A similar biotransformation route has been reported for the pyrrole containing anti-inflammatory drug prinomide (**33**) (Hawkins, 1989). *In vivo* studies in primates with radiolabelled drug led to the formation of the corresponding pyrrolinone **34** which was chemically unstable. The major and urinary metabolites of this drug were the corresponding spiro-lactams **36** and **37** which are likely to be formed via cyclization of the corresponding hydroxypyrrolinone **35** as shown in Figure 7.5.

More recently the authors have attempted to characterize the fate of the initially formed pyrrolinones in greater detail with the aid of the 5-acetoxypyrrole derivative **38**. This compound was prepared by treatment of 5-hydroxycotinine with acetic anhydride. At low pH, **38** undergoes stoichiometric hydrolysis to yield 5-hydroxycotinine, one of the metabolites isolated from incubations of β-nicotyrine under acid work-up conditions. At neutral pH, a mixture of 5-hydroxycotinine and the 5-hydroxypyrrolinone is produced. Under basic conditions a mixture of the pyrrolinones **26** and **27** is formed rapidly and then undergoes oxidation to the 5-hydroxypyrrolinone and a variety of undetermined decomposition products. Thus, the behaviour of this synthetic, blocked analogue of **29** parallels that of the hydroxypyrrole/pyrrolinone system generated metabolically from β-nicotyrine. Additional support for this general conclusion has been obtained from ESR studies. Incubations of the 5-acetoxypyrrole derivative under neutral conditions led to a carbon centred radical that was trapped by N-*tert*-butyl-α-phenylnitrone. The spectrum (Figure 7.6) displays a doublet of triplets, consistent with a trapped *tert*-carbon centred radical species (**39**) in which the radical is split by the nitrogen quadrapole and the β-proton dipole into the six lines observed (Janzen and Haire, 1990).

These *in vitro* studies have now been supplemented with *in vivo* studies, also in the rabbit. Twenty-four hour urine samples obtained from rabbits administered i.p. 50 mg/kg β-nicotyrine contained very little unchanged drug. Small amounts of 5-hydroxycotinine and the 5-hydroxypyrrolinone were detected with the aid of HPLC–diode array analysis. By far the principal metabolite present in the urine had the HPLC retention time and diode array UV spectrum shown in Figure 7.7. These characteristics are identical with those of synthetic *trans*-3-hydroxycotinine (**11**). This was an unexpected finding and therefore additional evidence to support this structure assignment was pursued. GC-EIMS ion chromatographic analysis of the urine extract obtained from the treated rabbits showed a single peak when monitored at m/z 192, the parent ion corresponding to a monohydroxylated cotinine deri-

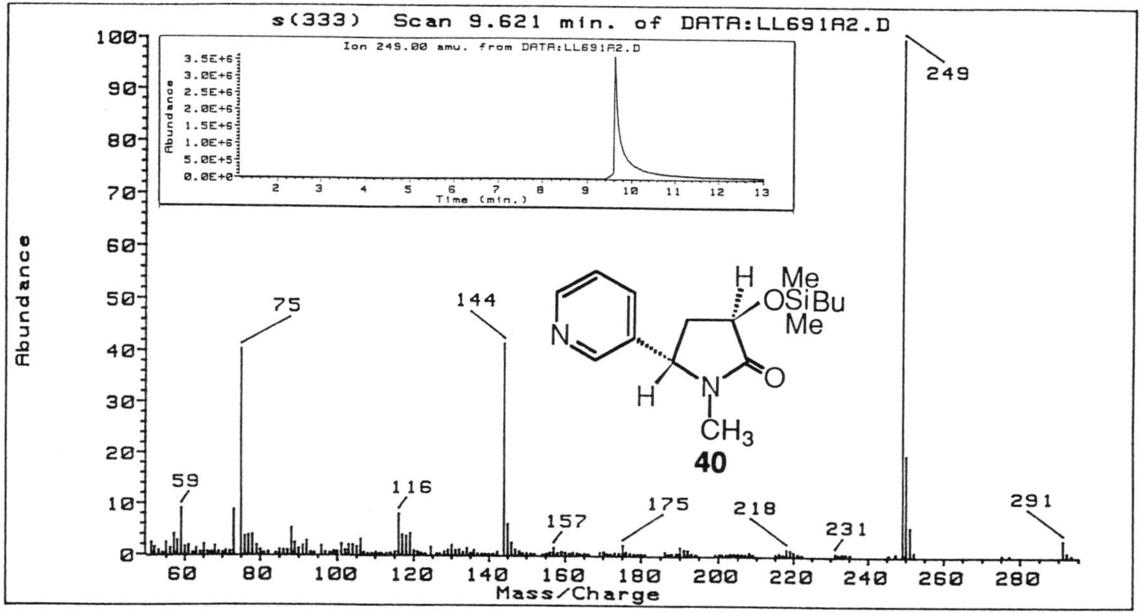

Figure 7.9 GC–EI mass spectrum of the β-nicotyrine derived metabolite as the corresponding *tert*-butyl dimethylsilyl derivative which is identical to that of the *trans*-3-hydroxycotinine derivative. The inset shows the ion chromatogram monitored at m/z 249, the base peak for the compound.

vative. The retention time and mass spectrum of this material were identical to those of authentic *trans*-3-hydroxycotinine (Figure 7.8) and were different from those of 5-hydroxycotinine. Finally, the authors analysed the corresponding O-*tert*-butyldimethylsilyl derivative **40** by GC-EIMS. Once again the retention time of the ion chromatogram (monitored at m/z 249, the base peak of **40**) and mass spectral features were identical to those of the synthetic standard (Figure 7.9).

The results of the *in vivo* studies led the authors to propose the metabolic pathway for β-nicotyrine summarized in Figure 7.10. The cytochrome P450 catalysed oxidation of β-nicotyrine generates the mixture of pyrrolinones **26** and **27** which then undergoes autoxidation to yield the 5-hydroxypyrrolinone **30**. Hydration of the 4-pyrrolinone isomer **27** gives 5-hydroxycotinine (**17**). Although without authentic standards it is not possible to rule out the isomeric 4-hydroxycotinine analogue **41**, the results of the present studies are interpreted as strong evidence that β-nicotyrine is also transformed *in vivo* to 3-hydroxycotinine (**11**). The diastereochemical features of the metabolite remain to be determined. Attempts to identify this compound in the rabbit liver microsomal preparations and in the various hydrolytic mixtures of the 5-acetoxypyrrole compound failed. Therefore, it appears that metabolic 3-hydroxycotinine is not derived from the pyrrolinone metabolites **26** or **27**. To what extent the 3-hydroxycotinine found in the urine of smokers may be derived from β-nicotyrine remains to be determined.

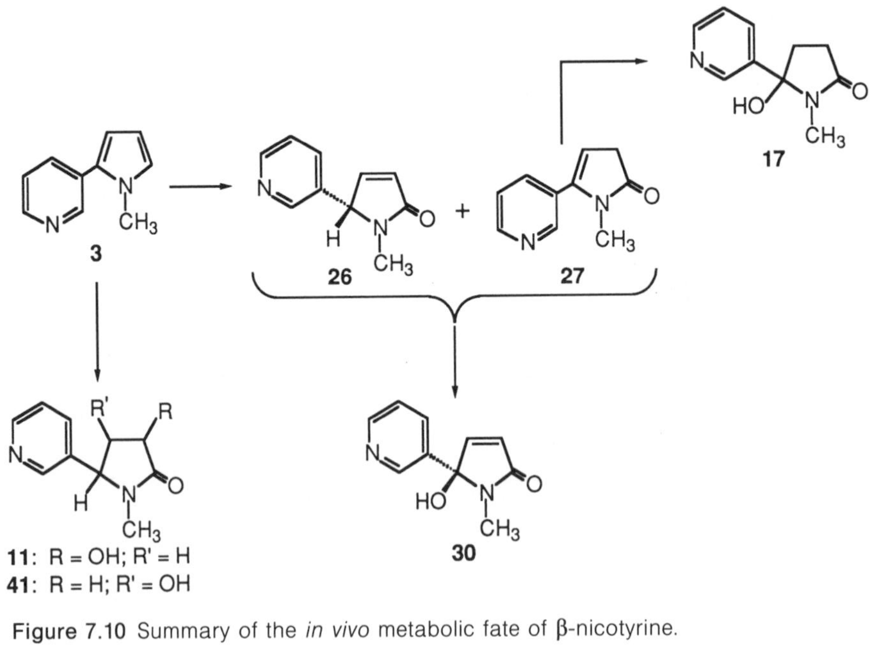

Figure 7.10 Summary of the *in vivo* metabolic fate of β-nicotyrine.

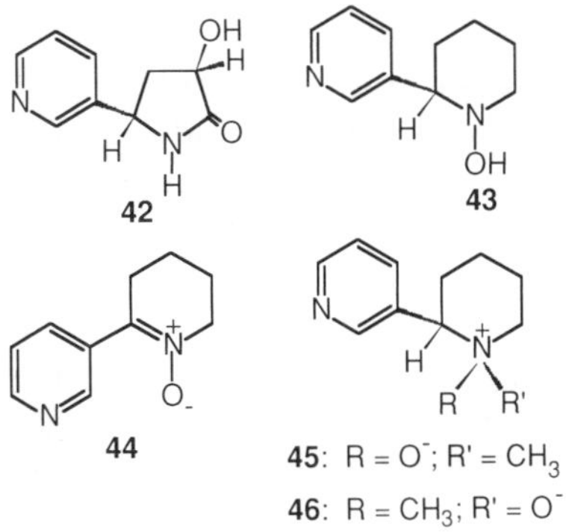

Figure 7.11 Structures of metabolites derived from nornicotine, anabasine and N-methylanabasine.

7.4 OTHER MINOR TOBACCO ALKALOIDS

With the exception of nornicotine (**4**), which is viewed as a major tobacco alkaloid (Zhang *et al.*, 1990), very few studies have been reported on the other minor tobacco alkaloids. As mentioned in the discussion on the metabolism of cotinine, desmethylcotinine (**20**) (Wada *et al.*, 1961) is likely to be derived from nornicotine as well as from cotinine. It is of interest that 5-hydroxydesmethylcotinine (**21**), which may arise from metabolic oxidation of **20**, has been reported as a metabolite of nicotine in the rat (Kyerematen *et al.*, 1988) while the corresponding 3-hydroxy isomer (**42**, see Figure 7.11) has not been detected. It will be recalled that *trans*-3-hydroxycotinine (**11**) is the principal metabolite of nicotine excreted by humans. Myosmine (**5**), which is derived metabolically from nornicotine in tobacco plants (Leete and Chedekel, 1974), apparently has not been detected as a human metabolite of nornicotine.

In vitro studies in rodents on the metabolism of anabasine (**6**) have established its conversion to the corresponding N-hydroxy derivative **43** (Figure 7.11) which subsequently is oxidized to the nitrone **44** (Beckett *et al.*, 1972). N-Methylanabasine (7) is reported to undergo N-oxidation to form the diastereomeric N′-oxides **45** and **46** but apparently does not undergo α carbon oxidation as is the case with nicotine (Jenner and Gorrod, 1973). For a more detailed discussion of the early reports on the metabolism of these compounds, the reader is referred to a review by Gorrod and Jenner (Gorrod and Jenner, 1975). It should be apparent from the brevity of these comments that, with the exception of cotinine and β-nicotyrine, very little is known about the metabolic fates of the minor tobacco alkaloids.

ACKNOWLEDGEMENT

Support for unpublished studies described in this chapter from the Harvey W. Peters Center for the Study of Parkinson's Disease and Central Nervous System Disorders is gratefully acknowledged.

REFERENCES

Andersen, R.A., Fleming, P.D., Burton, H.R. *et al.* (1991) Nitrosated, acylated, and oxidized pyridine alkaloids during storage of smokeless tobaccos: Effects of moisture, temperature, and their interactions. *J. Agric. Food Chem.*, **39**, 1280–87.

Baker, J.T. and Sifniades, S. (1979) Synthesis and properties of pyrrolin-2-ones. *J. Org. Chem.*, **44**, 2798–800.

Batsch, H. and Montesano, R. (1984) Relevance of nitrosamines to human cancer. *Carcinogenesis*, **5**, 1381–93.

Beckett, A.H., Gorrod, J.W. and Jenner, P. (1972) A possible relation between pKa1 and lipid solubility and the amounts excreted in urine of some tobacco alkaloids given to man. *J. Pharm. Pharmacol.*, **23**, 55S–61S.

Benowitz, N.L. and Jacob, III, P. (1991) Nicotine metabolism in humans. *Clin. Pharmacol. Ther.*, **50(4)**, 462–3.

Bowman, E.R. and McKennis, Jr., H. (1962) Studies on the metabolism of (−)-cotinine in the human. *J. Pharmacol. Exp. Ther.* **135**, 306–11.

Bowman, E.R., Hansson, E., Turnbull, L.B. *et al.* (1964) Disposition and fate of (−)-cotinine-H^3 in the mouse. *J. Pharmacol. Exp. Ther.*, **143**, 301–8.

Brandange, S. and Lindblom, L. (1979) The enzyme 'aldehyde oxidase' is an iminium oxidase. *Biochem. Biophys. Res. Comm.*, **91**, 991–6.

Caldwell, W.S., Green, J.M., Byrd, G.D. *et al.* (1992) Characterization of the glucuronide conjugate of cotinine – A previously unidentified major metabolite of nicotine in smoker's urine. *Chem. Res. in Toxicol*, **5**, 280–5.

Clark, M.S.G., Rand, M.J. and Vanov, S. (1965) Comparison of pharmacological activity of nicotine and related alkaloids occurring in cigarette smoke. *Arch. Int. Pharmacodyn.*, **156(2)**, 363–78.

Cundy, K.C. and Crooks, P.A. (1987) Biotransformation of primary nicotine metabolites II. Metabolism of [^{3}H]-S-(−)-cotinine in the guinea pig: determination of *in vivo* urinary metabolites by high-performance liquid radiochromatography. *Xenobiotica*, **17(7)**, 785–92.

Dagne, E. and Castagnoli, Jr., N. (1972a) Structure of hydroxycotinine, a nicotine metabolite. *J. Med. Chem.* **15**, 356–60.

Dagne, E. and Castagnoli, Jr., N. (1972b) Cotinine N-oxide, a new metabolite of nicotine. *J. Med. Chem.*, **15**, 840–1.

Dagne, E., Gruenke, L. and Castagnoli, Jr. N. (1974) Deuterium isotope effects in the *in vivo* metabolism of cotinine. *J. Med. Chem.*, **17**, 1330–3.

Desai, D.H. and Amin, S. (1991) Synthesis of a hapten to be used in development of immunoassays for trans-3′-hydroxycotinine, a major metabolite of cotinine. *Chem. Res. in Toxicol*, **4**, 524–7.

Djordjevic, M.V., Brynnemann, K.D. and Hoffmann, D. (1989) Identification and analysis of a nicotine-derived N-nitrosamine acid and other nitrosamino acids in tobacco. *Carcinogenesis*, **10(9)**, 1725–31.

Foth, H., Aubrecht, J., Horne, M. *et al.* (1992) Increased cotinine elimination and cotinine-N-oxide formation by phenobarbital induction in rat and mouse. *Clinical Investigator*, **70**, 175–81.

Frankenburg, W.G. and Vaitekunas, A.A. (1957) The chemistry of tobacco fermentation. I. Conversion of the alkaloids. D. Identification of cotinine in fermented leaves. *J. Am. Chem. Soc.*, **79**, 149–51.

Gorrod, J.W. and Hibberd, A.R. (1982) The metabolism of nicotine $\Delta^{1'(5')}$ iminium ion, *in vivo* and *in vitro*. *Eur. J. Drug Metab. Pharmacokin.*, **7**, 293–8.

Gorrod, J.W. and Jenner, P. (1975) The metabolism of tobacco alkaloids, in *Essays in Toxicology*, Vol. 6, Academic Press, New York, pp. 35–78.

Harke, H.-P., Schüller, D., Frahm, B., *et al.* (1974) Demethylation of nicotine and cotinine in pigs. *Res. Comm. Chem. Path. Pharmacol.*, **9(4)**, 595–9.

Hawkins, D.R. (Ed.) (1989) *Biotransformation. A survey of the biotransformations of drugs and chemicals in animals*, Vol. 2, The Royal Society of Chemistry, Cambridge (England), pp. 310–11.

Hibberd, A.R. and Gorrod, J.W. (1981) Nicotine $\Delta^{1'(5')}$ iminium ion: a reactive intermediate in nicotine metabolism. *Adv. Exp. Med. Biol.*, **136(Part B)**, 1121–31.

Hibberd, A.R. and Gorrod, J.W. (1985) Comparative N-oxidation of nicotine and cotinine by hepatic microsomes, in *Biological Oxidation of Nitrogen in Organic Molecules*, (Eds J.W. Gorrod and L.A. Damani) Ellis Horwood, Chichester (England), pp. 246–50.

Hill, D.L., Laster, Jr., W.R. and Struck, R.F. (1972) Enzymatic metabolism of cyclophosphamide and nicotine and production of a toxic cyclophosphamide metabolite. *Cancer Res.*, **32**, 658–65.

Hoffman, D., Harly, N.H., Fisenne, I. *et al.* (1990) Carcinogenic agents in snuff. *J. Nat. Cancer Inst.*, **76(3)**, 435–7.

Jacob, III, P., Benowitz, N.L. and Shulgin, A.T. (1988) Recent studies of nicotine metabolism in humans. *Pharmacol., Biochem. and Behavior*, **30**, 249–53.

Jacob, III, P., Shulgin, A.T. and Benowitz, N.L. (1990) Synthesis of (3'*R*, 5'*S*)-*trans*-3'-hydroxycotinine, a major metabolite of nicotine. Metabolic formation of 3'-hydroxycotinine in humans is highly stereoselective. *J. Med. Chem.*, **33**, 1888–91.

Janzen, E.G. and Haire, D.L. (1990) Two decades of spin trapping, in *Advances in Free Radical Chemistry*, Vol. 1, (Ed. D. Tanner) JAI Press Inc., Greenwich, Connecticut, pp. 253–95.

Jenner, P. and Gorrod, J.W. (1973) Comparative *in vitro* hepatic metabolism of some tertiary N-methyl tobacco alkaloids in various species. *Res. Comm. Chem. Pathol. Pharmacol.*, **6(3)**, 829–43.

Kyerematen, G.A., Morgan, M.L., Chattopadhyay, B. *et al.*· (1991a) Nicotine metabolism in humans. *Clin. Pharmacol. Ther.*, **50(4)**, 462–3.

Kyerematen, G.A. and Vesell, E.S. (1991b) Metabolism of nicotine. *Drug Metabolism Reviews*, **23(1&2)**, 3–41.

Kyerematen, G.A., Morgan, M. and Chattopadhyay, B. *et al.* (1990) Disposition of nicotine and eight metabolites in smokers and nonsmokers: Identification in smokers of two metabolites that are longer lived than cotinine. *Clin. Pharmacol. Ther.*, **48(6)**, 641–51.

Kyerematen, G.A., Taylor, L.H., deBethizy, J.D. *et al.* (1987) Radiometric-high-performance liquid chromatographic assay for nicotine and twelve of its metabolites. *J. Chromatog.*, **419**, 191–203.

Kyerematen, G.A., Taylor, L.H., deBethizy, J.D. *et al.* (1988) Pharmacokinetics of nicotine and 12 metabolites in the rat. *Drug Metabol. Dispos.*, **16**, 125–9.

Leete, E. and Chedekel, M.R. (1974) Metabolism of nicotine in *Nicotiana glauca*. *Phytochemistry*, **13**, 1853–9.

McKennis, Jr., H., Schwartz, S.L., Turnbull, L.B. *et al.* (1962) The corrected structure of ketoamide arising from the metabolism of (−)-nicotine. *J. Am. Chem. Soc.*, **84**, 4598–9.

McKennis, Jr., H., Turnbull, L.B., Bowman, E.R. *et al.* (1959) Demethylation of cotinine *in vivo*. *J. Am. Chem. Soc.*, **81**, 3951–4.

Murphy, P.J. (1973) Enzymatic oxidation of nicotine to nicotine-$\Delta^{1'(5')}$ iminium ion. A newly discovered intermediate in the metabolism of nicotine. *J. Biol. Chem.*, **248**, 2796–800.

Neurath, G.B., Dunger, M. Orth, D. *et al.* (1987) *Trans*-3′-hydroxycotinine as a main metabolite in urine of smokers. *Int. Arch. Occup. Environ. Health*, **59**, 199–201.

Nguyen, T.-L., Dagne, E., Gruenke, L. *et al.* (1981) The tautomeric structures of 5-hydroxycotinine, a secondary mammalian metabolite of nicotine. *J. Org. Chem.*, **46**, 758–60.

O'Doherty, S., Revans, A., Smith, C.L. *et al.* (1988) Determination of *cis*- and *trans*-3-hydroxycotinine by high performance liquid chromatography. *J. High Resol. Chromatog. & Chromatog. Comm.*, **11**, 723–5.

Obach, R.S. and Van Vunakis, H. (1990) Nicotinamide adenine dinucleotide (NAD)-dependent oxidation of nicotine-$\Delta^{1'(5')}$-iminium ion to cotinine by rabbit liver microsomes. *Biochem. Pharmacol.*, **39(1)**, R1–R4.

Ortiz de Montellano, P.R. (Ed.) (1986) *Cytochrome P450*, Plenum Press, New York.

Peterson, L.A. and Castagnoli, Jr., N. (1988) Regio- and stereochemical studies on the α-carbon oxidation of (*S*)-nicotine by cyctochrome P450 model systems. *J. Med. Chem.*, **31**, 637–40.

Rashid, M.A., Nikolin, B. and Nikolin, A. (1983) Investigation of some analytical, biochemical and toxicological characteristics of hydroxycotinine. *Folia Medica*, **18**, 17–27.

Schmeltz, I. and Hoffmann, D. (1977) Nitrogen-containing compounds in tobacco and tobacco smoke. *Chem. Rev.*, **77(3)**, 295–311.

Schumacher, J.N., Green, C.R., Best, F.W. *et al.* (1977) Smoke composition. An extensive investigation of the water-soluble portion of cigarette smoke. *J. Agricultural Food Chem.*, **25(2)**, 310–5.

Shigenaga, M. (1989a) Studies on the metabolism and bioactivation of (*S*)-nicotine and β-nicotyrine. Ph.D. Thesis, University of California, San Francisco.

Shigenaga, M.K., Kim, B.H., Caldera-Munoz, P. *et al.* (1989b) Liver and lung microsomal metabolism of the tobacco alkaloid β-nicotyrine. *Chem. I Res. Toxicol.*, **2**, 282–7.

Shulgin, A.T., Jacob, III, P., Benowitz, N.L. *et al.* (1987) Identification and quantitative analysis of cotinine-N-oxide in human urine. *J. Chromatog., Biomed. Applications*, **423**, 365–72.

Stålhandske, T. (1970) The metabolism of nicotine and cotinine by mouse liver preparation. *Acta Physiol. Scand.*, **78**, 236–48.

Stålhandske, T. and Slanina, P. (1982) Nicotyrine inhibits *in vivo* metabolism of nicotine without increasing its toxicity. *Toxicol. Appl. Pharmacol.*, **65**, 366–72.

Voncken, P., Schepers, G. and Schafer, K.-H. (1989) Capillary gas chromatographic determination of *trans*-3′-hydroxycotinine simultaneously with nicotine and cotinine in urine and blood samples. *J. Chromatog.*, **479**, 410–8.

Wada, E. Bowman, E., Turnbull, L.B. *et al.* (1961) Norcotinine (desmethylcotinine) as a urinary metabolite of nornicotine. *J. Med. Pharmacol. Chem.*, **4(1)**, 21–30.

Yamada, S., Sakai, T. and Ohashi, M. (1986) A model reaction of nicotine metabolism-photo-induced electron transfer oxidation of nicotine. *Photomedicine & Photobiology*, **8(2)**, 19–20.

Zhang, Y., Jacob, III, P. and Benowitz, N.D. (1990) Determination of nornicotine in smokers' urine by gas chromatography following reductive alkylation to N′-propylnornicotine. *J. Chromatog.*, **525**, 349–57.

Nicotine and metabolites: analysis and levels in body fluids

8

M. Curvall and E. Kazemi Vala

8.1 INTRODUCTION

Nicotine exposure during smoking, snuff usage and exposure to environmental tobacco smoke has been estimated by measuring nicotine and cotinine in body fluids. The best estimate of nicotine intake could be obtained from metabolic data. In recent years, several studies have been performed to determine the rate of metabolism, elucidation of the structure of the metabolites, quantitative assessment of the metabolic pathways and determination of urinary excretion profiles in tobacco users. These studies call for sensitive and reliable methods for the quantification of nicotine and its metabolites in different body fluids.

Both immunoassays and chromatographic methods have been developed for the analysis of nicotine and metabolites. The chromatographic determination of nonpolar metabolites of nicotine is usually carried out by high-resolution gas chromatography (HRGC) with selective detectors such as nitrogen–phosphorus (NPD) or mass spectrometric (MS) detectors. Polar metabolites, conjugates and quarternary metabolites are either converted to volatile derivatives before gas chromatographic analysis or analysed by high-performance liquid chromatography (HPLC) with UV, radiometric or MS detection (Figure 8.1).

Twenty or more metabolites of nicotine have been isolated and identified in urine from different species. Recent studies have shown that the main metabolites of nicotine found in urine after intravenous adminis-

Nicotine and Related Alkaloids: Absorption, distribution, metabolism and excretion. Edited by J.W. Gorrod and J. Wahren. Published in 1993 by Chapman & Hall, London. ISBN 0 412 55740 1

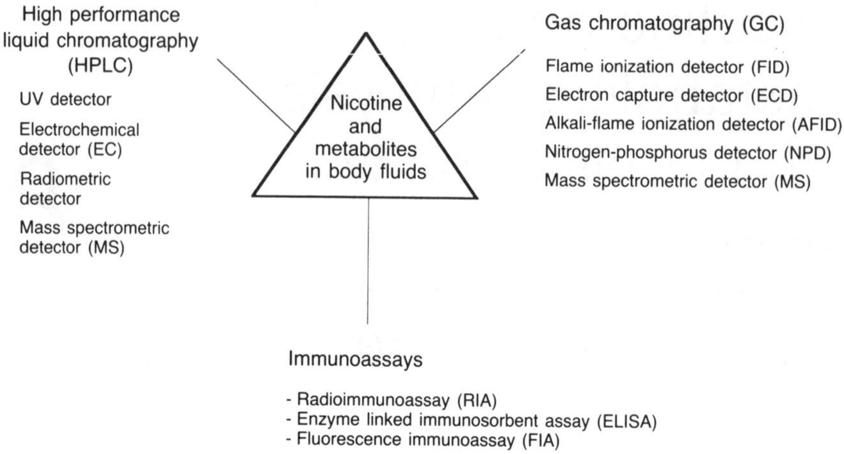

Figure 8.1 Bioanalytical methods for the analysis of nicotine and metabolites.

tration of nicotine to human and in urine of tobacco users are cotinine, *trans*-3′-hydroxycotinine, glucuronic acid derivatives of nicotine, cotinine and 3′-hydroxycotinine, nicotine-1′-N-oxide, cotinine-1-N-oxide, nornicotine, norcotinine and N-methyl-nicotinium ions (Figure 8.2).

In this review, the analysis of concentrations of nicotine and metabolites in body fluids of humans exposed to nicotine is summarized.

8.2 NICOTINE AND COTININE

Nicotine is an alkaloid and a tertiary amine and comprises a pyridine and a pyrrolidine ring. It is a weak base with a pKa value of 7.9 and is soluble both in water and in lipids. Nicotine is excreted partially unchanged by the kidney, but largely in the form of 20 or more different metabolites, which contain an intact pyridine ring. The major primary metabolite of nicotine in most species is cotinine. The conversion of nicotine to cotinine is a two-step reaction, which involves both a cytochrome P450 and an aldehyde oxidase enzyme system. Nicotine is absorbed through the mucous membranes of the mouth and the bronchial tree as well as across the alveolar capillary membrane. The extent of mucal absorption varies with the pH, such that nicotine is well absorbed in the mouth from alkaline (cigar) smoke and buffered moist snuff or chewing gum, but little is absorbed from acidic (cigarette) mainstream smoke (Armitage *et al.*, 1978).

There is a wide individual variation in blood, saliva and urine nicotine levels among smokers just after smoking a cigarette. Russell *et al.* (1980) have found that blood nicotine concentrations in 330 smokers varied from 4 to 72 ng/ml (average 33 ng/ml). There was a low correlation

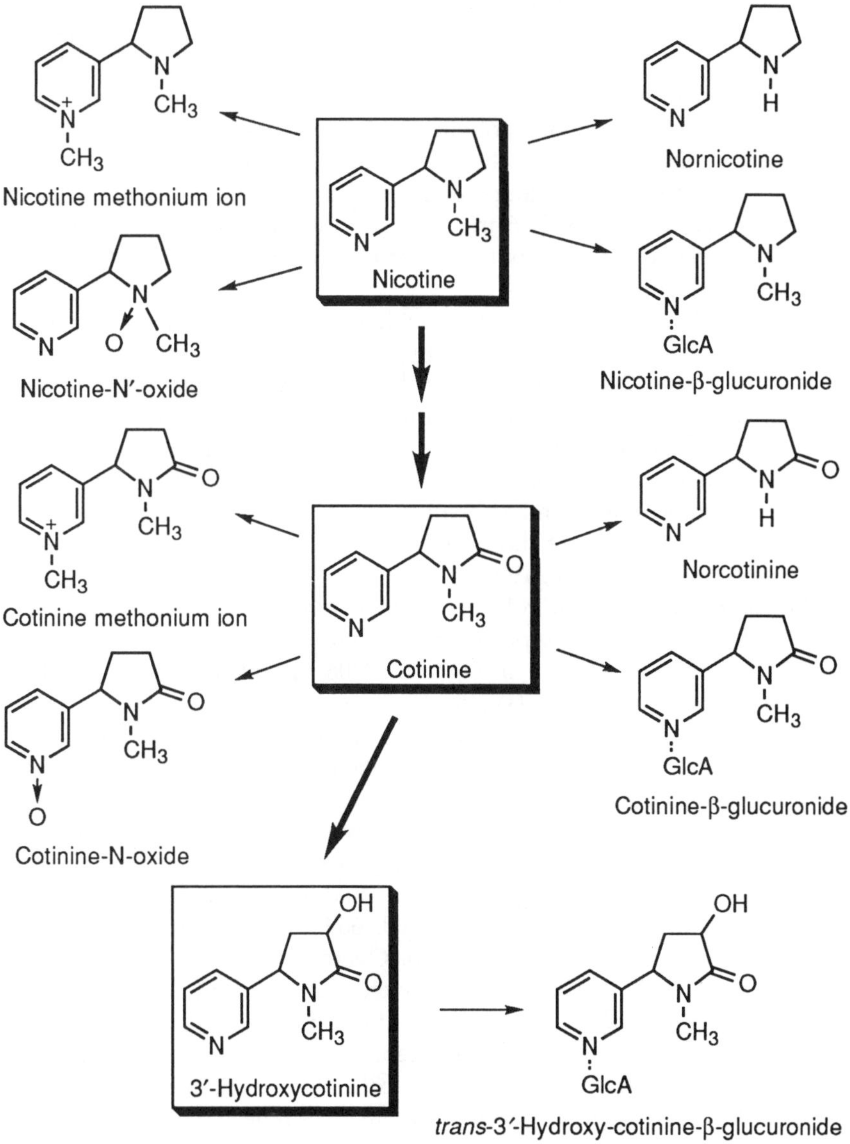

Figure 8.2 Urinary metabolites of nicotine and cotinine.

between blood nicotine concentrations and nicotine yields of cigarettes. Similar results were obtained by Herning *et al.* (1983), who found that only 25% of the individual differences in blood nicotine levels were attributable to the machine-determined nicotine yield, while 50 to 60% were attributable to individual differences in smoking behaviour.

Plasma, saliva and urinary levels of nicotine poorly reflect the actual intake of nicotine during cigarette smoking. The level of daily intake of nicotine is better estimated from metabolic clearance data obtained after intravenous infusion of nicotine in combination with blood and urinary nicotine concentration data obtained over 24 h when the subjects are exposed to tobacco. Using this estimation, Benowitz and Jacob (1984) have found that the average daily intake of nicotine in 22 subjects smoking cigarettes was 37.6 mg with a wide range of 10.5 to 78.6 mg. Intake correlated strongly with cigarettes smoked per day but not with machine-determined yields.

Cotinine has a longer elimination half-life than nicotine (15 h compared with 2 h) and is therefore more suitable for estimating nicotine exposure during the preceding few days (Benowitz *et al.*, 1983). Studies in which nicotine exposure was mimicked by an intravenous infusion of nicotine to smokers (Galeazzi *et al.*, 1985) and nonsmokers (Curvall *et al.*, 1990), have shown that the steady-state levels of plasma or saliva cotinine significantly correlate with nicotine intake. Saliva cotinine concentrations closely reflect those in plasma with a mean saliva to plasma ratio of 1.2 (Curvall *et al.*, 1990).

Mean serum levels of cotinine in men who smoked cigarettes (n = 50), cigars (n = 70) and pipes (n = 56) were found to be higher in pipe smokers (389 ng/ml) than in cigarette (306 ng/ml) and cigar smokers (121 ng/ml), respectively (Wald *et al.*, 1981). Morning plasma, saliva and urine cotinine concentrations were studied in two groups of smokers; nine active smokers who smoked less than ten cigarettes per day and 40 smokers who smoked more than 40 cigarettes per day. Although there was a relatively high degree of intersubject variability of cotinine levels among subjects claiming to smoke the same number of cigarettes per day, the light smokers had considerably lower mean levels in serum, saliva and urine (78.0, 66.9, 646.8 ng/ml) than heavy smokers (301.2, 283.7, 1100.7 ng/ml). Overall, only a slight correlation was obtained between cigarettes smoked per day and urine cotinine/creatinine (Wall *et al.*, 1988). Determination of cotinine concentrations in saliva and plasma is valid to differentiate smokers from nonsmokers (Pojer *et al.*, 1984) and as salivary collection is preferable to venepuncture in most field situations, salivary cotinine is considered to be the marker of choice in evaluating smoking prevalence (Haley *et al.*, 1983; Giusto and Eckhard, 1986; Carey and Abrams, 1988; McNeill *et al*, 1987; Abrams *et al.*, 1987; Van Vunakis *et al.*, 1989).

At pH 7.4 only 24% of nicotine is un-ionized and this can cross lipophilic cell membranes. As moist snuff and nicotine gum is buffered to an alkaline pH it is easily absorbed through the mucous membranes. Absorption of nicotine from cigarette smoke is facilitated by the huge alveolar surface area and after rapid absorption through the lungs, blood concentrations of nicotine rise quickly and reach a maximum level at the completion of cigarette smoking. In the case of smokeless tobacco and nicotine gum there is a prolonged absorption after termination of chewing and snuff taking. Nicotine is well absorbed from smokeless tobacco. Peak concentrations of nicotine in plasma after single doses of oral snuff or chewing tobacco are similar to, but are more sustained than, those seen after smoking a cigarette (Benowitz *et al.*, 1988).

In a cross-over study by Benowitz *et al.* (1989), in which cigarette smokers in a balanced order used cigarettes, oral snuff, chewing tobacco and abstained from tobacco, the exposure to nicotine was similar in peak levels and circadian pattern while smoking cigarettes, consuming oral snuff or chewing tobacco, although the average exposure measured as plasma AUC tended to be less with the use of smokeless tobacco.

When the steady-state levels of nicotine and cotinine were measured in different body fluids of 34 users of Swedish moist oral snuff (snus), it was found that the average levels of nicotine and cotinine in plasma were 20.8 and 304 ng/ml, respectively. In saliva, the mean steady-state value for cotinine was 1.8 times higher than in plasma (531 ng/ml) and in urine means of steady-state values for nicotine and cotinine were 1.05 and 2.49 μg/ml, respectively. No correlations were found between the consumption parameters (intake per day, number of times per day or habit duration) and the nicotine and cotinine concentrations in any of the body fluids.

The best estimate of the nicotine intake is obtained by measuring the amount of nicotine and all metabolites excreted in urine after nicotine exposure. At steady-state, the rate of excretion reflects the generation rate. In a study comprising 94 tobacco users, 45 moist snuff (snus) users, nine chewers and 40 smokers, it was found that the average doses of nicotine estimated from the total amount of nicotine and metabolites excreted during 24 h, were, 40, 60 and 33 mg per day for moist snuff users, chewers and smokers, respectively. The urinary excretion profiles did not differ between these groups of tobacco users (Curvall *et al.*, to be published).

While the absorption of nicotine from oral snuff via the mucous membrane is relatively slow and dependent on basic pH of the product, the absorption of nicotine from nasal snuff is rapid giving plasma concentrations within 10 min comparable to those obtained after terminating a cigarette (Russell *et al.*, 1981). In daily users of nasal snuff (n = 11) the average plasma nicotine peak concentrations averaged 36 ng/ml and cotinine 412 ng/ml.

With ageing, environmental tobacco smoke (ETS) becomes less acidic and pH may rise to pH 7.5, which enables buccal and nasal absorption by the nonsmoker. The presence of nicotine and its metabolites in body fluids is highly specific to environmental tobacco smoke exposure. As nicotine is relatively rapidly distributed to tissues and is rapidly metabolized it reflects recent or acute exposure to ETS, while cotinine, which is distributed to a much lesser extent than nicotine and has a longer elimination half-life, better reflects long-term exposure to cigarette smoke. Urine and saliva are particularly suited for epidemiological investigations, since they do not require invasive sampling techniques. Large increases in salivary nicotine levels have been observed in nonsmokers shortly after exposure to ETS (Jarvis *et al.*, 1984; Feyerabend *et al.*, 1982), probably as a result of a fast and direct gas to liquid transfer of nicotine in the mouth. Plasma nicotine concentrations from natural exposure do not differentiate self-reported exposed and nonexposed nonsmokers (Jarvis *et al.*, 1984). The concentrations in urine, on the other hand, discriminate between exposed and nonexposed nonsmokers and the average nicotine concentration in 188 urban nonsmokers was found to be 10.8 ng/ml which was 0.7% of the average of 1471 ng/ml in a sample of 229 cigarette smokers (Russell *et al.*, 1986).

Jarvis *et al.* (1985) related saliva cotinine concentrations in 569 nonsmoking schoolchildren to the smoking of their parents. When neither parent smoked the average saliva cotinine level was 0.44 ng/ml and when both parents smoked the average level was 3.38 ng/ml, roughly 1% of the average seen in smokers. Greenberg *et al.* (1984) measured the concentration of nicotine and cotinine in saliva and urine of 32 infants with household exposure to tobacco smoke and 19 unexposed infants. The concentrations of urine cotinine, which proved to be the best indicator for chronic exposure, were significantly higher in the exposed (median 225.3 ng/mg creatinine) than in the unexposed group (median 4 ng/mg creatinine). Reports of recent exposure to ETS and urinary cotinine levels were obtained on 663 never- and ex-smokers. Findings demonstrated that exposure to ETS is extremely prevalent as 91% had detectable cotinine levels. The cotinine levels ranged from 0 to 85 ng/ml with a mean of 8.84 ng/ml and 92% of the values were less than 20 ng/ml (Cummings *et al.*, 1990). In a comparison of different tests for distinguishing smokers from nonsmokers, it was found that the concentration of cotinine, whether measured in plasma, saliva or urine was the best indicator of smoking with sensitivity of 96–97% and a specificity of 99–100%. The average cotinine concentrations in plasma, saliva and urine were 1.5, 1.7 and 4.8 ng/ml for true nonsmokers ($n = 100$), while the corresponding values for cigarette smokers (n = 75) were 294, 330 and 1448 ng/ml (Jarvis *et al.*, 1987).

A wide range of nicotine concentrations (<20–512 ng/ml) in the milk of nursing mothers has been demonstrated (Luck and Nau, 1987). There is a linear correlation between both nicotine and cotinine concentrations in serum (or plasma) and milk of nursing smokers. The milk to serum concentration ratios are 2.9 for nicotine and 0.8 for cotinine, while the milk to plasma concentration ratios are 2.9 for nicotine and 1.2 for cotinine (Luck and Nau, 1984; Dahlström *et al.*, 1990).

Analysis of milk samples collected from all nursing periods within a 24 h interval from three groups of smoking mothers, who smoked 1–10, 11–20 and 21–40 cigarettes per day, respectively, showed average nicotine concentrations of 18, 28 and 48 ng/ml and average cotinine concentrations of 76, 125 and 230 ng/ml. Cotinine but not nicotine was significantly correlated with the number of cigarettes smoked by the mother (Luck and Nau, 1987). Urine cotinine excretion was found to be significantly greater in neonates of smoking mothers than in those whose mothers did not smoke. The initial median cotinine/creatinine ratio was 1233 ng/mg (14 to 3891 ng/mg) in 11 neonates of smokers and 14.5 ng/mg (0 to 40.6 ng/mg) in neonates of nonsmokers. Among the neonates of smokers there was no clear relationship between newborn urine cotinine/creatinine ratio and maternal cigarette consumption during 24 h before delivery (Etzel *et al.*, 1985). Similar cotinine/creatinine ratios (41–3980 ng/mg) were found in 22 breast fed children aged 3.7 d, but in this group of subjects the maternal smoking during pregnancy correlated with the urinary cotinine concentration of the infants (Dahlström *et al.*, 1990).

Nicotine and cotinine are transferred to the human foetus, placenta and amniotic fluid of smoking mothers. Nicotine concentrations in the placentas (range 3.3–28 ng/g), in amniotic fluid (range 1.5–23 ng/ml) and in foetal serum (range 0.5–25 ng/ml) were all higher than the corresponding maternal serum values, while cotinine concentrations in placental tissue (range 10–131 ng/g), amniotic fluid (range 5–188 ng/ml) and foetal serum (range 15–233 ng/ml) were lower than or similar to corresponding maternal serum levels (Luck *et al.*, 1985).

Nicotine and cotinine have also been found in breast fluid of non-lactating women. Thirty minutes after smoking two cigarettes, similar concentrations of nicotine were found in plasma and breast fluid, while the concentration of cotinine in breast fluid was half that in plasma (Hill and Wynder, 1979).

There are several forms for nicotine replacement therapy, such as transdermal administration in solution, oral administration in the form of chewing gum and nasal administration in solution. Nicotine polacrilex gum is widely used as substitution therapy during cigarette smoking cessation. Plasma levels of nicotine have been measured in abstinent cigarette smokers while chewing nicotine gum (Russell *et al.*, 1976; McNabb *et al.*, 1982). Plasma levels while chewing 2 mg nicotine gum were substantially lower (11.8 ng/ml) than, and while chewing 4 mg gum (23.2 ng/ml) they were similar to, levels of cigarette smokers (18.3 ng/ml) (McNabb *et al.*, 1982). Estimation of nicotine intake by measuring circadian blood nicotine concentrations together with the calculation of extraction efficiency of nicotine from gum by smokers switched to 2 or 4 mg nicotine gum, showed that nicotine levels and intake were much lower while chewing gum than during *ad libitum* smoking (Benowitz *et al.*, 1987).

Administration of 2 mg nicotine nasal solution (NNS) hourly, eight times a day to five subjects gave an average increase in plasma peak concentration of 16.3 ng/ml, which is higher than is normally obtained from 2 mg nicotine chewing gum (West *et al.*, 1984). Nicotine is readily absorbed through intact skin. When 30 mg transdermal nicotine patches were applied once daily for 24 h over a 7 d period to 24 male smokers, mean plasma nicotine concentrations during days 3–7 were in the region of 10–15 ng/ml, while the cotinine concentrations tended to increase over the study period and had a C_{max} value of 184 ng/ml at day 7 (Mulligan *et al.*, 1990). The responses of smokers and nonsmokers to transdermal nicotine patches over 24 h in three groups of subjects was studied by Srivastava *et al.* (1991). Either 15 or 30 mg patches were applied to two groups of nonsmokers and a 30 mg patch to a group of smokers. There was a higher absorption rate of transdermal nicotine in nonsmokers than in smokers and the average plasma nicotine C_{max} values were 13.9 ng/ml for smokers and 18.4 ng/ml and 7.9 ng/ml for nonsmokers dosed with 30 and 15 mg nicotine patches, respectively. Application once daily for 5 d of a nicotine user-activated transdermal therapeutic system (Nicotine UATTS), which was designed to deliver nicotine at a steady-state rate of 75 μg cm^2/h proved to establish steady-state conditions for nicotine (15.1 ng/ml) at day 3. Predose levels of cotinine on days 3–5 were about 200 ng/ml and accumulated during the day (Rose *et al.*, 1991).

8.2.1 CHROMATOGRAPHIC ANALYSIS

Although nicotine was one of the first alkaloids to be analysed by gas chromatography, it has proved difficult to analyse by gas chromatography at low concentrations owing to its adsorptive properties. The low concentrations found in nonsmokers exposed to environmental tobacco smoke together with the minute amounts of body fluid available, demand sensitive methods for the determination of nicotine and its metabolites. The specificity and sensitivity of the gas chromatographic methods depend largely on the sample size, the choice of internal standard, the work-up procedure, the gas chromatographic separation technique and the detector characteristics. The first gas chromatographic analysis of nicotine in urine described by Beckett and Triggs (1966) has been modified and completed by several workers. The early quantitative analyses were performed by packed column gas chromatography and flame ionization (FID) detection and involved large sample volumes and tedious clean-up procedures, which included a steam distillation step (Falkman *et al.*, 1975; Burrows *et al.*, 1971; Schievelbein and Grundke, 1968).

Sample volume is only a minor consideration when the matrix is urine. However, when the matrix is other body fluids such as plasma, saliva and mother's milk the sample volume could be limited, particularly in pharmacokinetic and disposition studies in which multiple samples are taken from the same subject. Improved detector sensitivity by use of NPD and MS detectors and introduction of simple microextraction procedures allow analysis of sample sizes down to 100 μl (Curvall *et al.*, 1982; Feyerabend and Russell, 1979; Feyerabend and Russell, 1990).

As the physico-chemical properties of nicotine and cotinine differ widely, separate dedicated internal standards have to be used. The internal standards, quinoline for nicotine and lidocaine for cotinine have been used in several studies. Large variations in the peak area ratios of nicotine to quinoline and cotinine to lidocaine are obtained, which is probably due to selective losses during work-up and/or during gas chromatography. As the structural analogues N-methylanabasine or N-ethylnornicotine for nicotine, and N-ethylnorcotinine or N-(2-methoxyethyl)-norcotinine for cotinine have similar extractability, volatility, affinity to glass and metal surfaces and gas chromatographic properties as those of nicotine and cotinine, high sensitivity is achieved.

Nicotine and cotinine are extracted from the matrix either by liquid-liquid extraction at alkaline pH or by solid phase extraction (Stehlik *et al.*, 1982; Teeuwen *et al.*, 1989). For simultaneous extraction of the two alkaloids chlorinated solvents such as dichloromethane have proved to give the best recoveries.

In most methods which report analysis of both nicotine and cotinine the two amines are eluted in parallel chromatographic runs using two separate chromatographs (Beckett and Triggs, 1966; Hengen and Hengen, 1978; Jacob *et al.*, 1981; Curvall *et al.*, 1982; Thompson *et al.*, 1982). Except for the method by Curvall *et al.*, these involve separate extraction procedures for nicotine and cotinine, so that actually these assays are composed of two analytical methods. Both simultaneous extraction and chromatography is accomplished by Davis (1986), Teeuwen *et al.* (1989), Kogan *et al.* (1981), Stehlik *et al.* (1982), Feyerabend and Russell (1990) and Voncken *et al.* (1989) (Table 8.1).

Table 8.1 Bioanalytical methods for the determination of nicotine and cotinine in human body fluids

Reference	Sample (ml)	Internal standard	Sample work-up	Recovery (%)	Determination	Calib. curve (ng/ml)	Quant. limit (ng/ml)	Precision (CV%)
Nicotine								
Blache *et al.* (1984)	Plasma (1–3)	Quinoline	Liquid–liquid extraction	83.2	GC–FID	0–100	–	10.5 (2.9 ng/ml)
Chien *et al.* (1988)	Plasma (0.1)	–	Liquid–liquid extraction	>85	HPLC–EC Reversed-phase	20–100	20	20.2 (20 ng/ml)
Degen and Schneider (1991)	Plasma (1–3)	Quinoline	Liquid–liquid extraction	97.5	GC–NPD	5–133	~2	17 (6–22 ng/ml)
Dow and Hall (1978)	Plasma (3)	Quinoline	Liquid–liquid extraction	–	GC–MS (SIM)	25–100	–	6.5 (10 ng/ml)
Falkman *et al.* (1975)	Blood (10)	Quinoline	Steam distillation Liquid–liquid extraction	>79.8	GC–FID	0–500	–	10.5 (10 ng/ml)
Feyerabend *et al.* (1975)	Plasma (3)	Quinoline	Liquid–liquid extraction	75	GC–NPD	5–50	0.1	5.0 (5 ng/ml)
Feyerabend and Russell (1979)	Biological fluid (1)	Quinoline	Liquid–liquid extraction	93	GC–NPD	5–100	0.1	–
	Biological fluid (0.1)					2–80	1	6.7 (2 ng/ml)
Feyerabend and Russell (1980a)	Biological fluid (3)	Isoquinoline	Liquid–liquid extraction	–	GC–NPD	5–100	0.1	2.0 (5 ng/ml)
Hartvig *et al.* (1979)	Plasma (1)	N-Propyl-nornicotine	Liquid–liquid extraction trichloroethyl carbamate derivatization	–	GC–ECD	–	10	8.8 (30 ng/ml)
Jones *et al.* (1982)	Plasma (1)	Nicotine-d_2	Liquid–liquid extraction	–	GC–MS (SIM)	5–100	–	3.6 (50 ng/ml)
Cotinine								
Daenens *et al.* (1985)	Plasma/ Urine (1)	3-[$^3H_3$]-Methylcotinine	Liquid–liquid extraction	–	GC–MS (SIM)	50–500	–	2.9 (200 ng/ml)
Feyerabend and Russell (1980b)	Plasma/ Saliva (1)	Pheniramine	Liquid–liquid extraction	90	GC–NPD	25–1000	1.0	1.8 (25–1000 ng/ml)
Feyerabend *et al.* (1986)	Plasma (1) Saliva (1) Urine (1)	Lignocaine	Liquid–liquid extraction	–	GC–NPD	0.25–10.0 0.25–10.0 0.25–100.0	0.1 0.1 0.1	7.7 (0.25–2.0 ng/ml)
Machacek and Jiang (1986)	Plasma (1)/ Saliva (1)	2-Phenylimidazole	Liquid–liquid extraction	>92.3	HPLC–UV Reversed-phase Ion pair	0–10000 (H_2O)	–	4.4 (150 ng/ml)
Norbury *et al.* (1987)	Plasma (0.2)	Methyprylone	Liquid–liquid extraction	111	GC–MS (SIM)	15–150	–	16 (40 ng/ml)
*Nicotine and cotinine**								
Barlow *et al.* (1987b)	Urine (0.5)	–	DETBA-derivatization	>91 >84	HPLC–UV Reversed-phase	0–20000 (H_2O) 0–20000 (H_2O)	– –	9.0 (smokers' urine) 8.4 (smokers' urine)
Curvall *et al.* (1982)	Plasma (0.1–1)	N-Methylanabasine N-Ethylnorcotinine	Liquid–liquid extraction	83 96	GC–NPD	5–100 5–500	<1 <1	8.6 (5 ng/ml) 13.6 (5 ng/ml)

Davis (1986)	Plasma (1)	N-Ethylnornicotine N-Ethylnorcotinine	Liquid–liquid extraction	102.2 98.8	GC–NPD	1–100 4–400	1.0 4.0	– –
Feyerabend and Russell (1990)	Biological fluid (0.1)	5-Methylcotinine	Liquid–liquid extraction	–	GC–NPD	1–100 1–1000	0.1 0.1	13.0 (1.0 ng/ml) 7.7 (1.0 ng/ml)
Hengen and Hengen (1978)	Plasma (1)	Modaline Lidocaine	Liquid–liquid extraction	80 95	GC–NPD	10–50 50–200	– –	14 (10 ng/ml) 7.0 (50 ng/ml)
Horstmann (1985)	Urine (4)	Amphetamine	Liquid–liquid extraction	94 51	HPLC–UV Straight-phase	0.1–10000 0.1–10000	50 50	<5.0 (0.1–10000 ng/ml) <5.0 (0.1–10000 ng/ml)
Jacob *et al.* (1981)	Plasma (1)	N-Ethylnornicotine N-(2-methoxy-ethyl)norcotinine	Liquid–liquid extraction	–	GC–NPD	5–50 5–600	1.0 5.0	4.9 (5 ng/ml) 4.2 (5 ng/ml)
Jacob *et al.* (1991)	Plasma (1)	Nicotine-d_4 Cotinine-d_4	Liquid–liquid extraction	– –	GC–MS (SIM)	1–100 10–1000	1.0 10.0	4.0 (1–40 ng/ml) 6.0 (10–400 ng/ml;)
Kogan *et al.* (1981)	Plasma (1)	Ketamine Ketamine	Liquid–liquid extraction	– –	GC–NPD	5–25 50–250	2 5	14.3 (5–25 ng/ml) 9.3 (50–250 ng/ml)
Maskarinec *et al.* (1978)	Plasma (1–5) Urine (20)	–	XAD-2 clean-up	>85.7 >86.4	HPLC–UV Straight-phase	10–10000 –	– –	<5.0 (10–10000 ng/ml)
McManus *et al.* (1990)	Urine (1)	Cotinine-d_3	Filtration by centrifugation	– –	LC–MS (CI/SIM) Thermospray Reversed-phase	80–8000 20–8000	80 20	6.0 (5000 ng/ml) 6.0 (5000 ng/ml)
Parviainen and Barlow (1988)	Urine (1)	–	DETBA derivatization Liquid–liquid extraction	>73 >75	HPLC–UV Reversed-phase	0–90 (H_2O) 0–100 (H_2O)	– –	14.3 (6.4 ng/ml) 13.1 (5.6 ng/ml)
Stehlik *et al.* (1982)	Plasma (2)/ Urine (4)	Quinoline Lidocaine	Solid-phase/ liquid–liquid extraction	>80 >80	GC–FID	0–250 (plasma)/ 0–15 000 (urine)	5 (plasma)/ 30 (urine)	– –
Teeuwen *et al.* (1989)	Plasma (1)/ Urine (0.1)	β-Nicotyrine Nikethamide	Solid-phase extraction	>75 >98	GC–NPD	0–60 plasma; 50–1000 urine 5–400 plasma; 50–1000 urine	1; 50 (plasma; urine) 5; 50 (plasma; urine)	10 (>2; 50 plasma; urine) 10 (>5; 50 plasma; urine)
Thompson *et al.* (1982)	Tissue (1 g)	Methylanabasine/ Nicotine-d_3 Cotinine-d_3	Liquid–liquid extraction	81 87	GC–NPD/ GC–MS (SIM) GC–MS	5–1000 ng/g 5–1000 ng/g	2.0 ng/g 2.0 ng/g	21 (10 ng/g) 25 (10 ng/g)
Verebey *et al.* (1982)	Plasma (0.5)	Ketamine	Liquid–liquid extraction	– –	GC–NPD	4–40 40–400	– –	9.8 (20 ng/ml) 5.6 (456 ng/ml)
Voncken *et al.* (1989)	Plasma/ Urine (1)	N-Ethylnornicotine N-Ethylnorcotinine	Liquid–liquid extraction	>77 >69	GC–MS (SIM)	10–1320 10–900	3 plasma; 2 urine 10 plasma; 10 urine	9.5 (10 ng/ml, plasma) 4.0 (200 ng/ml, urine) 2.3 (200 ng/ml, plasma) 1.3 (500 ng/ml, urine)
Watson (1977)	Urine (3)	Desmethyl-imipramine	Liquid–liquid extraction	>98 >98	HPLC–UV Straight-phase	100–10000 –	– –	7.3 (smokers' urine) 9.0 (smokers' urine)

*Nicotine (upper line) and cotinine (lower line).

Gas chromatographic separation of nicotine and/or cotinine, standards and endogenous compounds may be achieved with packed or with capillary open tubular columns. Simple one-step extraction of the analytes from the matrix require refined separation techniques in the analytical step. By using high-resolution capillary gas chromatography involving splitless injection, with temperature-programmed capillary gas chromatography with NPD detection, optimal sensitivity and resolution of the peaks could be achieved.

A rapid method has previously been described by Feyerabend and Russell (1990) in which nicotine and cotinine are simultaneously determined using a single extraction step without any further concentration, purification or evaporation steps. The same internal standard, 5-methylcotinine, is used for both analytes, only one calibration curve is used for both high and low levels of nicotine and cotinine and only 100 μl sample volume is required. Due to a simple work-up procedure, short retention times in the chromatographic analysis and automatic injections onto the chromatograph, a large number of samples could be analysed per day.

HPLC methods for quantification of nicotine and cotinine in body fluids of smokers have been reported by several groups (Table 8.1). These methods, which involve a simple liquid–liquid extraction or precolumn extraction followed by liquid column chromatography with UV or electrochemical detection allow rapid and simultaneous determination of the two analytes. Those methods using straight-phase silica column chromatography have low sensitivity due to poor resolution and adsorption effects and could accordingly only be used for analysis of nicotine and cotinine in smokers' urine (Watson, 1977; Horstmann, 1985). Better resolution of the analytes and higher sensitivity is obtained by usage of reversed-phase ion pair liquid chromatography (Maskarinec *et al.,* 1978; Hariharan *et al.,* 1988; Machacek and Jiang, 1986). Thermospray LC–MS has been described for the simultaneous determination of nicotine and 17 of its metabolites in urine (McManus *et al.,* 1990). After addition of the internal standard, cotinine-d_3, and clean-up by centrifugal filtration, the analytes were separated by reversed-phase HPLC and then introduced into the mass spectrometer via a thermospray interface. Chemical ionization of nicotine and its metabolites was achieved by post-column addition of ammonium acetate buffer with the filament of the ion source turned off. Selected ion monitoring–mass spectrometry (MS–SIM) was used to quantify each metabolite found in urine of smokers as its unique protonated molecular ion (M+H). These methods allow simultaneous determination of nicotine and cotinine in concentrations usually found in plasma and saliva of smokers and in urine of nonsmokers but are not sensitive enough for determination of these analytes in plasma and saliva of nonsmokers exposed to ETS.

Routine assessment of smoking status and cigarette smoke intake have been performed by spectrophotometric determination of coloured barbituric acid and diethylthiobarbituric acid urinary derivatives of nicotine metabolites (Kolonen and Puhakainen 1991; Peach *et al.*, 1985; Barlow *et al.*, 1987a; Puhakainen *et al.*, 1987). Nicotine and cotinine together with other metabolites have been individually determined by an HPLC method involving pre-column derivatization with barbituric acid followed by isocratic reversed-phase liquid column chromatography and UV detection (Barlow *et al.*, 1987b; Parviainen and Barlow, 1988; Parvainen *et al.*, 1990; Moore *et al.*, 1990).

8.2.2 IMMUNOASSAY

The first conventional radioimmunoassay (RIA) for analysis of nicotine was developed by Langone *et al.* (1973). Antibodies are produced by coupling hapten derivatives to macromolecules, which are injected into an animal. Known amount of labelled antigen and the antibody are in equilibrium with the antigen–antibody complex. To this mixture is added unlabelled antigen and the analyte (nicotine or cotinine). Labelled and unlabelled nicotine compete for the limited number of binding sites at the antibody and the extent of competition serves as a basis for the quantitative assay of the analyte. The measured radioactivity in the immune precipitate is inversely related to the concentration of the analyte in the test sample (Van Vunakis and Levine, 1974). These methods using polyclonal antibodies have limited sensitivity due to interference from the matrix when concentrated body fluids such as saliva and urine are used. In saliva or urine, variations in pH and ionic strength can lead to perturbation of the antigen–antibody reaction resulting in non-specific inhibition. Standard curves are obtained by estimating the percentage of tracer bound in the presence of known amounts of test compound. Standard curves are generally shallow and it is difficult to achive complete inhibition of immune binding with cotinine when a ^{125}I-labelled tyramine derivative is used as a tracer (Bjercke *et al.*, 1987).

Recently, stereospecific monoclonal antibody (McAb) for *S*-(−)-nicotine and *S*-(−)-cotinine have been prepared and used to develop a fluid phase RIA with ^{125}I or ^{3}H-labelled tracers as well as enzyme-linked immunosorbent assays (ELISA) and a fluorescence immunoassay (FIA) in microtitre plate format (Bjercke *et al.*, 1986; Bjerke *et al.*, 1987). In contrast to the conventional fluid phase RIAs the McAb-based assays are carried out in a solid phase system in which a cotinine–polylysine conjugate is passively adsorbed to the surface of plastic microtitre plates. Immobilized nicotine and cotinine and fluid phase nicotine and cotinine in the test sample compete with a variety of enzyme-labelled anti-immunoglobulin reagents. McAbs are homogenous in binding and specificity.

Assays with McAb compared with RIAs using rabbit antisera have been found to be more sensitive, have steeper standard curves and the antigen–antibody reactions completely inhibited by (−)cotinine even when the ^{125}I-labelled tyramine derivative is used (Bjercke *et al.*, 1987). Langone and Bjercke (1989) have developed an assay which relies on the ability of cotinine to inhibit binding between a monoclonal anti-cotinine antibody and a second McAb specific for the antigen combining region. By using this method, preparation of labelled hapten derivatives or macromolecular conjugates for solid phase assays becomes unnecessary. A good correlation between this method and the cotinine ELISA method developed by Bjercke *et al.* (1986) was demonstrated when levels in diluted nonsmokers' saliva spiked with cotinine and in diluted saliva from 35 smokers were monitored.

RIAs have been used for screening nicotine and cotinine levels in tobacco users. However, more sensitive immunoassays are needed to reproducibly quantify the low levels of nicotine and cotinine found in nonsmokers exposed to ETS or during post distributional pharmacokinetic studies. Various body fluids like urine, plasma (or serum), saliva, amniotic, celebrospinal, breast and spinal fluids have been analysed. The sample volumes in immunoassays are only 1 to 10% of those used in chromatographic methods. The limit of quantification is about ten times higher for immunoassays than for the most sensitive chromatographic methods (Tables 8.1 and 8.2). The concentrations of nicotine and cotinine in serum determined by RIA and GC have been found to agree, although somewhat higher by RIA (Jacob *et al.*, 1981; Langone and van Vunakis, 1982). When cotinine concentrations in serum from smokers were analysed both by RIA and by GC, the results from RIA were 30–50% higher than from GC and no correlation was found between the two methods when samples with cotinine levels below 7 ng/ml were analysed. GC and RIA methods have been compared in an interlaboratory study, in which nicotine and cotinine levels in plasma and urine were analysed. The urinary cotinine concentrations differed widely from laboratory to laboratory, which can be due to the poor specificity of some RIA methods used (Biber *et al.*, 1987). High cross-reactivity due to the presence of compounds closely related to nicotine and cotinine leads to an overestimation of the cotinine concentration (Richie *et al.*, 1990).

8.3 3′-HYDROXYCOTININE

In the early 1960s McKennis and co-workers isolated hydroxycotinine from the urine of smokers and a nonsmoker dosed with cotinine orally

Table 8.2 Immunoassays for determination of nicotine and cotinine in human body fluids

Reference	Sample (μl)	Sample work-up	Recovery (%)	Hapten derivative	Determination	Calib. curve (ng/ml)	Quant. limit (ng/ml)	Precision (CV%)
Nicotine								
Castro *et al.* (1979)	Serum (50)	–	–	N-Succinyl-6-amino-nicotine	Automated RIA (^{125}I)	10–50	10	6.9 (20 ng/ml)
Castro *et al.* (1980)	Plasma (50)	1 h incubation 24 h incubation	–	(*R,S*)-6-(p-Amino-benzamido)-nicotine	RIA (^{125}I)	10–150	10	4.6 (10–150 ng/ml) 2.9 (10–150 ng/ml)
Castro and Monji (1986)	Plasma (50)	–	–	(*R,S*)-6-(p-Amino-capramido)-nicotine	EIA/ Fluorescence	6–200	6	5.7 (6 ng/ml)
Cotinine								
Knight *et al.* (1985)	Serum (100)/ Urine (10)	–	–	*trans*-4′-Carboxycotinine	RIA (^{125}I)	90–500	–	10–15 (smokers' serum)
Langone and Bjercke (1989)	Saliva (10)	–	–	Anticotinine F(ab′)$_2$	ELISA	50–500	4	8.2 (5 ng/ml)
Matsukura *et al.* (1979)	Urine (100)	Liquid–liquid extraction	81	1-2(β-Aminoethyl)-cotinine	RIA (^{125}I)	100–1000	2	17 (10 ng/ml)
*Nicotine and cotinine**								
Bjercke *et al.* (1986, 1987)	Serum (100) Saliva (100) Urine (100)	–	–	3′-Hydroxymethyl-nicotine *trans*-4′-Carboxycotinine	RIA (^{125}I, ^{3}H)	5–500	10 (saliva) 5 (saliva)	13.3 (10 ng/ml) 9.4 (5 ng/ml)
	Serum (100) Saliva (100) Urine (100)	Micro-titre plate assay	–	3′-Hydroxymethyl-nicotine *trans*-4′-Carboxycotinine	ELISA (McAb, ^{3}H) FIA	5–500	10 (saliva) 5 (saliva)	13.3 (10 ng/ml) 9.4 (5 ng/ml)
Hill and Wynder (1979)	Breast fluid	Liquid–liquid extraction	97 85	*trans*-3′-Succinyl-methylnicotine *trans*-4′-Carboxycotinine	RIA (^{125}I)	–	–	10 (smokers' breast fluid)
Langone *et al.* (1973, 1982)	Serum (200–500) Plasma (200–500) Saliva (20)	–	–	*trans*-3′-Succinyl-methylnicotine *trans*-4′-Carboxycotinine	RIA (^{125}I, ^{3}H)	–	5 (plasma) 2 (urine) 10–15 (saliva)	–

*Nicotine (upper line) and cotinine (lower line)

(McKennis *et al.*, 1962). Hydroxycotinine was also found to be a metabolite of nicotine and cotinine in other species such as dogs and rats (Voncken *et al.*, 1989). The exact structure, stereochemistry and urinary concentrations were not established at that time. About ten years later, Dagne and Castagnoli (1972a) showed the exact structure to be *trans*-1-methyl-3-(*R*)-hydroxy-5-(*S*)-3-pyridyl-2-pyrrolidone and also described a synthesis. Kyerematen *et al.* (1988) have shown that 4.5% of the radiolabelled dose of nicotine given to rats was recovered as 3′-hydroxycotinine, when about 70% of the total amount of a radiolabelled compound was traced. Recently, *trans*-3′-hydroxycotinine was quantified and recognized as a major nicotine metabolite in the urine of smokers (Neurath *et al.*, 1987b). The amounts of five urinary metabolites relative to nicotine, in nine cigarette smokers smoking cigarettes with 1.35 mg nicotine/cigarette, were found to be, nicotine 1.0; cotinine 0.9; nicotine-1′-N-oxide 0.2; nornicotine 0.1; N-methylnicotinium ions 0.1; and 3′-hydroxycotinine 3.5. It was also found that the average serum level of 3′-hydroxycotinine of these smokers was 69 ± 91 ng/ml (Neurath *et al.*, 1987b). Quantification of nicotine and seven of its main metabolites in urine collected over 7 d after intravenous infusion of nicotine to abstinent tobacco users have shown that 3-hydroxycotinine accounts for 36% and together with the phase 2 metabolite, *trans*-3′-hydroxycotinine glucuronic acid, for 48% of the total dose (Curvall *et al.*, 1991).

Only a few bioanalytical methods for the determination of 3′-hydroxycotinine in urine and blood have been published (Table 8.3). In an investigation of the *in vivo* biotransformation of (G-^{3}H)-*S*-cotinine in the guinea pig, Cundy and Crooks (1987) have used a combination of reversed-phase and cation-exchange HPLC for the identification and quantification of 3′-hydroxycotinine together with other urinary metabolites of *S*-(−)-cotinine. Analysis of urine samples from guinea pigs after i.p. administration of (G-^{3}H)-*S*-(−)-cotinine showed that 3′-hydroxycotinine is also the major urinary metabolite in guinea pigs as it accounted for as much as 56% of the dose given, when about 96% of the radioactive dose was traced.

Kyerematen *et al.* (1987) have developed a radiometric high-performance liquid chromatographic assay for simultaneous determination of concentrations of nicotine and twelve urinary metabolites. By using a cyano reversed-phase column system and optimizing pH, ionic strength and solvent composition of the binary gradient mobile phase a good separation and short retention times were obtained.

The thermospray liquid chromatography–mass spectrometry method (see section 8.2), using reversed-phase column chromatography in the separation step and quantification by selected ion monitoring–mass spectrometry, has been used to quantify 3′-hydroxycotinine in smokers' urine in the range 20–8000 ng/ml.

Table 8.3 Bioanalytical methods for the determination of *trans*-3′-hydroxycotinine in human body fluids

Reference	Sample (ml)	Internal standard	Sample work-up	Recovery (%)	Determination	Calib. curve (ng/ml)	Quant. limit (ng/ml)	Precision (CV%)
McManus et al. (1990)	Urine (1)	Cotinine-d_3	Filtration by centrifugation	–	LC–MS (CI/SIM) Thermospray Reversed-phase	20–8000	20	6.0 (5000 ng/ml)
Moore *et al.* (1990)	Urine (1)	N-Acetyl-nornicotine	DETBA derivatization C_{18} Sep-Pak clean-up	–	Automated HPLC–UV, Reversed-phase	50–2000	50	<3.0 (100–1000 ng/ml)
Neurath and Pein (1987a)	Plasma (1)	–	Liquid–liquid extraction	75.2	GC–ECD	20–200	1	–
	Urine (1)	–	HFBA derivatization	88.8		5000–10000	5	–
O'Doherty *et al.* (1988)	Urine (10)	–	DETBA derivatization C_{18} Sep-Pak clean-up	>85	HPLC–UV Reversed-phase	0–5000	–	–
Voncken *et al.* (1989)	Plasma (1)	N-Ethylnornicotine	Liquid–liquid extraction	49	GC–MS (SIM)	20–3400	50 (GC–MS)	3.0 (200 ng/ml)
	Urine (1)	N-Ethylnorcotinine		82	GC–NPD GC–MS (SIM)		50 (GC–NPD) 10 (GC–MS)	7.3 (280 ng/ml)

An improved colorimetric HPLC assay has been developed for the simultaneous determination of nicotine, cotinine and 3′-hydroxycotinine in human urine (Moore *et al.*, 1990). This method involves derivatization of the analytes with diethylthiobarbituric acid, sample clean-up on a C_{18} Sep-Pak cartridge and separation and quantification of an aliquot on a reversed-phase HPLC system connected to a UV detector. By automation of the technique no degradation of the analyte derivatives occurred as the time between formation of the complex and injection onto the HPLC column was minimized.

Generally gas chromatographic methods are not suitable for the quantitative analysis of highly polar biotransformation products. Neurath and Pein (1987a) have developed a method for the determination of 3′-hydroxycotinine in plasma and urine, which comprises extraction with dichloromethane at alkaline pH, derivatization of an aliquot of the extract with heptafluorobutyric acid anhydride, quantitative analysis by packed column chromatography and electron capture detection. A more convenient method by which nicotine, cotinine and 3′-hydroxycotinine could be determined simultaneously from a single urine or blood sample in one GC analysis without previous derivatization was established by Voncken *et al.* (1989). After addition of the internal standards, N-ethylnornicotine and N-ethylnorcotinine, the analytes were extracted in a single step using dichloromethane and saturated aqueous potassium carbonate solution. The quantitative analyses were performed by gas chromatography using either an NPD detector to determine 3′-hydroxycotinine alone in urine or a mass selective detector for the simultaneous determination of nicotine, cotinine and 3′-hydroxycotinine in urine and blood. The quantification limit for urine samples was 10 ng/ml when a mass spectrometer was used as a detector. This method has previously been used to identify and quantify *cis*-3′-hydroxycotinine (Voncken *et al.*, 1990) as an *S*-(−)-nicotine metabolite in the urine of rats and hamsters dosed with nicotine and in the urine of smokers (about 0.1%).

8.4 GLUCURONIC ACID CONJUGATES OF NICOTINE, COTININE AND *trans*-3′-HYDROXYCOTININE

The phase 1 metabolites of nicotine account for about 65% of the dose of nicotine. Enzymatic hydrolysis of urine samples from tobacco users have shown that nicotine, cotinine and *trans*-3′-hydroxycotinine are excreted as glucuronic acid derivatives (Curvall *et al.*, 1991). Formation of phase 2 metabolites is an important pathway in the metabolism of nicotine, as the glucuronides account for close to 35% of the total amount of the nicotine dose excreted in the urine after intravenous administration of nicotine to abstinent tobacco users. These metabolites are found in urine but not in blood or saliva of tobacco users. In 94 tobacco users, the average amounts in urine of these glucuronides as a percentage of the total dose excreted during 24 h, were found to be 3, 9 and 23% for nicotine–GlcA, cotinine–GlcA and 3′-hydroxycotinine–GlcA, respectively. The ratio of the glucuronide over parent compound was 0.5, 1.2 and 0.6. These results are in agreement with the findings in a study by Byrd *et al.* (1992), which comprised 11 smokers.

At present, there is no direct method for the analysis of the glucuronides, which is why they are analysed, after enzymatic hydrolysis, as free nicotine, cotinine and 3′-hydroxycotinine. Urine samples are divided into four portions, the first one is analysed for nicotine and cotinine, the second for 3′-hydroxycotinine using HRGC with capillary column separation and NPD detection. The third and fourth portions, are analysed after enzymatic hydrolysis with β-glucuronidase for the total (free and conjugated) amounts of nicotine, cotinine and 3′-hydroxycotinine. The amounts of conjugated alkaloids are calculated from these results. Good precision of these assays is obtained by use of optimal conditions in the enzymatic hydrolysis step (Curvall *et al.*, 1991). The coefficients of variation for the determination of nicotine–GlcA, cotinine–GlcA and 3′-hydroxycotinine–GlcA were found to be 5.7, 2.8 and 6.6% at urinary concentrations of 2.1, 5.3 and 10.5 μg/ml.

8.5 N-OXIDES OF NICOTINE AND COTININE

Two diastereomers, *cis*- and *trans*-nicotine-1′-N-oxide have been identified as natural products in leaf of *Nicotiana tabacum* (Phillipson and Handa, 1978). Over 20 years ago, Booth and Boyland (1970) showed the presence of *cis* and *trans* isomers of nicotine-1′-N-oxide in smokers' urine. N-Oxidation of nicotine is catalysed by a microsomal flavoprotein and gives two diastereomers of nicotine-1′-N-oxide while N-oxidation of cotinine is catalysed by a cytochrome P450 enzyme system and gives cotinine-1-N-oxide. The nicotine-N-oxides are converted by reduction back to nicotine (Dajani *et al.*, 1975a; Dajani *et al.*, 1975b), a step that has been shown to be stereoselective in some species (Sepkovic *et al.*, 1986). The reduction of nicotine-1′-N-oxide to nicotine is dependent on route of administration (Beckett *et al.*, 1970). Approximately twice as much nicotine was recovered in the urine after oral administration of the N-oxide (about 40%), than when the corresponding dose of nicotine-1′-N-oxide was administered rectally, while after intravenous administration the 1′-N-oxide is quantitatively recovered. Beckett *et al.* (1971) showed that 24 h urinary excretion of nicotine-1′-N-oxide by smokers was in the range 0.11 to 2.78 mg and the ratio of cotinine/nicotine-1′-N-oxide was in the range 0.75 to 3.8.

Cotinine-1-N-oxide, which was first isolated and characterized by

Dagne and Castagnoli (1972b) from urine of a monkey after intravenous administration of cotinine, has also been found in urine of tobacco users. Average steady-state concentrations of nicotine-1′-N-oxide and cotinine-1-N-oxide in 24 h urine from 25 cigarette smokers were 574 and 370 ng/ml giving a total excretion of 1.08 and 0.70 mg respectively, during 24 h.

Of the urinary excretion profile of 96 tobacco users (22 smokers, 45 snuff users and 9 chewers) nicotine-1′-N-oxide and cotinine-1-N-oxide accounted for 7% each of the total amount of nicotine and metabolites excreted during 24 h. As there is no difference in metabolic profile between these groups of tobacco users, the route of administration does not seem to affect the recovery of the N-oxides. These findings are in agreement with the findings by Beckett *et al.* (1971), who found that the recovery of nicotine-1′-N-oxide was identical whether nicotine was administered orally or intravenously. They also showed that the urinary excretion was independent of pH and urinary flow.

N-Oxidation of nicotine results in a marked decrease in pKa of the parent amine. Nicotine-1′-N-oxide has a considerable dipole moment, displays high polarity and low volatility. As it is extremely soluble in water but not in lipids, it does not partition into organic solvents. In addition, nicotine-1′-N-oxide is thermolabile and easily undergoes thermal rearrangements. These characteristics are also true for cotinine-1-N-oxide, but it is more easily extracted into organic solvents. Mainly because of these properties and due to the fact that the levels in body fluids are low, it is difficult to isolate, extract and quantify these nicotine metabolites.

Bioanalysis of nicotine-1′-N-oxide has hitherto been accomplished by gas chromatography in two different ways (Table 8.4). One way involves removal of the parent base by extraction, reduction of the N-oxide with a reducing reagent ($TiCl_3$) (Beckett *et al.*, 1971) and analysis of the formed nicotine and cotinine by commonly used gas chromatographic methods with NPD detection. Stehlik *et al.* (1982) has developed a method which includes a better separation of the N-oxides and the parent amines by solid-phase extraction (SP), followed by *in situ* reduction with SO_2 and quantification by gas chromatography with FID detection. As the extraction of nicotine and cotinine prior to the reduction step results in unavoidable losses of cotinine, better precision in the determination is achieved when two separate analyses are performed, one for nicotine and cotinine and one for the total amount of N-oxides and parent compounds. The second way involves thermal conversion of nicotine-1′-N-oxide by Meisenheimer rearrangement to a volatile derivative (Jacob *et al.*, 1986). This method, which only allows analysis of the nicotine-1′-N-oxide and not cotinine-N-oxide, involves a tedious sample work-up procedure with SP extraction, derivatization, a two-stage

Table 8.4 Bioanalytical methods for the determination of nicotine-1′-N-oxides and cotinine-1′-N-oxide in human body fluids

Reference	Sample (ml)	Internal standard	Sample work-up	Recovery (%)	Determination	Calib. curve (ng/ml)	Quant. limit (ng/ml)	Precision (CV%)
Nicotine-1′-N-oxide								
Beckett *et al.* (1971)	Urine (4)	Phendimetrazine	Liquid–liquid extraction $TiCl_3$ reduction	–	GC–FID	100–1000	–	9 (100–1000 ng/ml)
Jacob *et al.* (1986)	Urine (2)	5-Methylnicotine	Liquid–liquid extraction Solid-phase extraction Derivatization	–	Automated GC–NPD	10–4000	–	3.5 (100 ng/ml)
McManus *et al.* (1990)	Urine (1)	Cotinine-d_3	Liquid–liquid extraction Filtration by centrifugation	–	LC–MS (CI–SIM) Thermospray Reversed phase	80–8000	90	–
Cotinine-N-oxide								
Shulgin *et al.* (1987)	Urine (2)	N′-Methyl-N′-(3-Pyridylmethyl)-acetamide-N-oxide-propionamide-N-oxide	Solid-phase extraction	–	HPLC–UV Reversed-phase	50–2000	10	10.0 (100 ng/ml) 18.4 (100 ng/ml)
*Nicotine-1′-N-oxide and cotinine-1-N-oxide**								
Stehlik *et al.* (1982)	Plasma (2)/ Urine (4)	Phendimetrazine-N-oxide Phendimetrazine	Solid-phase extraction SO_2 reduction	30 (plasma) 60 (urine)	GC–FID	0–250 (plasma) 0–1500 (urine)	15 (plasma) 100 (urine)	– –

*NNO (upper line) and CNO (lower line)

liquid–liquid extraction and quantification using automated gas chromatographic analysis with NPD detection (Table 8.4).

None of the gas chromatographic methods allows separation of the *cis* and *trans* isomers of nicotine-1′-N-oxide. Brandänge *et al.* (1977) have developed a method which not only allows the separation of the *cis* and *trans* isomers of nicotine-1′-N-oxide, but all five possible N-oxides of nicotine. A method for the simultaneous determination of the N-oxides of nicotine and cotinine has been developed by Kyerematen *et al.* (1987). This method, which allows determination of the N-oxides in plasma and urine samples, comprises SP extraction on silica, separation by reversed-phase chromatography and monitoring with a UV detector. The collected fractions were quantified by a scintillation counter. A sensitive method for the determination of cotinine-1-N-oxide, which involves reversed-phase silica column chromatography and UV detection, has been developed by Shulgin *et al.* (1987). Nicotine-1′-N-oxide in urine can be accurately determined in the range 80–8000 ng/ml by use of the rapid thermospray liquid chromatography method (see section 8.2), involving sample work-up by centrifugal filtration, separation by reversed-phase liquid chromatography and quantification by selective ion mass spectrometry (McManus *et al.*, 1990).

8.6 N-METHYLATED DERIVATIVES OF NICOTINE AND COTININE

In 1963, McKennis *et al.*, described the isolation and identification of nicotine-isomethonium ion in the urine of dogs after intravenous administration of nicotine, and cotinine methonium ion in the urine of dogs and man after oral administration of cotinine. Cundy *et al.* (1984b, 1985a, 1985b) have shown by studies on guinea pigs dosed with ^{14}C-labelled (±)-nicotine, that *in vivo* N-methylation is stereospecific as only *R*-(+)-nicotine is methylated and the *S*-(−)-enantiomer acts as an inhibitor. Besides the N-methylnicotinium and N-methylcotininium ions, the presence of N-methylnornicotinium ion and two diastereomers of N-methyl-N′-oxo-nicotinium ions has been demonstrated in urine of guinea pigs (Cundy *et al.*, 1985c; Pool and Crooks, 1985; Pool *et al.*, 1986). *In vivo* biotransformation studies of *R*-(+)-^{14}C-labelled methylnicotinium ion in guinea pigs showed that 60.9% is excreted unchanged in 24 h urine. N-Methylcotininium ion, N-methylnornicotinium ion and the N-methyl-N′-oxo-nicotinium ions accounted for 17.5, 8.0 and 9.0% of the total dose, respectively (Pool and Crooks, 1985). There is a great species variation in the formation of methonium ions of nicotine and cotinine. No N-methylation products were found in the 24 h urine of *S*-(−)-[^{3}H-N′-CH_3]-nicotine treated guinea pigs, hamsters, rats and rabbits, while two N-methylated metabolites, N-methylnicotinium ion

(4%) and N-methylcotininium ion (2%), were observed in guinea pigs after administration of ^{3}H-labelled *R*-(+)-nicotine (Nwosu and Crooks, 1988). Since the methonium ions of nicotine and cotinine are highly polar and extremely water soluble, there are great difficulties in quantitating minute amounts of these compounds in body fluids and tissues.

Analysis of N-methylated urinary metabolites in animals administered radiolabelled nicotine has been performed by liquid radiochromatographic methods, which involve separation on ion-exchange columns and quantification by radiometric detectors. The metabolites are identified by their ability to co-migrate with nonlabelled standards, and are monitored by UV detection. Quantification is achieved by measuring the radioactivity of each metabolite identified by UV detection (Cundy *et al.*, 1984a, 1984b, 1985b, 1985c, 1987). Although 4–12% of the naturally occurring *S*-(−)-nicotine is racemized to *R*-(+)-nicotine during cigarette smoking (Klus and Kuhn, 1977), the N-methylation pathway of nicotine seems to be of minor importance. The amount of N-methylnicotinium ions excreted in the urine by smokers is about 10% of the amount of nicotine excreted (Neurath *et al.*, 1987a).

8.7 NOR- DERIVATIVES OF NICOTINE AND COTININE

Nornicotine is a major tobacco alkaloid and comprises up to 20% of the total alkaloids in cigarette tobacco. It has pharmacological and toxicological actions similar to those of nicotine (Zhang *et al.*, 1990). Since nornicotine is present in tobacco it has been questioned whether it is a true human metabolite of nicotine. Recent studies have shown that nornicotine is formed by oxidative dealkylation of nicotine both in man and in animals (Neurath *et al.*, 1991; Kyerematen *et al.*, 1988). In a study by Cundy and Crooks (1984b), radiolabelled nicotine was injected intraperitoneally to guinea pigs. Only small amounts of nornicotine (1.6%) were detected in the total 24 h urine void, when 77% of the radioactivity was traced. After single intra-arterial doses of labelled nicotine to rats, it was found that nornicotine accounted for 8% of the total recovery of the administered radioactivity, when the total activity traced was 70% of the administered dose. In a group of 25 smokers, the steady-state concentrations of nornicotine in urine were found to be in the range 6 to 207 ng/ml giving a total excretion of 12 to 361 μg over 24 h.

High-performance liquid chromatographic methods have been developed to detect and quantitate labelled nornicotine in urine of rats (Kyerematen *et al.*, 1987) and guinea pigs (Cundy *et al.*, 1984a) after administration of labelled nicotine. The method by Cundy and Crooks, which was developed to determine N-methylated metabolites of nicotine, involves separation of the analytes by cation-exchange liquid-chromatograph and radiometric detection (see section 8.6). In order to simultaneously determine nicotine and 12 metabolites in biological fluids, Kyerematen *et al.* (1987) have developed a radiometric high-performance liquid chromatographic assay. The separation of urinary metabolites was performed using reversed-phase liquid chromatography and good resolution, short retention times and high sensitivity were obtained by optimizing pH, ionic strength and solvent composition of the mobile phase. McManus *et al.* (1990) have analysed norcotinine in smokers' urine using the thermospray liquid chromatography–mass spectrometry method described earlier. Linear response was obtained for norcotinine in the range 20 to 8000 ng/ml (Table 8.5).

Only a few gas chromatographic methods for the analysis of nornicotine have been published to date (Table 8.5). Neurath *et al.* (1991) developed a gas chromatographic method for the analysis of nornicotine in urine. After extraction into organic solvent and derivatization to heptanoates, nornicotine and the internal standard (anabasine) are quantified using a gas chromatograph equipped with a packed column and an NPD detector. This method has allowed the analysis of nornicotine in urine samples from six smokers who received nicotine and cotinine intravenously on separate occasions. The results from this study indicate that nornicotine and norcotinine are minor metabolites of nicotine as each of them constitutes less than 1% of the nicotine dose.

In order to investigate the pharmacokinetics and metabolic origin of nornicotine in humans, a rapid and sensitive gas chromatographic assay was developed by Zhang *et al.* (1990). The N′-alkylderivatives of nornicotine were found to be stable in acid-base partitioning steps during clean-up and to have good chromatographic properties. The nornicotine together with the internal standard (5-methylnornicotine) was, after basic extraction into organic solvent, converted to N′-propylderivatives by reaction with propionaldehyde under acidic conditions. The quantitative determinations were performed on a gas chromatograph equipped with capillary column and NPD detector. Recovery, precision and quantitation limit for the method is given in Table 8.5.

8.8 SUMMARY

Measurements of nicotine and its metabolites in different body fluids have become an important component in studies of the nicotine disposition in tobacco users and pharmacokinetic studies of nicotine and its metabolites, epidemiological studies on nonsmokers exposed to environmental tobacco smoke, behavioural studies on the role of nicotine in smoking and clinical studies of nicotine replacements.

A great many chromatographic methods and immunoassays have

Table 8.5 Bioanalytical methods for the determination of nornicotine and norcotinine in human urine

Reference	Sample (ml)	Internal standard	Sample work-up	Recovery (%)	Determination	Calib. curve (ng/ml)	Quant. limit (ng/ml)	Precision (CV%)
Nornicotine								
Neurath *et al.* (1991)	Urine (1)	Anabasine	Heptanoate derivatization Liquid–liquid extraction	–	GC–NPD	5–40	–	–
Zhang *et al.* (1990)	Urine (1)	5-Methylnornicotine	Liquid–liquid extraction N-propyl derivatization	>90	GC–NPD	2–1500	2	4.3 (20 ng/ml)
Norcotinine								
McManus *et al.* (1990)	Urine (1)	Cotinine-d_3	Filtration by centrifugation	–	LC–MS (CI/SIM) Thermospray Reversed-phase	20–8000	20	6.0 (5000 ng/ml)

been established for the determination of nicotine and cotinine in body fluids of smokers. Due to the high specificity and sensitivity, high-resolution gas chromatography has proved to be the best method for determination of such low levels of nicotine and cotinine as are usually found in nonsmokers exposed to ETS. HPLC and GC methods have recently been established for the determination of 3′-hydroxycotinine in body fluids at levels usually found in smokers. The N-oxides, glucuronides, N-methylated derivatives and the nor- derivatives in urine of tobacco users are either determined by indirect GC methods or by HPLC, which do not allow quantification at low concentrations.

Until now the exposure to environmental tobacco smoke has been estimated from nicotine and cotinine concentrations in body fluids. The best estimate of the nicotine dose is, however, obtained if the total amounts of nicotine and its metabolites are quantified in urine. As there is only one method for the simultaneous determination of nicotine and its metabolites and as none of the methods described above, except the HR–GC methods for determination of nicotine and cotinine, could be used to quantify these alkaloids at low nanogram levels, more precise and sensitive methods have to be developed, which allow simultaneous determination of these compounds at levels usually found in nonsmokers exposed to ETS.

REFERENCES

Abrams, D.B., Follick, M.J., Biener. L. *et al.* (1987) Saliva cotinine as a measure of smoking status in field settings. *Am. J. Public Health*, **77 (7)**, 846–8.

Armitage, A., Dollery, C., Houseman, T. *et al* (1978) Absorption of nicotine from small cigars. *Clin. Pharmacol. Ther.*, **23 (2)**, 143–51.

Barlow, R.D., Stone, R.B., Wald, N.J. *et al.* (1987a) The direct barbituric acid assay for nicotine metabolites in urine: a simple colorimetric test for the routine assessment of smoking status and cigarette smoke intake. *Clin. Chim. Acta*, **165**, 45–52.

Barlow, R.D., Thompson, P.A. and Stone, R.B. (1987b) Simultaneous determination of nicotine, cotinine and five additional nicotine metabolites in the urine of smokers using pre-column derivatisation and high-performance liquid chromatography. *J. Chromatog.*, **419**, 375–80.

Beckett, A.H. and Triggs, E.J. (1966) Determination of nicotine and its metabolite, cotinine, in urine by gas chromatography. *Nature*, **211**, 1415–7.

Beckett, A.H., Gorrod, J.W., and Jenner, J. (1970) Absorption of (−)-nicotine-1′-N-oxide in man and its reduction in the gastrointestinal tract. *J. Pharm. Pharmacol.*, **22**, 722–3.

Beckett, A.H., Gorrod, J.W. and Jenner, P. (1971) The analysis of nicotine-1′-N-oxide in urine, in the presence of nicotine and cotinine, and its application to the study of *in vivo* nicotine metabolism in man. *J. Pharm. Pharmac*, **23**, 55S–66S.

Benowitz, N.L., Hall, S.M., Herning, R.I. *et al.* (1983) Smokers of low-yield cigarettes do not consume less nicotine. *New Engl. J. Med.*, **309 (3)**, 139–42.

Benowitz, N.L. and Jacob III, P. (1984) Daily intake of nicotine during cigarette smoking. *Clin. Pharmacol. Ther.*, **35 (4)**, 499–504.

Benowitz, N.L., Jacob III, P. and Savanapridi, D. (1987) Determinants of nicotine intake while chewing nicotine polacrolex gum. *Clin. Pharmacol. Ther.*, **41**, 467–73.

Benowitz, N.L., Porchet, H., Sheiner, L. *et al.* (1988) Nicotine absorption and cardiovascular effects with smokeless tobacco use: Comparison with cigarettes and nicotine gum. *Clin. Pharmacol. Ther.*, **44**, 23–6.

Benowitz, N.L., Jacob III, P. and Yu, L. (1989) Daily use of smokeless tobacco: systemic effects. *Ann. Int. Med.*, **111**, 112–6.

Biber, A., Scherer, G., Hoepfner, I. *et al.* (1987) Determination of nicotine and cotinine in human serum and urine: An interlaboratory study. *Toxicol. Lett.*, **35**, 45–52.

Bjercke, R.J., Cook, G., Rychlik, N. *et al.* (1986) Stereospecific monoclonal antibodies to nicotine and cotinine and their use in enzyme-linked immunosorbent assays. *J. Immun. Methods*, **90**, 203–13.

Bjercke, R.J., Cook, G. and Langone, J.J. (1987) Comparison of monoclonal and polyclonal antibodies to cotinine in nonisotopic and isotopic immunoassays. *J. Immun. Methods*, **96**, 239–46.

Blache, D., Thevenon, C., Ciavatti, M. *et al.* (1984) A sensitive method for the routine determination of plasma nicotine by flame ionization gas–liquid chromatography. *Anal. Biochem.*, **143**, 316–9.

Booth, J. and Boyland, E. (1970) The metabolism of nicotine into two optically active stereoisomers of nicotine-1′-oxide by animal tissues *in vitro* and by cigarette smokers. *Biochem. Pharmacol.*, **19**, 733–42.

Brandänge, S., Lindblom, L. and Samuelsson, D. (1977) Stereochemistry of the oxidation of nicotine to its 1′-N-oxides. The action of tungstate (VI) and molybdate (VI). *Acta Chem. Scand.*, **B31**, 907–22.

Burrows, I.E., Corp, P.J., Jackson, G.C. *et. al.* (1971) The Determination of nicotine in human blood by gas–liquid chromatography. *Analyst*, **96**, 81–4.

Byrd, G.D., Chang, K-M., Greene, J.M. *et al.* (1992) Evidence for urinary excretion of glucuronide conjugates of nicotine, cotinine, and *trans*-3′-hydroxycotinine in smokers. *Drug Metab. Disp.*, **20 (2)**, 192–5.

Carey, K. B. and Abrams, D. B. (1988) Properties of saliva cotinine in young adult light smokers. *Am. J. Public Health*, **78 (7)**, 842–3.

Castro, A., Monji, N., Malkus, H. *et al* (1979) Automated radioimmunoassay of nicotine. *Clin. Chim. Acta*, **95**, 473–81.

Castro, A., Monji, N., Ali, H. *et al.* (1980) Nicotine antibodies: comparison of ligand specificities of antibodies produced against two nicotine conjugates. *Eur. J. Biochem.*, **104**, 331–40.

Castro, A. and Monji, N. (1986) Nicotine enzyme immunoassay. *Res. Comm. Chem. Path. Pharmacol.*, **51 (3)**, 393–404.

Chien, C-Y., Diana, J.N. and Crooks, P.A. (1988) High-performance liquid chromatography with electrochemical detection for the determination of nicotine in plasma. *J. Pharm. Sci.*, **77 (3)**, 277–9.

Cummings, K.M., Markello, S.J., Mahoney, M. *et al.* (1990) Measurement of current exposure to environmental tobacco smoke. *Arch. Environ. Health*, **45 (21)**, 74–8.

Cundy, K.C. and Crooks, P.A. (1984a) High performance liquid chromatographic method for the determination of N-methylated metabolites of nicotine. *J. Chromatog. Biomed. App.*, **306**, 291–301.

Cundy, K.C., Godin, C.S. and Crooks, P.A. (1984b) Evidence of stereospecificity in the *in vivo* methylation of [^{14}C] ±-nicotine in the guinea pig. *Drug Metab. Disp.*, **12**, 755–9.

Cundy, K.C., Crooks, P.A. and Godin, C.S. (1985a) Remarkable substrate–inhibitor properties of nicotine enantiomers towards a guinea pig lung aromatic azaheterocycle N-methyltransferase. *Biochem. Biophys. Res. Comm.*, **128**, 312–6.

Cundy, K.C., Godin, C.S. and Crooks, P.A. (1985b) Stereospecific *in vitro* N-methylation of nicotine in guinea pig tissues by an *S*-adenosyl-methionine-dependent N-methyltransferase. *Biochem. Pharmacol.*, **34**, 281–4.

Cundy, K.C., Sato, M. and Crooks, P.A. (1985c) Stereospecific *in vivo* N-methylation of nicotine in the Guinea pig. *Drug Metab. Disp.*, **13 (2)**, 175–85.

Cundy, K.C. and Crooks, P.A. (1987) Biotransformation of primary nicotine metabolites II. Metabolism of [3H]-*S*-(−)-cotinine in the guinea pig: determination of *in vivo* urinary metabolites by high-performance liquid-radiochromatography. *Xenobiotica*, **17 (7)**, 785–92.

Curvall, M., Kazemi-Vala, E. and Enzell, C.R. (1982) Simultaneous determination of nicotine and cotinine in plasma using capillary column gas chromatography with nitrogen-sensitive detection. *J. Chromat.*, **232**, 283–93.

Curvall, M., Kazemi Vala, E., Enzell, C.R. *et al.* (1990) Simulation and evaluation of nicotine intake during passive smoking: Cotinine measurements in body fluids of nonsmokers given intravenous infusions of nicotine. *Clin. Pharmacol. Ther.*, **47 (1)**, 42–9.

Curvall, M., Kazemi Vala, E. and Englund, G. (1991) Conjugation pathways in nicotine metabolism. *Advances in Pharmacological Sciences, Effects of Nicotine on Biological Systems* (Eds F. Adlkofer and K. Thurau), Birkhäuser Verlag, Basel, Boston, Berlin, 69–75.

Daenens, P., Laruelle, L. and Callewaert, K. (1985) Determination of cotinine in biological fluids by capillary gas chromatography–mass spectrometry–selected-ion monitoring. *J. Chromat.*, **342**, 79–87.

Dagne, E. and Castagnoli Jr, N. (1972a) Structure of hydroxycotinine, a nicotine metabolite. *J. Med. Chem.*, **15 (4)**, 356–60.

Dagne, E. and Castagnoli, N. (1972b) Cotinine N-oxide, a new metabolite of nicotine. *J. Med. Chem.*, **15 (8)**, 840–1.

Dahlström, A., Lundell, B., Curvall, M. *et al.* (1990) Nicotine and cotinine concentrations in the nursing mother and her infant. *Acta Paediatr. Scand.*, **79**, 142–7.

Dajani, R.M., Gorrod, J.W. and Beckett, A.H. (1975a) *In vitro* hepatic and extra-hepatic reduction of (−)-nicotine-1′-N-oxide in rats. *Biochem. Pharmacol.*, **24**, 109–17.

Dajani, R.M., Gorrod, J.W. and Beckett, A.H. (1975b) Reduction *in vivo* of nicotine-1′-N-oxide by germ-free and conventional rats. *Biochem. Pharmacol.*, **24**, 648–50.

Davis, R.A. (1986) The determination of nicotine and cotinine in plasma. *J. Chromat. Sci.*, **24**, 134–41.

Degen, P.H. and Schneider, W. (1991) Rapid and sensitive determination of low concentrations of nicotine in plasma by gas chromatography with nitrogen-specific detection. *J. Chromat.*, **563**, 193–8.

Dow, J. and Hall, K. (1978) Capillary-column combined gas chromatography–mass spectrometry method for the estimation of nicotine in plasma by selective ion monitoring. *J. Chromat.*, **153**, 521–5.

Etzel, R.A., Greenberg, R.A., Haley, N.J. *et al.* (1985) Clinical and laboratory observations. Urine cotinine excretion in neonates exposed to tobacco smoke products *in utero*. *J. Pediat.*, **107 (1)**, 146–8.

Falkman, S.E., Burrows, I.E., Lundgren, A. *et al.* (1975) A modified procedure for the determination of nicotine in blood. *Analyst*, **100**, 99–104.

Feyerabend, C., Levitt, T. and Russell, M.A.H. (1975) A rapid gas-liquid chromatographic estimation of nicotine in biological fluids. *J. Pharm. Pharmacol.*, **27**, 434–6.

Feyerabend, C. and Russell, M.A.H. (1979) Improved gas-chromatographic method and micro-extraction technique for the measurement of nicotine in biological fluids. *J. Pharm. Pharmacol.*, **31**, 73–6.

Feyerabend, C. and Russell, M.A.H. (1980a) Assay of nicotine in biological materials: sources of contamination and their elimination. *J. Pharm. Pharmacol.*, **32**, 178–81.

Feyerabend, C. and Russell, M.A.H. (1980b) Rapid gas–liquid chromatographic determination of cotinine in biological fluids. *Analyst*, **105**, 998–1001.

Feyerabend, C., Higenbottam, T. and Russell, M.A.H. (1982) Nicotine concentrations in urine and saliva of smokers and nonsmokers. *Br. Med. J.*, **284**, 1002–4.

Feyerabend, C., Bryant, A.E., Jarvis, M.J. *et al.* (1986) Determination of cotinine in biological fluids of nonsmokers by packed column gas-liquid chromatography. *J. Pharm. Pharmacol.*, **38**, 917–9.

Feyerabend, C. and Russell, M.A.H. (1990) A rapid gas–liquid chromatographic method for the determination of cotinine and nicotine in biological fluids. *J. Pharm. Pharmacol.*, **4**, 450–2.

Galeazzi, R.L., Daenens, P. and Gugger, M. (1985) Steady-state concentration of cotinine as a measure of nicotine-intake by smokers. *Eur. J. Clin. Pharmacol.*, **28**, 301–4.

Giusto, E.D. and Eckhard, I. (1986) Some properties of saliva cotinine measurements in indicating exposure to tobacco smoking. *Am. J. Public Health*, **76 (19)**, 1245–6

Greenberg, R.A., Haley, N.J., Etzel, R.A. *et al.* (1984) Measuring the exposure of infants to tobacco smoke. Nicotine and cotinine in urine and saliva. *New Eng. J. Med.*, **310 (17)**, 1075–8.

Haley, N.J., Axelrad, C.M. and Tilton, K.A. (1983) Validation of self-reported smoking behavior: biochemical analyses of cotinine and thiocyanate. *Am. J. Public Health*, **73 (10)**, 1204–7.

Hariharan, M., VanNoord, T. and Greden, J.F. (1988) A high-performance liquid-chromatographic method for routine simultaneous determination of nicotine and cotinine in plasma. *Clin. Chem.*, **34 (4)**, 724–9.

Hartvig, P., Ahnfelt, N-O., Hammarlund, M. *et al.* (1979) Analysis of nicotine as a trichloroethyl carbamate by gas chromatography with electron-capture detection. *J. Chromat.*, **173**, 127–38.

Hengen, N. and Hengen, M. (1978) Gas-liquid chromatographic determination of nicotine and cotinine in plasma. *Clin. Chem.*, **24 (1)**, 50–3.

Herning, R.I., Jones, R.T., Benowitz, N.L. *et al* (1983) How a cigarette is smoked determines blood nicotine levels. *Clin. Pharmacol. Ther.*, **33 (1)**, 84–90.

Hill, P. and Wynder, E.L. (1979) Nicotine and cotinine in breast fluid. *Cancer Lett.*, **6**, 251–4.

Horstmann, M. (1985) Simple high-performance liquid chromatographic method for rapid determination of nicotine and cotinine in urine. *J. Chromatog.*, **344**, 391–6.

Jacob, P., Wilson, M. and Benowitz, N.L. (1981) Improved gas chromatographic method for the determination of nicotine and cotinine in biologic fluids. *J. Chromatog.*, **222**, 61–70.

Jacob, III, P., Benowitz, N.L., Yu, L. *et al.* (1986) Determination of nicotine-N-oxide by gas chromatography following thermal conversion to 2-methyl-6-(3-pyridyl)tetrahydro-1,2-oxazine. *Anal. Chem.*, **58**, 2218–21.

Jacob, III, P., Yu, L., Wilson, M. *et al* (1991) Selected ion monitoring method for determination of nicotine, cotinine and deuterium-labelled analogs: Absence of an isotope effect in the clearance of (*S*)-nicotine-3-3′-d_2 in humans. *Biol. Mass Spec.*, **20**, 247–52.

Jarvis, M., Tunstall-Pedoe, H., Feyerabend, C. *et al.* (1984) Biochemical markers of smoke absorption and self reported exposure to passive smoking. *J. Epidem. Community Health*, **38**, 335–9.

Jarvis, M.J., Russel, M.A.H., Feyerabend, C. *et al.* (1985) Passive exposure to tobacco smoke: saliva cotinine concentrations in a representative population sample of non-smoking schoolchildren. *Br. Med. J.*, **291**, 927–9.

Jarvis, M.J., Tunstall-Pedoe, H., Feyerabend, C. *et al.* (1987) Comparison of tests used to distinguish smokers from nonsmokers. *Am. J. Public Health*, **77 (11)**, 1435–8.

Jones, D., Curvall, M., Abrahamsson, L. *et al.* (1982) Quantitative analysis of plasma nicotine using selected ion monitoring at high resolution. *Biomed. Mass Spectrom.*, **9 (12)**, 539–45.

Klus. H. and Kuhn, H. (1977) A study of the optical activity of smoke nicotines. *Fachliche Mitt. Oesterr. Tabakregie*, **17**, 331–6.

Knight, G.J., Wylie, P., Holman, M.S. *et al.* (1985) Improved ^{125}I radioimmunoassay for cotinine by selective removal of bridge antibodies. *Clin. Chem.* **31 (1)**, 118–21

Kogan, M.J., Verebey, K., Jaffee, J.H. *et al.* (1981) Simultaneous determination of nicotine and cotinine in human plasma by nitrogen detection gas–liquid chromatography. *J. Forensic Sci.*, **26 (1)**, 6–11.

Kolonen, S.A. and Puhakainen, V.J. (1991) Assessment of the automated colorimetric and the high-performance liquid chromatographic methods for nicotine intake by urine samples of smokers, smoking low- and medium-yield cigarettes. *Clin. Chim. Acta*, **196**, 159–66.

Kyerematen, G.A., Taylor, L.H., deBethizy, J.D. *et al.* (1987) Radiometric-high-performance liquid chromatographic assay for nicotine and twelve of its metabolites. *J. Chromatog.*, **419**, 191–203.

Kyerematen, G.A., Taylor, L.H., deBethizy, J.D. *et al.* (1988) Pharmacokinetics of nicotine and 12 metabolites in the rat. Application of a new radiometric high performance liquid chromatography assay. *Drug Metab. Disp.*, **16** (**1**), 125–9.

Langone, J.J., Gjika, H.B. and Van Vunakis, H. (1973) Nicotine and its metabolites. radioimmunoassays for nicotine and cotinine. *Biochemistry*, **12 (24)**, 5025–30.

Langone, J.J. and Van Vunakis, H. (1982) Radioimmunoassay of nicotine, cotinine, and γ-(3-pyridyl)-γ-oxo-N-methylbutyramide. *Methods in Enzymology*, **84**, 628–40.

Langone, J.J. and Bjercke, R.J. (1989) Idiotype-anti-idiotype hapten immunoassays: Assay for cotinine. *Anal. Biochem.*, **182**, 187–92.

Luck, W. and Nau, H. (1984) Nicotine and cotinine concentrations in serum and milk of nursing smokers. *Br. J. Clin. Pharmacol.*, **18**, 9–15.

Luck, W., Nau, H., Hansen, R. *et al.* (1985) Extent of nicotine and cotinine transfer to the human fetus, placenta and amniotic fluid of smoking mothers. *Dev. Pharmacol. Ther.*, **8**, 384–95.

Luck, W. and Nau, H. (1987) Nicotine and cotinine concentrations in the milk of smoking mothers: influence of cigarette consumption and diurnal variation. *Eur. J. Pediatr.*, **146**, 21–6.

Machacek, D.A. and Jiang, N-S. (1986) Quantification of cotinine in plasma and saliva by liquid chromatography. *Clin. Chem.* **32 (6)**, 979–82.

Maskarinec, M.P., Harvey, R.W. and Caton, J.E. (1978) A novel method for the isolation and quantitative analysis of nicotine and cotinine in biological fluids. *J. Anal. Toxicol.* **2**, 124–6.

Matsukura, S., Sakamoto, N., Seno, Y. *et al* (1979) Cotinine excretion and daily cigarette smoking in habituated smokers. *Clin. Pharmacol. Ther.*, **25**, 555–61.

McKennis, H., Turnbull, L.B. and Bowman, E.R. (1963) N-methylation of nicotine and cotinine *in vivo*. *J. Biol. Chem.*, **238 (2)**, 719–23.

McManus, K.T., deBethizy, J.D., Garteiz, D.A. *et al.* (1990) A new quantitative thermospray LC–MS method for nicotine and its metabolites in biological fluids. *J. Chromatog. Sci.*, **28**, 510–6.

McNabb, M.E., Ebert, R.V. and McCusker, K. (1982) Plasma nicotine levels produced by chewing nicotine gum. *JAMA*, **248 (7)**, 865–8.

McNeill, A.D., Jarvis, M.J., West, R. *et al.* (1987) Saliva cotinine as an indicator of cigarette smoking in adolescents. *Br. J. Addiction*, **82**, 1355–60.

Moore, J., Greenwood, M. and Sinclair, N. (1990) Automation of a high-performance liquid chromatographic assay for the determination of nicotine, cotinine and 3-hydroxycotinine in human urine. *J. Pharm. Biomed. Analysis*, **8 (8–12)**, 1051–4.

Mulligan, S.C., Masterson, J.G., Devane, J.G. *et al.* (1990) Clinical and pharmacokinetic properties of a transdermal nicotine patch. *Clin. Pharmacol. Ther.*, **47 (3)**, 331–7.

Neurath, G.B. and Pein, F.G. (1987a) Gas chromatographic determination of trans-3′-hydroxycotinine, a major metabolite of nicotine in smokers. *J. Chromatog.*, **415**, 400–6.

Neurath, G.B., Dünger, M., Orth, D. *et al.* (1987b) *Trans*-3′-hydroxycotinine as a main metabolite in urine of smokers. *Int. Arch. Occup. Environ. Health*, **59**, 199–201.

Neurath, G.B., Orth, D. and Pein, F.G. (1991) Detection of nornicotine in human urine after infusion of nicotine. *Advances in Pharmacological Sciences, Effects of Nicotine on Biological Systems* (Eds F. Adlkofer and K. Thurau), Birkhauser Verlag, Basel, Boston, Berlin, 45–9.

Norbury, C.G. (1987) Simplified method for the determination of plasma cotinine using gas chromatography–mass spectrometry. *J. Chromatog.*, **414**, 449–53.

Nwosu, C.G. and Crooks, P.A. (1988) Species variation and stereoselectivity in the metabolism of nicotine enantiomers. *Xenobiotica*, **18 (12)**, 1361–72.

O'Doherty, S., Revans, A., Smith, S.L. *et al.* (1988) Determination of *cis*- and *trans*-3-hydroxycotinine by high performance liquid chromatography. *J. High Resolution Chromatog. and Chromatog. Comm.*, **11**, 723–5.

Parviainen, M.T. and Barlow, R.D. (1988) Assessment of exposure to environmental tobacco smoke using a high-performance liquid chromatographic method for the simultaneous determination of nicotine and two of its metabolites in urine. *J. Chromatog.*, **431**, 216–21.

Parviainen, M.T., Puhakainen, E.V.J., Laatikainen, R. *et al.* (1990) Nicotine metabolites in the urine of smokers. *J. Chromatog.*, **525**, 193–202.

Peach, H., Ellard, G.A., Jenner, P.J. *et al.* (1985) A simple, inexpensive urine test of smoking. *Thorax*, **40**, 351–7.

Phillipson. J.D. and Handa, S.S. (1978) Alkaloid N-oxides in some species of *Solanaceae* and *Papaveracea*, in *Biological Oxidation of Nitrogen* (Ed. J.W. Gorrod), Elsevier/North Holland, Biomedical Press, 409–20.

Pojer, R., Whitfield, J.B., Poulos, V. *et al.* (1984) Carboxyhemoglobin, cotinine, and thiocyanate assay compared for distinguishing smokers from non-smokers. *Clin. Chem.*, **30 (8)**, 1377–80.

Pool, W.F. and Crooks, P.A. (1985) Biotransformation of primary nicotine metabolites. 1. *In vivo* metabolism of R-(+)-[^{14}C-N-CH_3] N-methylnicotinium ion in the Guinea Pig. *Drug Metab. Disp*, **13 (5)**, 578–81.

Pool, W.F., Houdi, A.A., Damani, L.A. *et al.* (1986) Isolation and characterization of N-methyl-N′-oxonicotinium ion, a new urinary metabolite of *R*-(+)-nicotine in the Guinea pig. *Drug Metab. Disp.*, **14**, 574–9.

Puhakainen, E.V.J., Barlow, R.D. and Salonen, J.T. (1987) An automated colorimetric assay for urine nicotine metabolites: a suitable alternative to cotinine assays for the assessment of smoking status. *Clin. Chim. Acta*, **170**, 255–62.

Richie, J.P., Leutzinger, Y., Axelrad, C.M. *et al.* (1990) Contribution of *trans*-3′-hydroxycotinine and glucuronide conjugates of nicotine metabolites to the measurement of cotinine by RIA. *Advances in Pharmacological Sciences. Effects of Nicotine on Biological Systems*, (Eds F. Adlkofer and K. Thurau), Birkhäuser Verlag, Basel, Boston, Berlin, pp. 77–81.

Ross, H.D., Chan, K.K.H., Piraino, A.J. *et al.* (1991) Pharmacokinetics of multiple daily transdermal doses of nicotine in healthy smokers. *Pharm. Res.*, **8 (3)**, 385–8.

Russell, M.A.H., Feyerabend, C. and Cole, P.V. (1976) Plasma nicotine levels after cigarette smoking and chewing nicotine gum. *Br. Med. J.*, **1**, 1043–6.

Russell, M.A.H., Jarvis, M., Iyer, R. *et al.* (1980) Relation of nicotine yield of cigarettes to blood nicotine concentrations in smokers. *Br. Med. J.*, **280**, 972–6.

Russell, M.A.H., Jarvis, M.J., Devitt, G. *et al.* (1981) Nicotine intake by snuff users. *Br. Med. J.*, **283**, 814–7.

Russell, M.A.H., Jarvis, M.J. and West, R.J. (1986) Use of urinary nicotine concentrations to estimate exposure and mortality from passive smoking in non-smokers. *Br. J. Addiction*, **81**, 275–81.

Schievelbein, H. and Grundke, K. (1968) Gas-chromatographische Methode zur Bestimmung von Nicotin in Blut und Geweben. *Z. Anal. Chem,.* **237**, 1–9.

Sepkovic, D.W., Haley, N.J., Axelrad, C.M. *et al.* (1986) Short-term studies on the *in vivo* metabolism on N-oxides of nicotine in rats. *J. Toxicol. Environ. Health*, **18**, 205–14.

Shulgin, A.T., Jacob III, P., Benowitz, N.L. *et al.* (1987) Identification and quantitative analysis of cotinine-N-oxide in human urine. *J. Chromatog.*, **423**, 365–72.

Srivastava, E.D., Russell, M.A.H., Feyerabend, C. *et al.* (1991) Sensitivity and tolerance to nicotine in smokers and nonsmokers. *Psychopharmacol.*, **105**, 63–8.

Stehlik G., Kainzbauer, J., Tausch, *et al.* (1982) Improved method for routine determination of nicotine and its main metabolites in biological fluids. *J. Chromatog.*, **232**, 295–303.

Teeuwen, H.W.A., Aalders, R.J.W., Van Rossum, J.M. *et al.* (1989) Simultaneous estimation of nicotine and cotinine levels in biological fluids using high-resolution capillary-column gas chromatography combined with solid phase extraction work-up. *Mol. Biol. Reports*, **13**, 165–75.

Thompson, J.A., Ho, M-S. and Petersen, D.R. (1982) Analyses of nicotine and cotinine in tissues by capillary gas chromatography and gas chromatographymass spectrometry. *J. Chromatog.*, **231**, 53–63.

Van Vunakis, H. and Levine, L. (1974) Use of double antibody technique and nitrocellular membrane filtration to separate free antigen from antibody-bound antigen in radioimmunoassay, in *Immunoassays for drugs subject to abuse* (Eds S.J. Mulé, I. Sunshine, M. Braude and R.E. Willette), CRC, Cleveland, Ohio, .23–35.

Van Vunakis, H., Tashkin, R.P., Rigas, B. *et al.* (1989) Relative sensitivity and specificity of salivary and serum cotinine in identifying tobacco-smoking status of self-reported nonsmokers and smokers of tobacco and/or marijuana. *Arch. Environ. Health*, **44(1)**, 53–8.

Verebey, K.G., DePace, A., Mulé, S.J. *et al.* (1982) A rapid, quantitative GLC method for the simultaneous determination of nicotine and cotinine. *J. Anal. Toxicol.*, **6**, 294–6.

Voncken, P., Schepers, G., Schäfer, K-H. (1989) Capillary gas chromatographic determination of *trans*-3′-hydroxycotinine simultaneously with nicotine and cotinine in urine and blood samples. *J. Chromatog.*, **479**, 410–18.

Voncken, P., Rustemeier, K. and Schepers, G., (1990) Identification of *cis*-3′-hydroxycotinine as a urinary nicotine metabolite. *Xenobiotica*, **20 (12)**, 1353–6.

Wald, N.J., Idle, M., Boreham, J. *et al.* (1981) Serum cotinine levels in pipe smokers: evidence against nicotine as cause of coronary heart disease. *Lancet*, **ii**, 775–7.

Wall, M.A., Johnson, J., Jacob, P. *et al.* (1988) Cotinine in the serum, saliva, and urine of nonsmokers, passive smokers, and active smokers. *Am. J. Public Health*, **78 (6)**, 699–701.

Watson, I.D. (1977) Rapid analysis of nicotine and cotinine in the urine of smokers by isocratic high-performance liquid chromatography. *J. Chromatog.*, **143**, 203–6.

West, R.J., Jarvis, M.J., Russell, M.A.H. *et al.* (1984) Plasma nicotine concentrations from repeated doses of nasal nicotine solution. *Br. J. Addiction*, **79**, 443–5.

Zhang Y., Jacob P. and Benowitz, N.L. (1990) Determination of nornicotine in smokers' urine by gas chromatography following reductive alkylation to N′-propylnornicotine. *J. Chromatog.*, **525**, 349–57.

Kinetics of nicotine and its metabolites in animals

9

J. Gabrielsson and M. Gumbleton

9.1 INTRODUCTION

The major interest in nicotine alkaloids derives from their carcinogenic effects and the dependence associated with tobacco smoking. Nicotine is a highly water soluble, strongly alkaline liquid with pKa of (-N= , -N<) 3.2 and 8.0, respectively. At physiological pH nicotine exists mainly in the form of the univalent nicotinium ion which appears, rather than the non-ionized base, to be the active form. Nicotine contains a chiral centre and is found naturally to exist mainly as the more potent (*S*) enantiomer. The quite complex pharmacology of nicotine can potentially give rise to clinically important pharmacodynamic drug interactions (Miller, 1990). However, tolerance to the pharmacological actions of nicotine can develop, complicating any relationship between pharmacological action and dose/systemic concentrations; the rapidity of tolerance development is dependent upon both the physical form in which nicotine is administered and the route of administration (Cohen and Roe, 1981; US Public Health Service, DHEW Publications Nos. (HSM) 72-7516, 1972 and (HSM) 71-7513, 1971; Benowitz, 1990).

In humans nicotine clearance is relatively high, with a plasma clearance ranging from approximately 1 l/min to 2.5 l/min, of which 85–95% is accounted for by metabolic clearance (Benowitz, 1986; Feyerabend *et al.* 1985; Kyerematen and Vessell, 1991). The fraction of nicotine in plasma that is bound to proteins is low and approximates to

Nicotine and Related Alkaloids: Absorption, distribution, metabolism and excretion. Edited by J.W. Gorrod and J. Wahren. Published in 1993 by Chapman & Hall, London. ISBN 0 412 55740 1

5%, with what must be a significantly greater degree of binding in tissues to account for a relatively large distribution volume averageing 1.7–3.0 l/kg. As a consequence of the magnitude of these primary pharmacokinetic parameters the elimination half-life for nicotine is quite short, in the range of 1 to 3 h. An additional terminal phase to the nicotine concentration–time profile has been suggested (Russell and Feyerabend, 1978; Rosenberg *et al.*, 1980), reflecting a more slowly equilibrating tissue compartment. In smokers this terminal phase has been estimated to have a half-life of only about 5 h, while in non-smokers it can be as long as 10 h (Kyerematen *et al.*, 1982), although obviously such values are critically dependent upon the precision of detection at the lower concentration limits of the assay. However, an additional terminal phase of 5–10 h should not contribute significantly to the $AUC_{0-\infty}$, and a clearance estimate based upon calculation of an $AUC_{0-\infty}$ incorporating a poor (if any) definition of this terminal phase will still remain a reasonable predictor for nicotine steady-state concentrations. Nicotine is extensively metabolized, mainly in the liver but also in the lungs and kidneys (Gorrod and Jenner, 1975; Turner *et al.*, 1975). In man up to eight metabolites of nicotine have thus far been isolated and identified (Kyerematen *et al.*, 1990), with six primary pathways recognized for nicotine metabolism in mammals: α-carbon oxidation of the pyrrolidine ring to form cotinine, N-oxidation of the pyrrolidine ring to form nicotine-N-oxide, N-methylation of the pyridine ring to form N-methylnicotinium and N-demethylation of the pyrrolidine ring to form nornicotine. The other two primary pathways for nicotine metabolism include the formation of nicotine $\Delta^{4',5'}$-enamine and of the nicotine glucuronide (see Figure 9.1; reproduced from Kyerematen and Vesell, 1991). α-Carbon oxidation to cotinine is quantitatively the major primary metabolic pathway, with cotinine itself being essentially devoid of nicotinic receptor agonist actions. The plasma clearance of cotinine in man is lower than that of nicotine, approximating 0.07 l/min of which 10–20% comprises renal excretion (Curvall *et al.*, 1990; Benowitz *et al.*, 1983; De Schepper *et al.*, 1978); under steady-state conditions the plasma concentration of cotinine can be up to ten-fold that of nicotine (Gritz *et al.*, 1981). The distribution volume for cotinine is in the range of 0.8 to 1.2 l/kg with a terminal half-life ranging from 12 to 20 h, the fraction of cotinine in plasma that is bound to proteins is less than 3%.

The aim of this chapter is to review the absorption and disposition characteristics of nicotine and cotinine with emphasis upon its pharmacokinetics in animals. In addition, the chapter will highlight the established technique of utilizing subcutaneously implanted mini-pumps for animal pharmacokinetic studies, a technique that enables the control of drug input-rate over an extended time period.

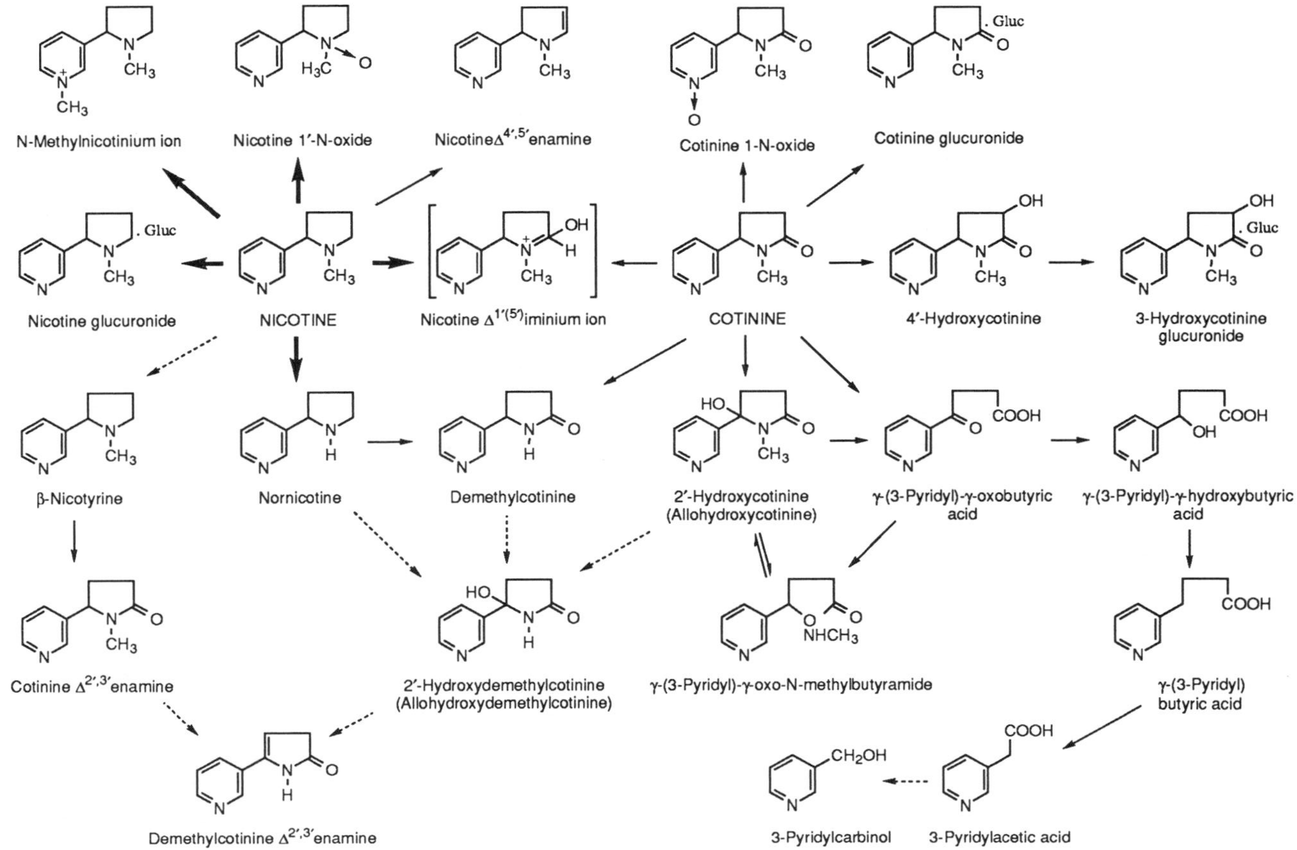

Figure 9.1 Pathways of nicotine metabolism in mammals. Solid arrows show established pathways, broken arrows indicate unconfirmed reactions. (Reproduced with minor modification from Kyerematen and Vesell, 1991.)

9.2 ABSORPTION KINETICS AND FIRST-PASS METABOLISM

The influence of pH on the absorption of weak acids or bases through lipid membranes will be dependent upon the pKa of the ionizable functionalities and the lipid solubility of the non-ionized species. For nicotine the equilibria will be shifted toward the non-ionizable species (i.e. favouring absorption across biological membranes) as its aqueous environment becomes more alkaline, i.e. at a pH less than 5, the diprotonated species of nicotine should predominate, while at pH 5 to 8 nicotine will exist primarily as the monoprotonated species. At a pH of greater than 9 nicotine will be greater than 90% non-ionized. That the non-ionized species is lipid soluble and readily permeates biological membranes (Russell and Feyerabend, 1978) is testified by the rapid systemic absorption of nicotine following inhalation (see below), and by its access to and metabolism within the endoplasmic recticulum of the cell. Nicotine is not only absorbed via the lungs, but also through the skin, the gastrointestinal tract, and buccal and nasal mucosae (Kyerematen and Vesell, 1991), in addition to undergoing reabsorption across the renal tubular membranes (Pillotti, 1980; Schechter *et al.*, 1977).

Cigarette smokers absorb about 90% of the nicotine content of the smoke that they inhale, although factors influencing the amount absorbed include the tobacco type, the moisture content of the tobacco, the pH of the membrane surfaces which the smoke comes into contact with and, of course, the depth of inhalation (Armitage *et al.*, 1975). In tobacco smoke aerosols, nicotine is predominantly in the particulate or solid phase, nevertheless equilibrium with aqueous phases and absorption across the epithelial tissues of the pulmonary airway into the systemic vasculature is recognized to be extremely rapid (Armitage *et al.*, 1975; Haines *et al.*, 1976) e.g. venous plasma concentrations of nicotine peak within 1 min of ceasing to inhale cigarette smoke. With regard to first-pass metabolism within the lung, in the isolated perfused dog lung about 30% of a vapourized administration of nicotine is metabolized within the lung to cotinine and nicotine-N-oxide before entering the pulmonary circulation (Turner *et al.*, 1975), first-pass metabolism approximating only 7% of the dose when nicotine is administered via the pulmonary artery (Beckett *et al.*, 1971). In microsomal preparations of rabbit lung, nicotine has been shown to be metabolized primarily to nicotine $\Delta^{1',5'}$-iminium ion (Williams *et al.*, 1990), an intermediate compound in the sequential metabolism to cotinine (Obach and van Vunakis, 1990). Studies with reconstituted purified P450 enzymes and the use of inhibitory antibodies (inhibiting metabolism of nicotine to the nicotine $\Delta^{1',5'}$-iminium ion to the extent of 95%) has implicated phenobarbital inducible cytochrome P450 form 2 (equivalent to updated 'protein name' cytochrome P450 IIB1; Neubert *et al.*, 1987) as the catalytic enzyme involved; nicotine was found to be a poor substrate for rabbit lung flavin-monooxygenase. The phenobarbital inducibility of nicotine metabolism in lung tissue has also been recently confirmed in the rat species (Foth *et al.*, 1991), where pretreatment with phenobarbital increased nicotine lung clearance two-fold; cotinine metabolism by lung tissue has also been found to be induced by phenobarbital pretreatment.

Nicotine 'chewing gum' and transdermal delivery systems are clinically important products, and have been developed as an adjunct to smoking cessation therapy. In a recent clinical study involving 11 subjects the average absolute transdermal bioavailability of nicotine was estimated at 82%, with 10% of the nicotine absorbed after removal of the delivery device, indicating that skin can act as a depot or reservoir for nicotine (Benowitz *et al.*, 1991). Although the buccal absorption of nicotine avoids hepatic first-pass metabolism the systemic nicotine bioavailability from nicotine gum has been estimated at only 30–40% (Benowitz *et al.*, 1987). Obviously this could reflect a number of confounding factors including expectoration, inadequate chewing technique and the swallowing of saliva containing high amounts of extracted nicotine. In this latter regard the extent of the intestinal bioavailability of nicotine may well be of importance. However, the first-pass bioavailability following oral administration has not been adaquately studied, although data obtained in two human subjects suggest an oral bioavailability of about 30% (Beckett *et al.*, 1971). Contrary to common expectations, which are based upon nicotine clearance values approximating hepatic plasma flow and an assumption that metabolic elimination is mainly hepatic in origin, recent data in the conscious rat indicated that nicotine can exhibit a high oral bioavailability (Foth *et al.*, 1990). Using an i.v. nicotine dose of 0.3 mg/kg, nicotine plasma clearance was calculated to approximate 19 ml/min (a value exceeding hepatic–splanchnic plasma flow*) with an oral bioavailability exceeding 90%. It should be noted that in this study dose independency in nicotine first-pass or systemic kinetics was not confirmed, a factor which could undermine the bioavailability assessments, although in the rat, linearity in the systemic kinetics of nicotine at i.v. doses of 0.1 mg/kg to 1.0 mg/kg has been demonstrated (Kyerematen *et al.*, 1988). Corroborative evidence for a low extraction ratio across rat liver has been provided by Rüdell *et al.* (1987), who reported a hepatic extraction ratio for nicotine of only 0.12 in an isolated perfused rat liver preparation. In this preparation phenobarbital pretreatment enhanced nicotine metabolism up to eight-fold, although an inducer of cytochrome P448, 5,6-benzoflavone, exerted only marginal influence.

*Estimates based on an average haematocrit of 0.55 and Gumbleton *et al.* (1990) referring to hepatosplanchnic blood flows in the conscious rat ranging from 65 to 90 ml/min per kg body weight.

9.3 DISPOSITION KINETICS

Nicotine is rapidly and extensively distributed throughout the body. Thirty to 60 min after intravenous administration of nicotine to rats, concentrations two to 15 times higher than those in plasma have been observed in a number of organs, i.e. adrenals, liver, brain, lung, heart, gastro-intestinal tissue, spleen, thymus and kidney in addition to the relatively lowly perfused (expressed as perfusion rate per gram of tissue) skeletal muscle (Schmiterlöw *et al.*, 1967; Stålhandske, 1967; Tsujimoto *et al.*, 1975). The extensive distribution of this highly lipid soluble alkaloid is reflected in reported distribution volumes of about 5 l/kg in the rat (Kyerematen *et al.*, 1988) and about 2.0 l/kg in macaques (Seaton *et al.*, 1991). It is also found in many body fluids including saliva and breast milk, the distribution balance reflecting the relative pH of blood and the particular fluid in question. High concentrations of nicotine have been observed in the salivary glands of monkeys and humans, the concentration ratio of nicotine in saliva:plasma generally exceeding 10 (Russell and Feyerabend, 1978). Nicotine has also been found to concentrate in breast milk, with milk:serum concentration ratios averageing 2.9 in a group of nursing smokers; the nursed infants' serum:maternal serum nicotine concentration ratio averageing 0.06 (Luck and Nau, 1984). In dogs the distribution of nicotine between blood and plasma is reflected by a concentration ratio (erythrocytes:plasma) of 0.8 (Schievelbein, 1982).

The kidney is the main organ for excretion of nicotine in man and other mammals (Turner, 1969), although the extent of renal excretion only approximates 10% of an i.v. dose. However, the renal excretion of nicotine displays pH dependency; at a urinary pH above 7 nicotine is readily reabsorbed through passive diffusional transfer across the renal tubule back into the renal tubular circulation, with the result that as little as 2% of a dose is excreted unchanged in urine. When urine is more acidic (pH less than 5) as much as 23% of the dose can be recovered in the urine. Such pH dependency in renal clearance reflecting, at alkaline pHs, an equilibrium shift toward the non-ionized lipophilic species with the potential for extensive passive reabsorption of the filtered drug load.

Secretion of nicotine into the bile contributes relatively insignificantly to overall clearance (Turner, 1969). In the rat 2.0–3.5% of injected radiolabelled (ring labelling) dose of nicotine has been found to be secreted into the bile within a 6 h collection period; autoradiographic analysis showed that the collected radioactivity comprised both nicotine and cotinine. After i.v. administration of ^{14}C-nicotine to cats, 0.5% of the radioactivity was found in the bile after a 4 h collection period. Following an i.p. injection of randomly ^{14}C-labelled nicotine into guinea pigs, 2% of the label could be recovered in the faeces after an 18 h collection period, and in the dog after i.v. injection of radiolabelled nicotine the faeces collected over the following 24 h contained about 5% radioactivity.

As mentioned earlier, nicotine is extensively metabolized, mainly in the liver but also in the lungs and kidneys, a major review of nicotine metabolism in mammals has recently been published by Kyerematen and Vesell (1991), and the interested reader is directed to this article. Only a brief summary of nicotine metabolism will be presented here. Figure 9.1 shows the pathways for nicotine metabolism in mammals and is reproduced from the scheme published by Kyerematen and Vesell (1991). In most mammalian species cotinine is the major primary metabolite of nicotine. The initial oxidation to the intermediate iminium ion is catalysed by cytochrome P450, with further oxidation to cotinine catalysed by cytosolic aldehyde dehydrogenases. In a reconstituted enzyme system (McCoy *et al.*, 1989) using a number of isolated and purified cytochrome P450s from rabbit liver, two isoenzymes showed particularly high rates for the C- and N-oxidation of nicotine, the phenobarbital inducible IIB1 and the constitutive IIC3 enzymes. Further hydroxylation with the formation of 3-hydroxycotinine from cotinine is also considered a significant pathway in the overall metabolism of nicotine. N-oxidation constitutes an important route for nicotine biotransformation, the metabolism of such tertiary amines generally catalysed almost exclusively by the flavin-monooxygenase (FMO) system. However, recent data indicate that at least in rabbit lung microsomes the formation of 1′-N-oxide from nicotine is in part catalysed by a cytochrome P450II enzyme. Nicotine 1′-N-oxide has been identified in the urine from the rat, cat and rabbit species, and *in vitro* following nicotine incubation with the livers and lungs from several animal species. Cotinine-N-oxide has also been detected in a variety of *in vitro* and *in vivo* experiments performed in dogs, rats and monkeys, cotinine-N-oxide formation being cytochrome P450 dependent. The N-oxidation products of nicotine are of interest in that their generation is implicated with the formation of carcinogenic nitrosamines. N-Demethylation is also recognized as a significant metabolic pathway for nicotine with the isolation of demethylcotinine in urine from a number of animal species following nicotine administration. Glucuronidation and N-methylation represent recognized conjugation pathways in the metabolism of nicotine.

There are a number of factors which can influence the metabolic transformation of nicotine, including species, age and sex differences. Detailed *in vitro* and *in vivo* kinetic measurements of the influence of sex and age have been undertaken. In each of four strains of rat (Kyerematen *et al.*, 1988) liver homogenates from males exhibited significantly higher rates than females with regard to both C- and N-oxidation of nicotine. Mature (100 day old) rats showed the largest sex difference, although a strain influence was also noticed, e.g. Long Evans species showed a large sex difference in both mature and young (40 day) rats, while for Wistar rats the sex difference was dramatically less for the young group. *In vivo*, the elimination half-lives for nicotine have been reported shorter in the male compared with the female rat (Kyerematen *et al.*, 1988), although the authors indicated that this was a result of a larger nicotine distribution volume in the female rat and not a reflection of a sex-dependent difference in clearance. Species differences in nicotine metabolism have also been demonstrated (Nwosu and Crooks, 1988; Kyerematen *et al.*, 1990): *in vivo* (Nwosu and Crooks, 1988) 4′-hydroxycotinine has been detected as the major urinary metabolite in hamsters, guinea pigs and humans, but in rats cotinine excretion is more significant; nicotine 1′-N-oxide is an important metabolite in guinea pigs and rats, but was not detected in hamsters and rabbits (Booth and Boyland, 1970). *In vitro* (Kyerematen *et al.*, 1990), hepatocytes isolated from guineas pigs, hamsters, mice, rats and humans have been shown to exhibit very large species differences in nicotine metabolism.

Although traces of nicotine 1′-N-oxide, cotinine-N-oxide and γ-3-(pyridyl)-methylaminobutyric acid have been isolated from rat plasma, analytical techniques of sufficient sensitivity have only allowed analysis of the systemic pharmacokinetics (based on vascular sampling) of nicotine and its major primary metabolite cotinine. At i.v. doses of 0.1 to 1 mg/kg the elimination half-life, volume of distribution and clearance of nicotine in the rat has been reported to approximate 1.0 h, to range from 5 to 5.7 l per kg body weight and 3.0 to 4.0 l/h per kg body weight, respectively (Foth *et al.*, 1990; Kyerematen *et al.*, 1988; Adir *et al.*, 1976). The nicotine clearance did not display any dose dependency (Kyerematen *et al.*, 1988) at i.v. doses of 0.1, 0.5 and 1.0 mg/kg, with dose normalized AUCs of 320, 348 and 267 ng/ml h per mg/kg dose, respectively. In the same study (Kyerematen *et al.*, 1988) plasma concentrations of the metabolite cotinine were measured, with an estimated cotinine elimination half-life ranging from 4.9–5.3 h. The dose normalized AUC values for cotinine were, respectively, 1.64, 1.15 and 1.07 mg/ml h per mg/kg nicotine dose administered, indicating that with an increase in dose there existed a tendency for either an increase in the fraction of nicotine clearance that comprised formation to cotinine, or a decrease in cotinine clearance. Gabrielsson and Bondesson (1987) have recently reported cotinine plasma clearance and distribution volume of about 0.04 l/h per kg body weight and 0.4 l/kg, respectively. Interestingly the data from that study (Gabrielsson and Bondesson, 1987) showed an increase in cotinine clearance with duration of infusion is consistent with the proposal that cotinine displays time dependent induction of its metabolism (Kyerematen *et al.*, 1983). In the male and female macaque nicotine plasma clearances of approximately 1 l/h per kg and 0.5 l/h per kg have been reported, clearly a sex dependent phenomenon and resembling that observed in humans. However, in contrast to the rat where the distribution volume is larger in the female than in the male, the distribution volumes in the macaque averaged 1.4 l/kg and 2.5 l/kg, respectively.

Devices (e.g. Alzet™ minipump) are now available that provide controlled drug input rate in experimental animals for prolonged periods of time. The use of subcutaneously implanted minipumps for animal pharmacokinetic studies allows the attainment of steady-state conditions, often necessary for studying parent compound/metabolite(s) tissue: blood equilibrium ratios. Indeed, application of such constant-rate input has been proposed as a modification to the standard protocol used in teratogenicity tests (Nau *et al.*, 1981; Gabrielsson, 1985). There are several advantages to the use of a constant-rate input minipump in steady-state pharmacokinetic studies. It allows the assessment of compounds with a short elimination half-life, with saturable elimination kinetics, with a narrow therapeutic index and with a steep concentration–response relationship; peaks and troughs of other more conventional multiple dosing procedures can be avoided; drug and metabolite concentrations in blood and plasma can be monitored conveniently and hepatosplanchnic first-pass metabolism can be avoided. Such a device has recently (J.L. Gabrielsson, unpublished) been implanted subcutaneously into both male and female (pregnant) rats in order to study the steady-state tissue partitioning of cotinine following nicotine infusion and cotinine infusion. The resulting tissue–arterial blood partition coefficients of cotinine in the male and female rats are shown in Tables 9.1 and 9.2, respectively. As summarized in Table 9.1, there is a significant difference in the cotinine partition coefficients for all tissues in male rats when comparing the results from the nicotine with those of the cotinine infusion. The individual tissue-to-blood partition coefficients following nicotine infusion were greater by two- to five- fold compared to corresponding values following cotinine infusion. The results demonstrated that there was a significant increase in the independent distribution volume of cotinine, from about 0.4 l/kg to about 1 l/kg (based on physiological considerations of partitioning) when the nicotine infusion was employed, as compared to the cotinine infusion. Speculation as to the basis for such an observation of increased cotinine distribution volume could include local peripheral metabolism of nicotine to cotinine; intra-cellular accumulation and subsequent sequential metabolism of the lipophilic nicotine to the hydrophilic cotinine molecule, i.e. intracellular trapping of cotinine; and increased distribution volume of the hydrophilic cotinine molecule as a

Table 9.1 Steady-state tissue-to-arterial blood partition coefficients of cotinine in the male rat following constant i.v. infusion of nicotine or cotinine

Organ-tissue	Nicotine infusion	Cotinine infusion
	mean ± S.D. ($n = 7$)	mean ± S.D. ($n = 7$)
Liver	1.76 ± 0.42	0.64 ± 0.12
Lung	1.83 ± 0.40	0.50 ± 0.08
Heart muscle	1.50 ± 0.39	0.55 ± 0.08
Kidneys	2.27 ± 0.52	0.99 ± 0.09
Intestine	1.68 ± 0.46	0.53 ± 0.09
Adipose tissue	0.39 ± 0.14	0.08 ± 0.02
Skeletal muscle	1.43 ± 0.37	0.51 ± 0.09
Brain	1.28 ± 0.31	0.48 ± 0.07

result of nicotine induced water retention arising as a result of stimulation of the release of arginine vasopressin. It is interesting to note that the equilibrium partitioning between organ/tissue was quantitatively and systematically smaller for the female group compared with the male group. Obviously the fact that the female population was pregnant may well be of importance with respect to this observation. Nevertheless a sex-dependent difference in nicotine distribution volume is observed in the rat (larger in the female rat than in the male), although the influence of sex upon the systemic kinetics of cotinine in the rat has not been addressed. Also of note is that in the female rat the method of input of cotinine (i.e. via nicotine infusion or cotinine infusion) did not appear to make any difference with regard to cotinine equilibrium distribution

Table 9.2 Steady-state tissue-to-arterial blood partition coefficients of cotinine in the female (pregnant) rat following constant i.v. infusion of nicotine or cotinine

Organ-tissue	Nicotine infusion	Cotinine infusion
	mean ± S.D. ($n = 4$)	mean ± S.D. ($n = 4$)
Liver	1.39 ± 0.15	1.61 ± 0.29
Lung	1.47 ± 0.10	1.41 ± 0.05
Heart muscle	1.23 ± 0.06	1.44 ± 0.11
Kidneys	1.91 ± 0.48	2.12 ± 0.15
Spleen	1.34 ± 0.18	1.41 ± 0.06
Intestine	1.26 ± 0.07	1.50 ± 0.03
Adipose tissue	0.13 ± 0.03	0.10 ± 0.05
Skeletal muscle	1.16 ± 0.18	1.36 ± 0.06
Brain	0.96 ± 0.04	1.17 ± 0.04
Foetal carcass	1.13 ± 0.25	1.32 ± 0.12

ratios. Once again the pregnant status of the female population is a co-variate in conjunction with sex.

A further observation to make on the data in Tables 9.1 and 9.2, is the low variability of cotinine tissue-to-blood partition coefficients between different tissues, regardless of the moiety (i.e. nicotine or cotinine) infused. This probably reflects the low protein binding of cotinine and its distribution that is essentially restricted to extracellular fluids. Using this information and previously published *in vitro* and *in vivo* data for cotinine kinetics in the rat, a physiological pharmacokinetic model for cotinine can be constructed (see Figure 9.2). In this figure brain tissue is included; adipose tissue is included as a deep storage compartment; hepatic, pulmonary and renal tissues are included because of their role in elimination; intestinal and skeletal muscle tissues are included because of their storage capacity; and finally heart muscle tissue are included with regard to the potential cardiotoxic effects of cotinine. Preliminary studies comparing the simulated kinetic behaviour based on the model (as in Figure 9.2) and experimental observations (unpublished) has allowed the authors to conclude that it is a suitable model for precise prediction of the disposition kinetics of cotinine in the rat following a single intravenous dose. The assumptions utilized in the model include:

- each tissue behaves as a well-stirred compartment;
- cotinine distribution is flow-limited;
- all kinetic processes are first order;
- equilibrium partition coefficients are concentration and time-independent;
- blood flow remains constant through the duration of study, i.e. cotinine has no haemodynamic actions.

The advantage of such a physiological model is that concentration–time profiles can be predicted in the organs and tissues of the body, i.e. sites not readily accessible for concentration sampling, but where pharmacological and toxicological actions are initiated. By mathematical scale-up of the kinetic parameters with respect to species physiology, animal physiological models can legitimately be extrapolated for use as predictive models in man.

In conclusion, this chapter has been intended to give a general overview of the kinetics of nicotine and its metabolites in animals. Although the disposition of nicotine clearly has implications in areas of nicotine tolerance development and carcinogenicity, relatively little is still known. Further advances in defining the quite complex metabolism and kinetics of nicotine and its metabolites will be dependent upon the development of sufficiently sensitive analytical techniques and the use of appropriate *in vitro* and *in vivo* animal models.

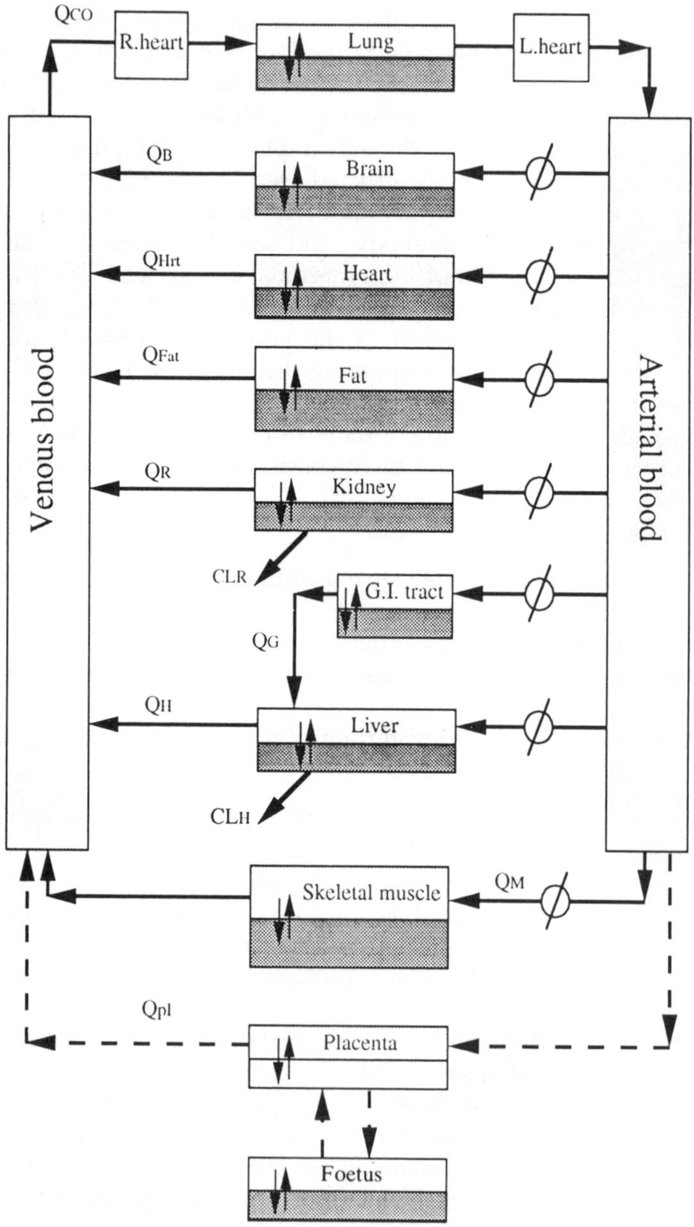

Figure 9.2 Physiological model for cotinine disposition in the rat.

REFERENCES

Adir, J., Miller, R.P. and Rotenberg, K. (1976) Disposition of nicotine after intravenous administration. *Res. Commun. Chem. Pathol. Pharmacol.* **13**, 173–83.

Armitage, A.K., Dollery, C.T., George, C.F. *et al.* (1976) Absorption and metabolism of nicotine from cigarettes. *Br. Med. J.* **4**, 313–6.

Beckett, A.H., Gorrod, J.W. and Jenner, P. (1971) The effect of smoking on nicotine metabolism *in vivo* in man. *J. Pharm. Pharmacol.* **23**, 55–61S.

Benowitz, N.L., Kuyt, F., Jacobs III, P. *et al.* (1983) Cotinine disposition and effects. *Clin. Pharmacol. Ther.* **34**, 604–11.

Benowitz, N.L. (1986) Clinical pharmacology of nicotine. *Ann. Rev. Med.* **37**, 21–32.

Benowitz, N.L., Jacobs III, P. and Savanapridi, C. (1987) *Clin. Pharmacol. Ther.* **46**, 467–73.

Benowitz, N.L. (1990) Pharmacokinetic considerations in understanding nicotine dependence. *Ciba Foundation Symposium* **152**, 186–209.

Benowitz, N.L., Chan, K., Denaro, C.P. *et al.* (1991) Stable isotope method for studying transdermal drug absorption: the nicotine patch. *Clin. Pharmacol. Ther.* **50**, 286–93.

Booth, J. and Boyland, E. (1970) Enzymatic oxidation of (−)-nicotine by guinea-pig tissues *in vitro*. *Biochem. Pharmacol.* **19**, 733–42.

Cohen, A.J. and Roe, F.J.C. (1981) Monographs on the pharmacology and toxicology of nicotine. Tobacco Advisory Council, Occasional Paper, London.

Curvall, M., Elwin, C.E., Vala-Kazemi, E. *et al.* (1990) The pharmacokinetics of nicotine in plasma and saliva from non-smoking healthy volunteers. *Eur. J. Clin. Pharmacol.* **38**, 281–7.

Feyerabend, C., Ings, R.M.J. and Russell, M.A.H. (1985) Nicotine pharmacokinetics and its application to intake for smoking. *Br. J. Clin. Pharmacol.* **19**, 239–47.

Foth, H., Walther, U.I., and Kahl, G.F. (1990) Increased hepatic nicotine elimination after phenobarbital induction in the conscious rat. *Toxicol. Appl. Pharmacol.* **105**, 382–92.

Foth, H., Looschen, H., Neurath, H. *et al.* (1991) Nicotine metabolism in isolated perfused lung and liver of phenobarbital and benzoflavone-treated rats. *Arch. Toxicol.* **65**, 68–72.

Gabrielsson, J. (1985) Modelling of drugs in pregnancy: Application of pharmacokinetic principles and physiological flow models. *Acta. Univ. Uppsala. Abstracts of Uppsala Dissertations*, Faculty of Pharmacy No 108, Uppsala, Sweden..

Gabrielsson, J., and Bondesson, U. (1987) Constant-rate infusion of nicotine and cotinine I: a physiological pharmacokinetic analysis of cotinine disposition and effects on clearance and distribution in the rat. *J. Pharmacokin. Biopharm.* **15**, 583–92.

Gorrod, J.W., and Jenner, P. (1975) The metabolism of tobacco alkaloids. *Essays in Toxicology*. **6**, 35–78.

Gritz, E.R., Baier-Weiss, V., Benowitz, N.L. *et al.* (1981) Plasma nicotine and cotinine concentrations in habitual smokeless tobacco users. *Clin. Pharmacol. Ther.* **30**, 201–9.

Gumbleton, M., Nicholls, P.J. and Taylor, G. (1990) Differential influence of laboratory anaesthetic regimens upon renal and hepatosplanchnic haemodynamics in the rat. *J. Pharm. Pharmacol.* **42**, 693–7.

Haines, C.F., Mahajan, D.K., Miljkovic, D. *et al.* (1976) Radioimmunoassay of plasma nicotine in habituated and naive smokers. *Clin. Pharmacol. Ther.* **16**, 1083–9.

Kyerematen, G.A., Taylor, L.H., de Bethizy, J.D. *et al.* (1988) Pharmacokinetics of nicotine and 12 metabolites in the rat: Application of a new radiometric high performance liquid chromatography assay. *Drug Metab. Dispos.* **16**, 125–9.

Kyerematen, G.A., Dvorchik, B.H. and Vesell, E.S. (1983) Influence of different forms of tobacco intake on nicotine elimination in man. *Pharmacology* **26**, 205.

Kyerematen, G.A., Damiano, M.D., Dvorchik, B.H. *et al.* (1982) *Clin. Pharmacol. Ther.* **32**, 769–80.

Kyerematen, G.A., Owens, G.F., Chattopadhyay, B. *et al.* (1988) Sexual dimorphism of nicotine metabolism and distribution in the rat. *Drug Metab. Dispos.* **16**, 823–30.

Kyerematen, G.A., Morgan, M., Warner, G. *et al.* (1990) Metabolism of nicotine by hepatocytes. *Biochem. Pharmacol.* **40**, 1747.

Kyerematen, G.A., Morgan, M.L., Chattopadhyay, B. *et al.* (1990) Disposition of nicotine and eight metabolites in smokers and nonsmokers; identification in smokers of two metabolites that are longer lived than cotinine. *Clin. Pharmacol. Ther.* **48**, 641–51.

Kyerematen, G.A. and Vessel, E.S. (1991) Metabolism of nicotine. *Drug Metab. Rev.* **23**, 3–41.

Luck, W. and Nau, H. (1984) Nicotine and cotinine in serum and milk of nursing mothers. *Br. J. Clin. Pharmacol.* **18**, 9–15.

Miller, A.G. (1990) Cigarettes and drug therapy: pharmacokinetic and pharmacodynamic considerations. *Clin. Pharm.* **9**, 125–35.

McCoy, G.D., Demarco, G.J. and Koop, G.R. (1989) Microsomal nicotine metabolism: A comparison of relative activities of six purified rabbit cytochrome P450 isoenzymes. *Biochem. Pharmacol.* **38**, 1185–92.

Nau, H., Zierer, R., Spielmann, H., Neubert, D. *et al.* (1981) A new model for embryotoxicity testing: teratogenicity and pharmacokinetics of valproic acid following constant-rate administration in the mouse using human therapeutic drug metabolite concentrations. *Life Sci.* **29**, 2803–14.

Neubert, D.W. *et al.* (1987) The P450 gene superfamily: Recommended nomenclature. *DNA* **6**, 1–11.

Nwosu, C.G. and Crooks, P.A. (1988) Species variation and stereoselectivity in the metabolism of nicotine enantiomers. *Xenobiotica.* **18**, 1361–72.

Obach, R.S. and van Vunakis, H. (1990) Radioimmunoassay of nicotine-$\Delta^{1'(5')}$-iminium ion, and intermediate formed during the metabolism of nicotine to cotinine. *Drug Metab. Dispos.* **18**, 508–13.

Pillotti, A. (1980) Symposium of the effects of nicotine on nervous function. *Acta. Physiol. Scand. Suppl.* **479**, 13–17.

Rosenberg, J., Benowitz, N.L., Jacob III, P. *et al.* (1980) Disposition kinetics and effects of intravenous nicotine. *Pharmac. Ther.* **28**, 517–22.

Rüdell, U., Foth, H., and Kahl, G.F. (1987) Eight-fold induction of nicotine elimination in perfused rat liver by pretreatment with phenobarbital. *Biochem. Biophys. Res. Commun.* **148**, 192–8.

Russell, M.A.H. and Feyerabend, C. (1978) Cigarette smoking: a dependence on high nicotine boli. *Drug Metab. Rev.* **8**, 29–57.

Schechter, S., Kozlowski, L.T. and Silverstein, B. (1977) Studies of the interaction of psychological and pharmacological determinants of smoking. *J. Exp. Psychol. Gen.* **106**, 3–11.
De Schepper, P.J., Van Hecken, A., Daenens, P. *et al.* (1978) Kinetics of cotinine after oral and intravenous administration to man. *Eur. J. Clin. Pharmacol.* **31**, 583–8.
Schievelbein, H. (1982) Nicotine resorption and fate. *Pharmac. Ther.* **18**, 233–47.
Schmiterlöw, C.G, Hansson, E., Andersson, G. *et al.* (1967) Distribution of nicotine in central nervous system. *Ann. New York Acad. Sci.* **142**, 2–14.
Seaton, M., Kyerematen, G.A., Morgan, M., *et al.* (1991) Nicotine metabolism in stumptailed macaques. *Drug Metab. Dispos.* **19**, 946–54.
Stålhandske, T. (1967) Effects of increased liver metabolism of nicotine on its uptake, elimination and toxicity in mice. *Acta Pharmacol. Toxicol.* **41**, 25S–9S.
Tsujimoto, A., Nakaghima, T., Tarino, S. *et al.* (1975) Tissue distribution of [^{3}H] nicotine in dogs and rhesus monkeys. *Toxicol. Appl. Pharmacol.* **32**, 21–31.
Turner, D.M. (1969) Metabolism of [^{14}C] nicotine in the cat. *Biochem. J.* **115**, 889.
Turner, D.M., Armitage, A.K., Briant, R.H. *et al.* (1975) Metabolism of nicotine by the isolated perfused dog lung. *Xenobiotica.* **5**, 539–51.
Williams, D.E, Shigenaga, M.K. and Castagnoli, N. (1990) The role of cytochrome P450 and flavin-containing monoxygenase in metabolism of (*S*)-nicotine by the rabbit lung. *Drug Metab. Dispos.* **18**, 418–28.
The health consequences of smoking: A report of the surgeon general. (1972) US Public Health Service, DHEW Publication No. (HSM) 72–7516.
The health consequences of smoking: A report of the surgeon general. (1971) US Public Health, DHEW Publication No. (HSM) 71–7513.

Pharmacokinetics of (*S*)-nicotine and metabolites in humans 10

P. Jacob III and N.L. Benowitz

10.1 INTRODUCTION

An understanding of the pharmacokinetics of nicotine and its metabolites is important in understanding the biological effects of nicotine and tobacco for various reasons. The rate of absorption and distribution profoundly affects the magnitude of pharmacological effect of nicotine (Benowitz *et al.*, 1990). The rate of metabolism of nicotine affects its pharmacology by determining the level of nicotine in the body with any given rate of consumption of tobacco. When elimination of nicotine is accelerated, people consume more tobacco, presumably to maintain a desired level of nicotine in the body (Benowitz *et al.*, 1990). Metabolites of nicotine may be pharmacologically active (Clark *et al.*, 1965) or may be toxic (Shigenaga *et al.*, 1988) and, therefore, it is important to know levels of human exposure. Metabolic data can be used to estimate nicotine intake and exposure to tobacco smoke (Benowitz, 1984; Benowitz and Jacob, 1984a).

An extensive review of the pharmacokinetics and pharmacodynamics of nicotine has been recently published (Benowitz *et al.*, 1990). This chapter will present recent research in which stable isotope methodology has been used to study quantitative aspects of nicotine metabolism. The authors will discuss a study to determine the fractional conversion of nicotine to its metabolite cotinine in smokers, a study comparing the pharmacokinetics of nicotine in smokers and nonsmokers, and studies of the bioavailability of nicotine administered by various routes.

Nicotine and Related Alkaloids: Absorption, distribution, metabolism and excretion. Edited by J.W. Gorrod and J. Wahren. Published in 1993 by Chapman & Hall, London. ISBN 0 412 55740 1

10.2 STABLE ISOTOPE METHODOLOGY

Stable isotope methodology is frequently used to study the metabolic disposition of drugs which are administered chronically, or to determine bioavailability (Baillie *et al.*, 1984). By administering the labelled drug intravenously and using mass spectrometry to distinguish labelled from natural drug, pharmacokinetic parameters can be determined in the presence of the natural drug and the absolute bioavailability of the natural drug can be measured.

People chronically self-administer nicotine in a dose that cannot be determined based on knowledge of the composition of the tobacco product. Much of the nicotine present in tobacco (Table 10.1) is not absorbed; losses by pyrolysis occur during smoking, in sidestream smoke, or, in the case of smokeless tobacco, expectoration. One approach to the study of nicotine pharmacokinetics has been to administer a known dose of nicotine during a period of tobacco abstinence (Benowitz *et al.*, 1982). However, this is not ideal, since smoking may influence the rate of nicotine metabolism (Beckett *et al.*, 1971; Kyerematen *et al.*, 1982; Lee *et al.*, 1987). For this reason, stable isotope methodology has been developed for quantitative studies of nicotine metabolic disposition (Benowitz *et al.*, 1991b). To carry out these studies,

Table 10.1 Nicotine content of tobacco products and medications

Product	Concentration of nicotine (mg/g tobacco)	Typical single dose (g tobacco)	Nicotine in single dose* (mg)	Nicotine in typical daily dose*
Cigarettes[a]	15.7 (13.3–26.9)	0.54	8.4	168 mg/ 20 cigs
Moist snuff[b,c]	10.5 (6.1–16.6)	1.4	14.5	157 mg/ 15 grams
Chewing tobacco[c,d,e]	16.8 (9.1–24.5)	7.9	133	1176 mg/ 70 grams
Pipe tobacco[e]	11.3			
Cigar tobacco[e]	13.9			
Nicotine gum			2 or 4	30 or 60 mg/ 15 pieces
Transdermal				8–114 mg/ patch

*Dose is amount of nicotine in product, not amount absorbed by user.
[a]Benowitz *et al.* (1983).
[b]Kozlowski *et al.* (1982).
[c]Gritz *et al.* (1981).
[d]Benowitz *et al.* (1988).
[e]Benowitz and Jacob, unpublished.

it was necessary to (1) synthesize suitable labelled analogues of nicotine and metabolites; (2) develop analytical methods capable of distinguishing and quantitating labelled and natural compounds; and (3) demonstrate bioequivalence of natural and labelled compounds.

10.3 SYNTHESIS OF LABELLED NICOTINE AND METABOLITES

Several requirements had to be met in choosing the particular stable isotope labelled analogues for our studies. The labelled analogues must have a sufficient mass enhancement to prevent interference by the small amounts of stable isotopes present in the natural drug (Beynon and Williams, 1963). Interference by naturally occurring isotopes in organic compounds is principally due to carbon-13, which amounts to about 1% of natural carbon. Thus, for nicotine and most metabolites which have ten carbon atoms, a mass enhancement of two reduces interference to negligible levels, unless concentrations of the natural compounds greatly exceed concentrations of the labelled analogues. Of the various stable isotopes, deuterium is by far the easiest and least expensive to incorporate into organic molecules. Since relatively large quantities of

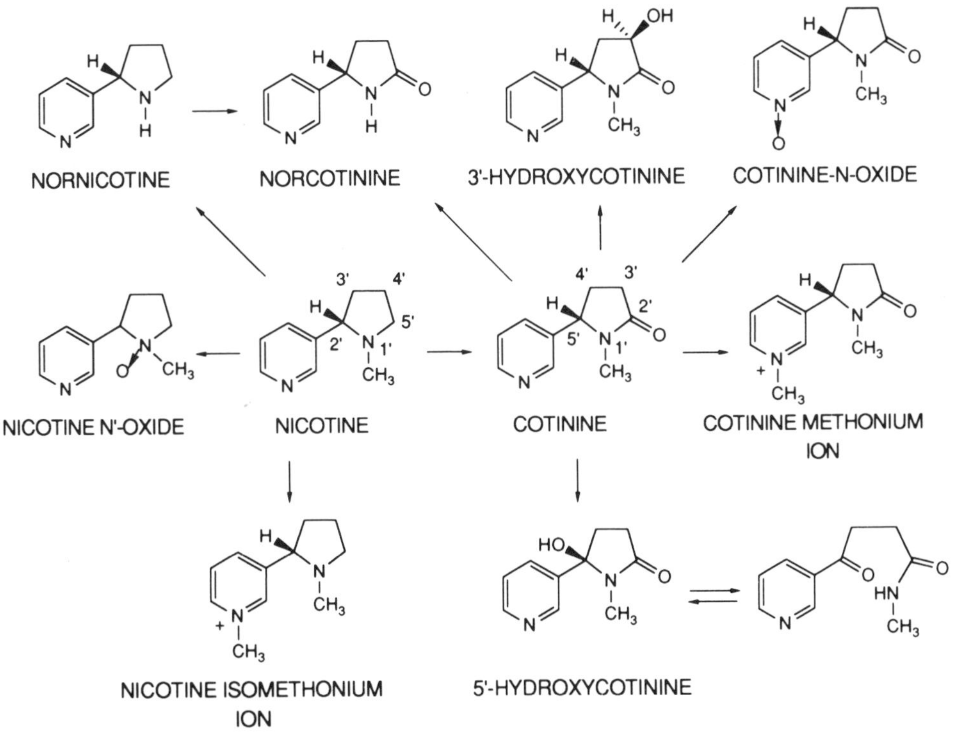

Figure 10.1 Nicotine metabolic pathways.

labelled nicotine were required, it was felt that it would not be feasible to use other isotopes, such as carbon-13 or nitrogen-15.

The label must be located on a part of the molecule that is not lost in formation of the metabolites of interest. The principal metabolites of nicotine result from oxidation of the pyrrolidine ring (Figure 10.1) (Gorrod, this volume). Consequently, these sites, especially the 4′ and 5′ carbon atoms of nicotine, would not be suitable for deuterium labelling, since the label would be lost in formation of the major metabolites cotinine and *trans*-3′-hydroxycotinine. In addition, deuterium located at a site of metabolism could significantly slow the rate of metabolism of nicotine (kinetic isotope effect), rendering the labelled analogue unsuitable for studies of the pharmacokinetics of the natural drug.

Another consideration is the analytical methodology to be employed. Generally, this will be mass spectrometry with selected ion monitoring. If a fragment ion rather than the molecular ion is monitored, the label must be located on this fragment.

Finally, if the desired labelled compound is not available commercially, it must be possible to synthesize sufficient quantities with the correct stereochemistry. The studies described in this chapter and studies now in progress involve over 200 human subjects. The doses of labelled nicotine and cotinine administered are not tracer doses, but are doses sufficient to produce nicotine blood levels comparable to those attained during smoking. Consequently, several grams of labelled nicotine and cotinine have had to be synthesized.

Stereochemistry is an important issue. Nicotine has one asymmetric centre and, as a result, exists as a pair of optical isomers (enantiomers) (Figure 10.2). Nicotine in tobacco is largely, if not entirely, the laevorotatory (*S*)-isomer. Small amounts of the (*R*)-isomer (up to 5% of the total nicotine) are found in tobacco smoke, presumably formed by racemization during combustion (Pool *et al.*, 1985). Pharmacological studies in animals and with *in vitro* preparations have shown that the (*S*)-isomer is more potent, with potencies five to 100 times that of the (*R*)-isomer, depending on the system (Ikushima *et al.*, 1982; Abood *et al.*, 1985; Sloan *et al.*, 1985; Risner *et al.*, 1988; Copeland *et al.*, 1991). In

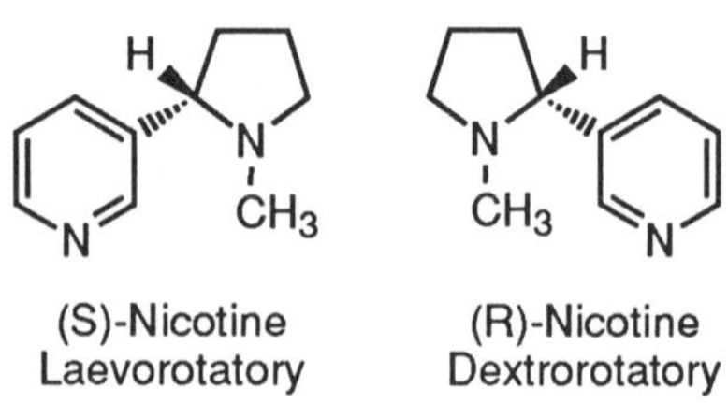

Figure 10.2 Nicotine enantiomers.

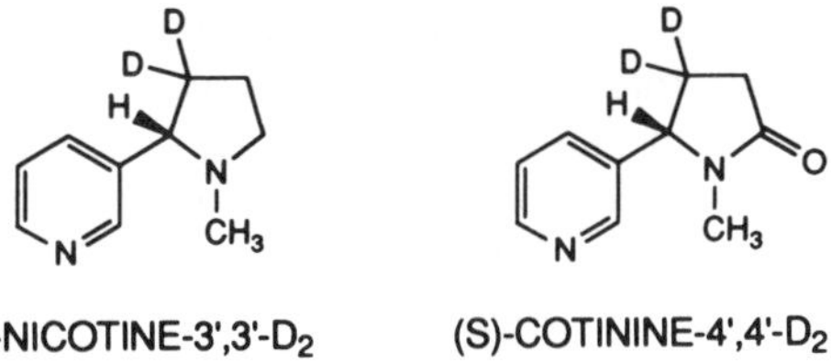

Figure 10.3 Structures of nicotine-d_2 and cotinine-d_2.

addition, it is known that the enantiomers are metabolized differently (Martin *et al.*, 1983; Jacob *et al.*, 1988a). Consequently, it is important to carry out metabolic studies with nicotine of defined stereochemistry, and to use pure (*S*)-nicotine if the results are to be relevant to understanding the metabolism of tobacco derived nicotine

A stable isotope labelled analogue of nicotine meeting the requirements discussed above is (*S*)-nicotine-3′,3′-d_2 (Figure 10.3) The 3′-position is remote from the major sites of metabolism, so the deuterium atoms located on the 3′-carbon will not be lost in the formation of major metabolites and would not be expected to lead to a kinetic isotope effect. An efficient synthesis of pure (*S*)-nicotine-d_2 has been developed (Jacob *et al.*, 1988c).

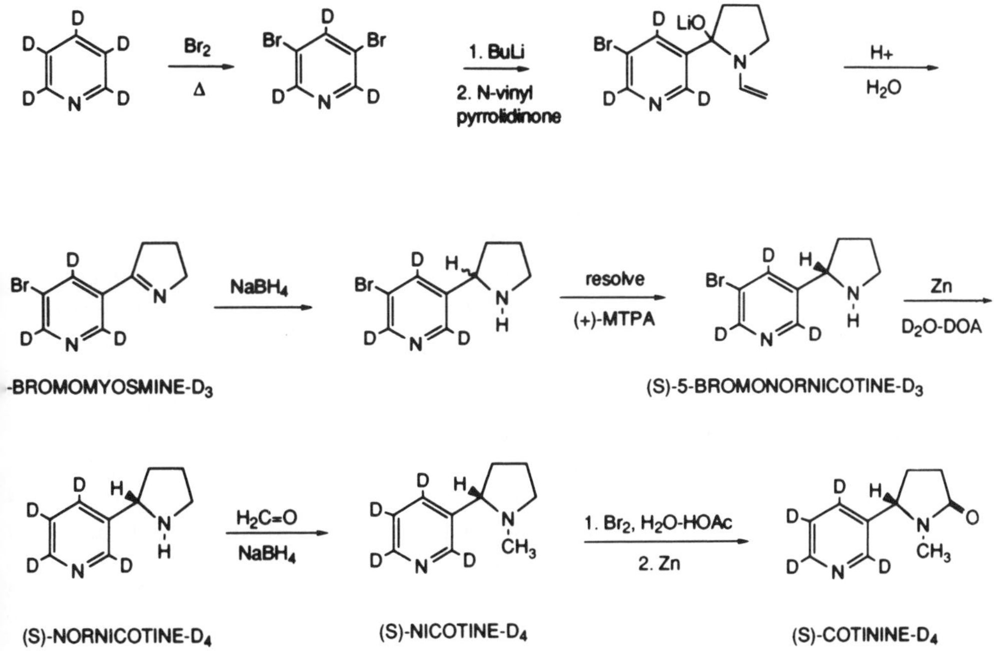

Figure 10.4 Synthesis of (*S*)-cotinine-2,4,5,6-d_4.

For some of the studies described below, analogues with a mass enhancement greater than two were required. Since the pyridine ring remains intact in all of the known nicotine metabolites, this site was chosen for incorporation of four deuterium atoms. The sequence shown in Figure 10.4 was used to synthesize (*S*)-cotinine-2,4,5,6-d_4.

10.4 ANALYTICAL METHODOLOGY

To distinguish labelled from natural compounds, mass spectrometry is required. Since the compounds of interest must be quantitated in a complex biological matrix, an efficient separation, such as capillary gas chromatography, prior to mass spectrometry is needed. For pharmacokinetic studies, the methods must be accurate and sensitive enough to measure low ng/ml concentrations in small plasma samples, as well as being practical for analysis of the many samples generated in large-scale clinical studies.

A method for quantitation of nicotine, cotinine and deuterium-labelled analogues in biological fluids has been developed (Jacob *et al.*, 1991). The method utilizes capillary gas chromatography–mass spectrometry (GC–MS) with selected ion monitoring, and is suitable for plasma or urine. Concentrations less than 1 ng/ml in 1 ml samples can be reliably quantitated. Methods for determination of *trans*-3′-hydroxycotinine (Jacob *et al.*, 1992), nicotine-N′-oxide (Jacob *et al.*, 1986) and deuterium-labelled analogues in biological fluids have also been developed.

10.5 PHARMACOKINETICS OF (*S*)-NICOTINE-3′,3′-d_2

A study was carried out to compare the pharmacokinetics of deuterium-labelled nicotine with natural nicotine (Jacob *et al.*, 1991). After overnight abstinence from smoking, five healthy male smokers were given an intravenous infusion of a 50:50 mixture of (*S*)-nicotine and (*S*)-nicotine-3′,3′-d_2. The dose (2 μg/kg/min) of total nicotine base infused for 90 min was chosen to produce blood levels typical of those found in heavy smokers. The plasma concentration–time curves are shown in Figure 10.5. Concentrations of nicotine and nicotine-d_2 followed similar profiles, although concentrations of natural nicotine were slightly higher, reflecting prior tobacco exposure. Pharmacokinetic parameters are presented in Table 10.2. Total clearance averaged about 1400 ml/min, and was virtually identical for labelled and natural nicotine. The terminal half-life ($t_{½}β$) was slightly longer for natural nicotine (168 vs. 149 min), possibly resulting from slow release of tobacco derived nicotine from deep tissue stores. Differences in pharmacokinetic parameters were not statistically significant.

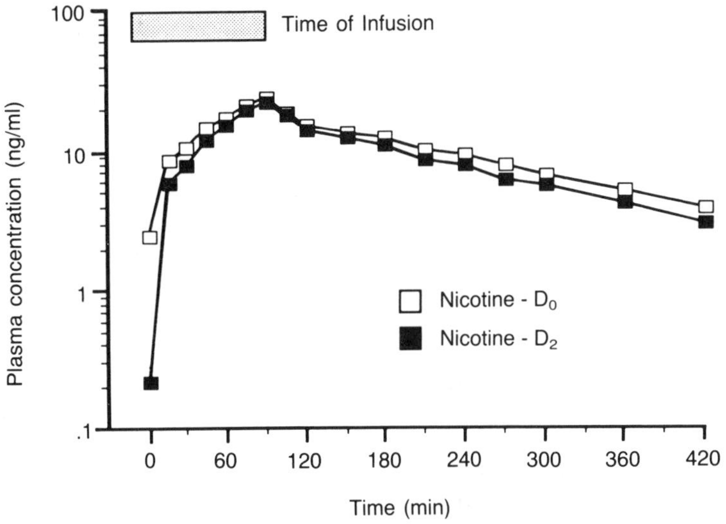

Figure 10.5 Mean plasma concentrations of (*S*)-nicotine and (*S*)-nicotine-3′,3′-d_2 in five human subjects during and after intravenous infusion of a 50:50 mixture of (*S*)-nicotine and (*S*)-nicotine-3′,3′-d_2. (From Jacob *et al.*, 1991, by permission.)

Table 10.2 Disposition kinetics of natural nicotine and (*S*)-nicotine-3′,3′-d_2 in five smokers[a]

Subject	Total Clearance (ml/min)		Half-life (min) ($t\text{-}\frac{1}{2}\beta$)		Volume of distribution (l) (VD_{ss})	
	Nicotine	Nicotine-d_2	Nicotine	Nicotine-d_2	Nicotine	Nicotine-d_2
1	1200	1210	215	178	319	264
2	1420	1500	167	122	317	255
3	1110	1210	198	165	303	274
4	1560	1610	143	139	304	298
5	1730	1570	119	144	273	294
Mean	1404	1420	168	149	303	277
S.D.	254	196	39	22	18	19

[a]Differences in pharmacokinetic parameters comparing natural and labelled nicotine were not significant by paired *t*-test.

10.6 PHARMACOKINETICS OF NICOTINE IN SMOKERS AND NONSMOKERS

Due to the ubiquitous presence of tobacco smoke in the environment, nonsmokers have measurable quantities of nicotine and its metabolites in their bodies (Jarvis *et al.*, 1984; Rylander *et al.*, 1989). Nonsmokers are also exposed to nicotine from ingestion of certain foods and beverages, for example, potatoes, tomatoes and tea (Davis *et al.*, 1991). Although the relative contributions of environmental tobacco smoke and dietary sources to nicotine and metabolite levels in biological fluids of nonsmokers remains to be determined, several studies have shown correlations between nicotine or cotinine concentrations with the degree of environmental tobacco smoke exposure (Jarvis *et al.*, 1984; Wall *et al.*, 1988; Jones *et al.*, 1991).

Biological fluid levels of nicotine and cotinine are frequently used as measures of exposure to tobacco smoke (Benowitz, 1984; Jarvis *et al.*, 1984; Stookey *et al.*, 1987; Cummings and Richard, 1988). To estimate exposure levels, extrapolations are sometimes made from cotinine levels of nonsmokers to the equivalent dose of nicotine or cigarettes consumed by smokers. An assumption of such an extrapolation is that smokers and nonsmokers metabolize nicotine and cotinine at the same rate.

Using stable isotope methodology, a study was carried out comparing the elimination kinetics of nicotine and cotinine in smokers and nonsmokers (Benowitz and Jacob, 1993). Eleven nonsmokers received a low dose (0.5 μg/kg/min) 30 min intravenous infusion of (*S*)-nicotine-3′,3′-d_2. Eleven smokers were administered the same dosing regimen and were allowed to smoke prior to and after the infusion. On a separate occasion, the smokers were administered a higher dose of (*S*)-nicotine-3′,3′-d_2 (2.0 μg/kg/min for 30 min) that produced nicotine levels comparable to those achieved by cigarette smoking. Following the infusion, nicotine concentrations in plasma were measured for 8 h, and cotinine concentrations were measured for up to 96 h (Figure 10.6). Pharmacokinetic parameters are presented in Table 10.3. Nicotine clearance was significantly greater in nonsmokers than in smokers, averageing 1320 ml/min compared with 1070 ml/min, for the low dose treatment. The terminal half-life of nicotine was shorter in nonsmokers, averageing 122 min compared with 157 min, but the difference was not significant. The half-life of cotinine tended to be shorter in nonsmokers, but the difference was not statistically significant.

The finding that nonsmokers metabolize nicotine more rapidly than smokers is consistent with results from a previous study in which smokers were found to metabolize nicotine more rapidly after seven days abstinence from tobacco than they did following overnight abstinence (Lee *et al.*, 1987). Possible explanations include inhibition of meta-

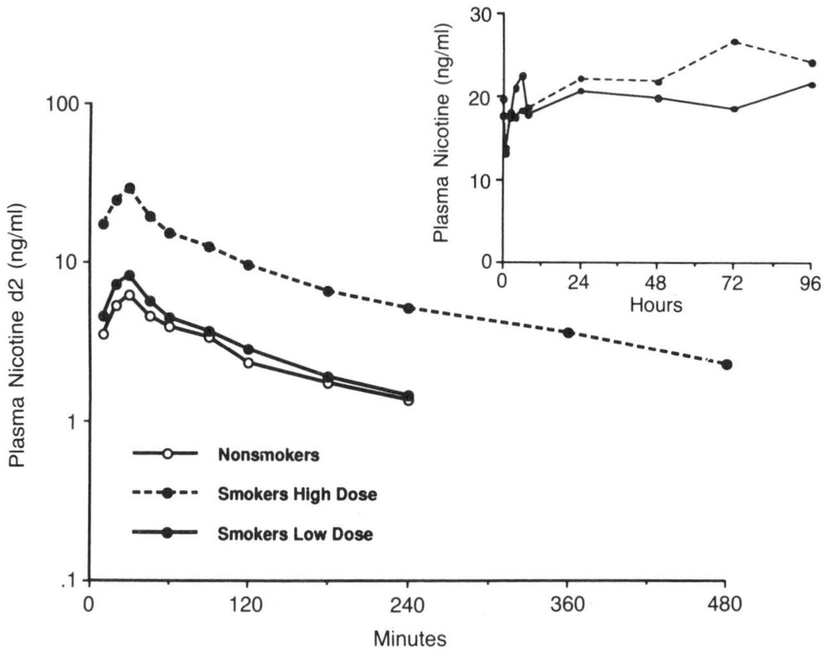

Figure 10.6 Mean plasma concentrations of (*S*)-nicotine-3′,3′-d_2 in 11 smokers and 11 nonsmokers following intravenous infusion. The low dose was 0.5 μg/kg/min for 30 min, the high dose was 2 μg/kg/min for 30 min. Concentrations of natural nicotine in the smokers' plasma are shown in the insert. (From Benowitz and Jacob, 1993.)

bolism by components of cigarette smoke such as carbon monoxide or hydrogen cyanide, or inhibition by nicotine itself or a metabolite. The findings contrast with those of Kyerematen *et al.* (1982), who reported faster clearance of nicotine in smokers. The reason for the discrepancy is not clear; however, in their study a very low dose of racemic ^{14}C-labelled nicotine was administered, and the difference could be explained by stereoselective metabolism of (*R*)-nicotine by a smoking induced isoform of cytochrome P450.

The finding that the half-life of cotinine is similar in smokers and nonsmokers is consistent with the results of Curvall *et al.* (1990), deShepper *et al.* (1987) and Jarvis *et al.* (1988). Curvall and deShepper infused cotinine intravenously into nonsmokers, and reported pharmacokinetic parameters similar to those reported by other investigators for smokers. Jarvis *et al.* found that the half-life of cotinine in nonsmokers following oral administration of nicotine was similar to the half-life of cotinine in smokers. In contrast, Kyerematen *et al.* (1982) reported a slightly shorter half-life of cotinine in smokers, which may have been due to the administration of racemic nicotine, as discussed above.

Table 10.3 Disposition kinetics of nicotine-3′,3′-d_2 in smokers and nonsmokers

	Infusion rate[a] (μg/kg/min)	Nicotine clearance (ml/min)	Nicotine V_{ss} (l)	Nicotine plasma half-life (h)	Cotinine plasma half-life (h)	Cotinine urine half-life (h)
Nonsmokers (N = 11)	0.5	1320 (940–1700)	185 (143–227)	2.0 (1.5–2.5)	21.0	15.2
Smokers (N = 11)	0.5	1070 (900–1250)	199 (150–249)	2.7 (1.8–3.5)	29.5[b]	17.8
Smokers (N = 11)	2.0	1170 (950–1400)	202 (165–239)	2.6 (2.0–3.1)	23.9[b]	14.9

[a]30 min infusion.
[b]Significant difference compared to other value.
() = 95% C.I.

Haley and co-workers (Haley *et al.*, 1989) reported a considerably shorter half-life of cotinine in smokers, 16.5 h compared with 27.3 h in nonsmokers. This could have been due to methodology. The half-life in nonsmokers was determined by measuring cotinine levels following environmental tobacco smoke exposure. Continued low level exposure to nicotine from environmental tobacco smoke or from dietary sources would have had the effect of prolonging the apparent half-life.

In summary, the authors' data do not support previous studies (Kyerematen *et al.*, 1982; Haley *et al.*, 1989) which concluded that smokers metabolize nicotine and/or cotinine faster than nonsmokers. In fact, the authors found slightly but significantly faster clearance of nicotine in nonsmokers. Our finding that the half-life of cotinine in nonsmokers is similar to the half-life of cotinine in smokers indicates that it is probably valid to use cotinine levels in biological fluids of nonsmokers to estimate nicotine exposure based on known pharmacokinetic parameters in smokers, provided that the fractional conversion of nicotine to cotinine in smokers and nonsmokers is similar.

10.7 BIOAVAILABILITY STUDIES

There has been considerable interest in determining nicotine intake for the purposes of assessing the relative health risks of various tobacco products. For example, the so-called low-yield cigarettes deliver less 'tar' and nicotine than other cigarettes as measured by smoking machines, but a number of studies have shown that machine determined delivery of nicotine does not correlate well with the amount of nicotine absorbed by smokers (Benowitz *et al.*, 1983; Ebert *et al.*, 1983; Benowitz and Jacob, 1984b; Feyerabend *et al.*, 1985; Benowitz *et al.*, 1986; Kozlowski *et al.*, 1988). Nicotine intake by nonsmokers is of interest for estimating passive exposure to tobacco smoke. The recent development of nicotine-containing medications has required methodology for determining bioavailability.

To determine the systemic dose, one must know the concentration of nicotine in blood (area under the plasma concentration time curve, AUC) and the clearance, CL:

$$\text{Dose} = \text{AUC} \times \text{CL}_{\text{total}}$$

Since the rate of nicotine metabolism is quite variable between individuals (Benowitz *et al.*, 1982), the clearance, as well as plasma levels (AUC), needs to be determined in each individual to obtain a good estimate of dose. It is advantageous to determine clearance under the conditions in which the drug is usually administered, since the drug itself or metabolites accumulated in the body may influence the rate of metabolism or, in the case of smoking, substances in tobacco smoke could induce or inhibit nicotine metabolism. The best way to do this is to administer the drug in its usual dosage form concurrently with an isotopically labelled drug administered intravenously to determine clearance.

The authors have used stable isotope methodology to determine nicotine intake during cigarette smoking (Benowitz *et al.*, 1991b) and bioavailability of orally (Benowitz *et al.*, 1991b) and transdermally (Benowitz *et al.*, 1991a, 1992) administered nicotine. Total clearance of nicotine was determined from the plasma AUC following intravenous infusion of (*S*)-nicotine-3′,3′-d_2, and used in conjunction with the AUC of natural nicotine to calculate the systemic dose:

$$CL_{total} = \frac{Dose_{nic\text{-}d_2}}{AUC_{nic\text{-}d_2}}$$

$$Dose_{nic} = AUC_{nic} \times CL_{total}$$

During a period of *ad libitum* cigarette smoking on a research ward, 13 subjects absorbed an average of 30.1 mg nicotine (range 14.5 to 78.9, S.D. 16.1) during a 24 h period (Table 10.4). The bioavailability of nicotine administered orally as the bitartrate salt in capsules was studied in seven subjects at doses ranging from 3 to 6 mg (Benowitz *et al.*, 1991b). At doses of 3 and 4 mg, oral nicotine was well tolerated and resulted in plasma levels of nicotine similar to those achieved after cigarette smoking. Oral bioavailability averaged 44% (Table 10.4), although there was considerable individual variability. This is some what higher than expected based on clearance data and hepatic blood flow, and suggests that there is some extrahepatic metabolism (Benowitz *et al.*, 1991b). It is known that nicotine is metabolized to some extent by the lung, and perhaps by other organs.

As an aid to smoking cessation, several pharmaceutical companies have developed systems for transdermal delivery of nicotine. These systems are designed to release nicotine over the course of 16 to 24 h

Table 10.4 Bioavailability of nicotine determined using stable isotope methodology

	Nicotine intake, mg/day (% bioavailability)			
	Cigarettes	Oral	Patch #1	Patch #2
Mean	30.1	(44%)	17 (77%)[a]	14.0 (99%)[a]
Range	14.5–78.9	(24–59%)	6–25	7.3–20.0
S.D.	16.1	(9%)	5.1	4.0
N	13	7	14	11

[a]Based on amount of nicotine released from transdermal patch

and produce peak blood levels about half of those typically found in heavy smokers. Prior to marketing, it was necessary to determine the bioavailability of nicotine from transdermal systems. Bioavailability data for two different systems (Benowitz *et al.*, 1991a; Benowitz *et al.*, 1992) determined using stable isotope methodology are given in Table 10.4. The two systems delivered systemic nicotine doses of 17.0 mg/24 h and 14.1 mg/16 h, respectively. The mean bioavailability based on the amount of nicotine released from the transdermal patches was high, averageing 77% and 99% for the two patches, although there was considerable variability among subjects. There is continuing interest in development of new products for delivery of nicotine, including lozenges and nasal sprays. Stable isotope methodology should be useful in studying the pharmacokinetics and bioavailability of nicotine from these new products.

10.8 URINARY EXCRETION OF NICOTINE METABOLITES

Nicotine is extensively metabolized, with only about 10% of the absorbed dose being excreted unchanged in urine (Jacob *et al.*, 1988b). The metabolic pathways of nicotine are discussed by Gorrod in this volume. To assess quantitative aspects of nicotine metabolism in smokers and to compare the metabolic profile of nicotine absorbed from tobacco smoke

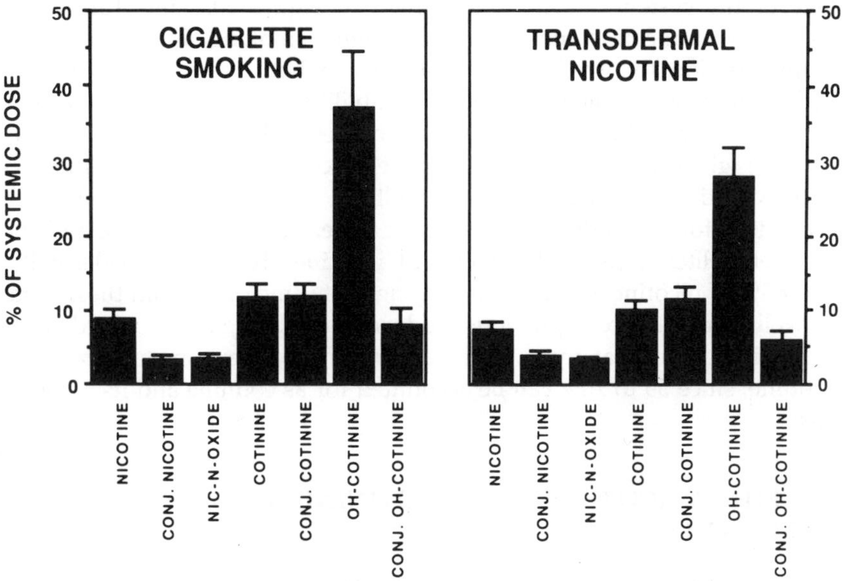

Figure 10.7 Nicotine and metabolites excreted in 24-h urine following smoking or transdermal nicotine administration.

Table 10.5 Urinary excretion of nicotine and metabolites following smoking or transdermal nicotine, as percentage of systemic dose

	Smoking (S.E.M.) (%)	Transdermal (S.E.M.) (%)
Nicotine	8.9 (1.3)	7.5 (0.9)
Nicotine conjugate	3.3 (0.6)	3.8 (0.7)
Nicotine N′-oxide	3.7 (0.4)	3.5 (0.2)
Cotinine	11.8 (1.8)	10.2 (1.1)
Cotinine conjugate	12.0 (1.7)	11.6 (1.6)
trans-3′-Hydroxycotinine	37.2 (7.4)	28.0 (4.0)
trans-3′-Hydrocycotinine conjugate	8.0 (2.3)	6.0 (1.3)

with the metabolic profile of transdermally administered nicotine, 24 h urinary excretion of nicotine and six metabolites were determined. This was done in 13 subjects using a crossover design in which the subject was studied smoking *ad libitum* at steady state, followed by a period during which transdermal nicotine administration had reached steady state. In each subject, the systemic dose of nicotine was calculated from circadian plasma nicotine levels used in conjunction with the clearance of (*S*)-nicotine-3′,3′-d_2 given intravenously, as described above.

Urinary excretion of nicotine and metabolites, expressed as a percentage of the systemic dose, is presented in Figure 10.7 and Table 10.5. Excretion of nicotine metabolites following smoking and transdermal administration followed similar patterns. In both cases, *trans*-3′-hydroxycotinine was the most abundant metabolite, accounting for 28 to 37% of the systemic dose. The other metabolites, which included cotinine, cotinine glucuronide, nicotine-N′-oxide, nicotine glucuronide, *trans*-3′-hydroxycotinine glucuronide, as well as unchanged nicotine, each accounted for 3 to 12% of the systemic dose. Together, nicotine and the six metabolites accounted for 71% of nicotine absorbed transdermally, and 85% of nicotine absorbed during cigarette smoking. From these data, as well as results from previous studies (Curvall *et al.*, 1991; Byrd *et al.*, 1992), it is clear that most of the absorbed nicotine is metabolized via cotinine, since 55 to 70% can be accounted for as cotinine and its metabolites.

10.9 PHARMACOKINETICS OF COTININE FORMATION AND ELIMINATION

The long plasma half-life of cotinine as compared to nicotine, and relatively high concentrations of cotinine in plasma of smokers has made the measurement of cotinine a valuable marker for exposure to tobacco smoke. To utilize cotinine levels to obtain an accurate estimate of nicotine exposure, it is necessary to know the fractional conversion of nicotine to cotinine and the rate of elimination of cotinine. As discussed previously in this chapter, it is advantageous to study the pharmacokinetics of nicotine and its metabolites during smoking. Consequently, the authors developed a dual stable isotope label technique for studying nicotine and cotinine elimination kinetics and the fractional conversion of nicotine to cotinine. This involves simultaneous administration of nicotine labelled with two deuterium atoms and cotinine labelled with four deuterium atoms, which makes it possible to distinguish cotinine generated from a known dose of nicotine (cotinine-d_2) from the cotinine given in known dose by infusion (cotinine-d_4, Figure 10.6) and from tobacco nicotine-derived cotinine (cotinine-d_0).

The experimental design is as follows. Subjects receive an intravenous infusion of an equal dose (2 μg/kg/min for 30 min) of (*S*)-nicotine-3′,3′-d_2 and (*S*)-cotinine-2,4,5,6-d_4. Blood samples are obtained at regular intervals during and following the infusion for analysis by GC–MS. Using a modification of a published procedure (Jacob *et al.*, 1991), concentrations of nicotine-d_2, cotinine-d_2, and cotinine-d_4 are determined. From the dose and concentrations of cotinine-d_4, clearance and half-life of cotinine are calculated. From the AUC of cotinine-d_2 generated from nicotine-d_2, the clearance of cotinine-d_4 and dose of nicotine-d_2 fractional conversion of nicotine to cotinine can be calculated.

$$CL_{cot} = \frac{Dose_{cot\text{-}d_4}}{AUC_{cot\text{-}d_4}}$$

$$f_{nic \rightarrow cot} = \frac{AUC_{cot\text{-}d_2} \times CL_{cot}}{Dose_{nic\text{-}d_2}}$$

The technique is illustrated in Figure 10.8, which shows plasma concentrations of nicotine-d_2, cotinine-d_2 and cotinine-d_4 during and after infusion of nicotine-d_2 and cotinine-d_4 in one subject. Some preliminary data from the studies are presented in Figure 10.8. In ten smokers, the clearance of nicotine-d_2 averaged 1270 ml/min with a range of 550 to 1860 ml/min. Cotinine-d_4 clearance averaged 52 ml/min with a range of 26 to 90 ml/min. Fractional conversion of nicotine to cotinine averaged 87% with a range of 71 to 97% (Figure 10.9).

The goal of the dual stable isotope studies is to characterize cotinine kinetics and fractional conversion of nicotine to cotinine in a large population of smokers in order to ascertain the variability among individuals including differences related to age, sex and ethnic group. These studies are in progress and will involve over 200 subjects.

Using fractional conversion, clearance data and plasma cotinine

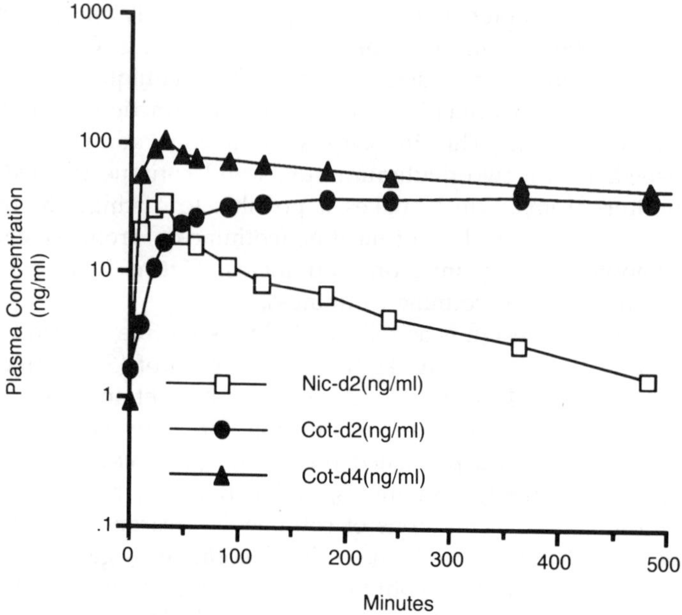

Figure 10.8 Plasma concentrations of nicotine-d_2, cotinine-d_2, and cotinine-d_4 in a smoker following intravenous infusion of a 50:50 mixture of nicotine-d_2 and cotinine-d_4.

levels, daily intake of nicotine from tobacco can be estimated. At steady state, the rate of cotinine elimination (COT_{elim}) is equal to the rate of generation (COT_{gen}), which is equal to the dose of nicotine multiplied by the fractional conversion:

$$COT_{elim} = COT_{gen} = f_{nic \rightarrow cot} \times Dose_{nic}$$

The rate of cotinine elimination is equal to the steady state plasma concentration ($P\text{-}COT_{ss}$) multiplied by cotinine clearance:

$$COT_{elim} = P\text{-}COT_{ss} \times CL_{cot}$$

Combining the two equations:

$$Dose_{nic} = \frac{P\text{-}COT_{ss} \times CL_{cot}}{f_{nic \rightarrow cot}}$$

For an 'average' smoker with plasma cotinine concentration of 300

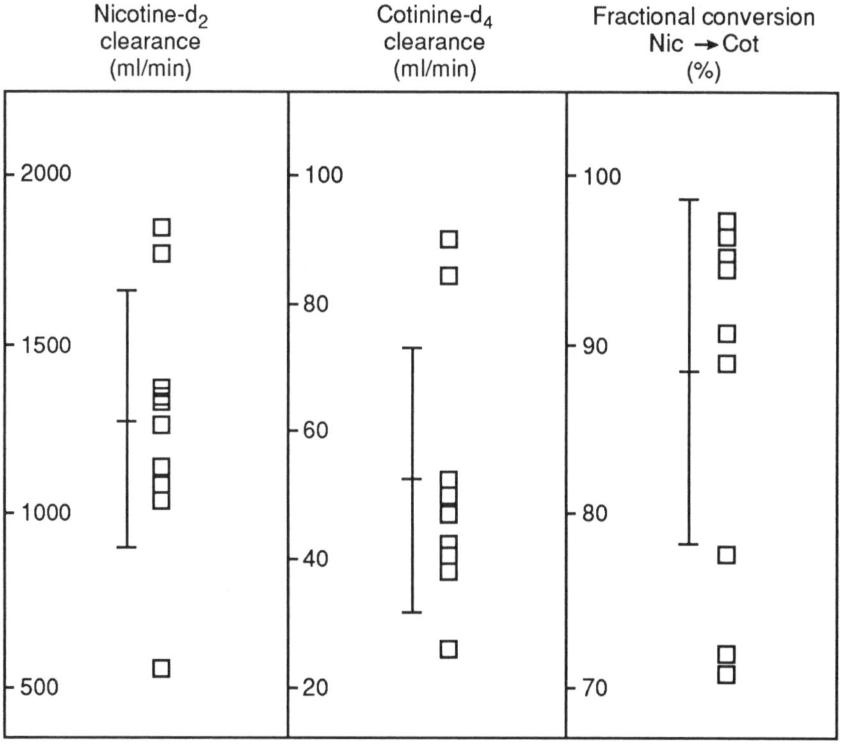

Figure 10.9 Nicotine and cotinine plasma clearance and fractional conversion of nicotine to cotinine in nine smokers, determined using stable isotope methodology. Bars indicate SEM.

ng/ml, cotinine clearance of 50 ml/min, and fractional conversion of 0.88:

$$\text{Dose}_{\text{nic}} = \frac{300 \text{ ng/ml} \times 50 \text{ ml/min} \times 1400 \text{ min/day}}{0.88} = 24.5 \text{ mg/day}$$

10.10 FUTURE STUDIES

Although considerable information on the pharmacokinetics of nicotine has been accumulated over the past few years, little is known about the pharmacokinetics of nicotine metabolites other than cotinine. Recent studies have shown that *trans*-3′–hydroxycotinine is the most abundant nicotine metabolite in smokers' urine (Neurath *et al.*, 1987; Jacob *et al.*, 1988b; Adlkofer *et al.* 1988) and, as such, its concentrations may be useful in estimating nicotine intake. In order to do so, it will be necessary to know the fractional conversion of nicotine to *trans*-3′-hydroxycotinine and the rate of clearance of *trans*-3′-hydroxycotinine. Studies of the metabolic disposition of *trans*-3′-hydroxycotinine are now in progress.

Recent studies have shown that glucuronide conjugates of nicotine, cotinine, and *trans*-3′-hydroxycotinine are significant urinary metabolites. Nothing is known about their pharmacokinetics. Since there is precedent for glucuronides having pharmacological activity (Bodd *et al.*, 1990), the possibility that nicotine glucuronide is active must be addressed. Conjugates may be subject to enterohepatic recirculation and, if this were true for nicotine glucuronide, it could represent a reservoir for sustained generation of nicotine in the body and play a role in maintaining tobacco dependency. For these reasons, studies of the pharmacokinetics of the glucuronide metabolites are planned.

10.11 CONCLUSION

During the past several years, considerable information on the pharmacokinetics and metabolism of nicotine has been published. Advances in analytical instrumentation and methodology have made it possible to characterize new metabolites and to carry out large-scale clinical studies in which low concentrations of nicotine and metabolites are measured in thousands of samples. Pharmacokinetic studies have been carried out in conjunction with studies of nicotine pharmacology to explore the relationship between metabolic disposition and effects (Benowitz *et al.*, 1990). Pharmacokinetic data have been used to estimate nicotine intake by smokers (Benowitz and Jacob, 1984a; Feyerabend *et al.*, 1985; Benowitz *et al.*, 1991b) in order to test the hypothesis that smokers regulate intake to achieve a desired level of nicotine in the body. Such studies have aided the understanding of tobacco dependence and the adverse health effects of smoking. Nicotine metabolite levels are used as markers for tobacco exposure (Benowitz, 1984; Jarvis *et al.*, 1984), and information on the pharmacokinetics of metabolites, in particular cotinine, has been used to estimate daily nicotine exposure. Studies comparing nicotine disposition in smokers and nonsmokers suggest that it is reasonable to use the large body of pharmacokinetic data on nicotine for smokers to estimate environmental tobacco smoke exposure in nonsmokers (Benowitz and Jacob, 1993).

In recent years, nicotine-containing pharmaceutical products have been developed for aiding smoking cessation. Techniques developed in the study of nicotine pharmacokinetics and metabolism, such as stable isotope methodology for determining bioavailability (Benowitz *et al.*, 1991b), have aided the development of these products (Benowitz *et al.*, 1991a, 1992). Pharmacokinetic data should be useful in optimizing dosing of nicotine substitutes in promoting smoking cessation for individual patients and in assessing smoking cessation treatment outcome.

ACKNOWLEDGEMENTS

The authors are grateful to Kaye Welch, Alexander Shulgin, Margaret Peng, Lisa Yu, Chin Savanapridi, Lila Glogowsky, Mario Cave, Lourdes Abayan, Clarissa Ramstead, Duan Min Jiang, Pat Buley, Irving Fong, Sandra Tinetti, Sylvia Wu, Lucy Wu and Gang Liang for their contributions to the research presented in this chapter. Research supported in part by USPHS Grants DA02277 and DA01696 and Grant # 1RT 169 from State of California Tobacco Related Disease Research Program.

REFERENCES

Abood, L.G., Grassi, S. and Noggle, H.D. (1985) Comparison of the binding of optically pure (−)- and (+)-[^{3}H]nicotine to rat brain membranes. *Neurochem. Res.*, **10**, 259–67.

Adlkofer, F., Scherer, G., Jarczyk, L. *et al.* (1988) Pharmacokinetics of 3′-hydroxycotinine, in *The Pharmacology of Nicotine* (Eds M.J. Rand and K. Thurau), ICSU Press, pp. 25–9.

Baillie, T.A., Rettenmeier, A.W., Peterson, L.A. *et al.* (1984) Stable isotopes in drug metabolism and disposition, in *Annual Reports in Medicinal Chemistry*, Academic Press, New York, p. 273.

Beckett, A.H., Gorrod, J.W. and Jenner, P. (1971) The analysis of nicotine-1′-N-oxide in urine, in the presence of nicotine and cotinine, and its application to the study of *in vivo* nicotine metabolism in man. *J. Pharm. Pharmacol.*, **23**, 55S–61S.

Benowitz, N.L. (1984) The use of biologic fluid samples in assessing smoke consumption, in *Measurement in the Analysis and Treatment of Smoking Behavior, NIDA Monograph #48* (Eds J. Grabowski and C.S. Bell), US Government Printing Office, Washington, D.C., pp. 6–26.

Benowitz, N.L., Chan, K., Denaro, C.P. *et al.* (1991a) Stable isotope method for studying transdermal drug absorption: The nicotine patch. *Clin. Pharmacol. Ther.*, **50**, 286–93.

Benowitz, N.L., Hall, S.M., Herning, R.I., *et al.* (1983) Smokers of low yield cigarettes do not consume less nicotine. *New Eng. J. Med.*, **309**, 139–42.

Benowitz, N.L. and Jacob, P., III. (1984a) Daily intake of nicotine during cigarette smoking. *Clin. Pharmacol. Ther.* **35**, 499–504.

Benowitz, N.L. and Jacob, P., III. (1984b) Nicotine and carbon monoxide intake from high- and low-yield cigarettes. *Clin. Pharmacol. Ther.*, **36**, 265–70.

Benowitz, N.L. and Jacob, P., III. (1993) Nicotine and cotinine elimination kinetics in smokers and nonsmokers. *Clin. Pharmacol. Ther.*, **53**, 316–23.

Benowitz, N.L., Jacob, P., III, Denaro, C. *et al.* (1991b) Stable isotope studies of nicotine kinetics and bioavailability. *Clin. Pharmacol. Ther.*, **49**, 270–7.

Benowitz, N.L., Jacob, P., III, Jones, R.T. *et al* (1982) Interindividual variability in the metabolism and cardiovascular effects of nicotine in man. *J. Pharmacol. Exp. Ther.*, **221**, 368–72.

Benowitz, N.L., Jacob, P., III, Olsson, P. *et al.* (1992) Intravenous nicotine retards transdermal absorption of nicotine: Evidence of blood flow-limited percutaneous absorption. *Clin. Pharmacol. Ther.*, **52**, 223–30.

Benowitz, N.L., Jacob, P., III, Yu, L., *et al.* (1986) Reduced tar, nicotine, and carbon monoxide exposure while smoking ultralow but not low-yield cigarettes. *JAMA*, **256**, 241–6.

Benowitz, N.L., Porchet, H. and Jacob, P., III. (1990) Pharmacokinetics, metabolism, and pharmacodynamics of nicotine, in *Nicotine Psychopharmacology: Molecular, Cellular and Behavioral Aspects* (Eds S. Wonnacott, M.A.H. Russell and I.P. Stolerman), Oxford University Press, Oxford, pp. 112–57.

Benowitz, N.L., Porchet, H., Sheiner, L. *et al.* (1988) Nicotine absorption and cardiovascular effects with smokeless tobacco use: Comparison with cigarettes and nicotine gum. *Clin. Pharmacol. Ther.*, **44**, 23–8.

Beynon, J.H. and Williams, A.E. (1963) *Mass and Abundance Tables for Use in Mass Spectrometry*, Elsevier, Amsterdam.

Bodd, E., Jacobsen, D., Lund, E. *et al.* (1990) Morphine-6-glucuronide might mediate the prolonged opioid effect of morphine in acute renal failure *Hum. Exp. Tox.*, **9**, 317–21.

Byrd, G.D., Chang, K., Greene, J.M. *et al.* (1992) Evidence for urinary excretion of glucuronide conjugates of nicotine, cotinine, and trans-3′-hydroxycotinine in smokers. *Drug Metab. Disp.*, **20**, 192–7.

Clark, M.S.G., Rand, M.J. and Vanov, S. (1965) Comparison of pharmacological activity of nicotine and related alkaloids occurring in cigarette smoke. *Arch. Int. Pharmacodyn.*, 363–79.

Copeland, J.R, Adem, A., Jacob, P., III *et al.* (1991) A comparison of the binding of nicotine and nornicotine stereoisomers to nicotinic binding sites in rat brain cortex. *Naunyn-Schmiedebergs Arch. Pharmacol.*, **343**, 123–7.

Cummings, S.R. and Richard, R.J. (1988) Optimum cutoff points for biochemical validation of smoking status. *Am. J. Public Health*, **78**, 574–5.

Curvall, M., Elwin, C.-E., Kazemi-Vala, E. *et al.* (1990) The pharmacokinetics of cotinine in plasma and saliva from nonsmoking healthy volunteers. *Eur. J. Clin. Pharmacol.*, **38**, 281–7.

Curvall, M., Kazemi Vala, E. and Englund, G. (1991) Conjugation pathways in nicotine metabolism, in *Effects of Nicotine on Biological Systems* (Eds F. Adlkofer and K. Thurau), Birkhauser-Verlag, Basel, pp. 69–75.

Davis, R.A., Stiles, M.F., deBethizy, J.D. *et al.* (1991) Dietary nicotine: A source of urinary cotinine. *Food and Chemical Toxicology*, **29**, 821–7.

deSchepper, P.J., Van Hecken, A. and Van Rossum, J.M. (1987) Kinetics of cotinine after oral and intravenous administration to man. *Eur. J. Clin. Pharmacol.* **31**, 583–8.

Ebert, R.V., McNabb, M.E., McCusker, K.T. *et al.* (1983) Amount of nicotine and carbon monoxide inhaled by smokers of low-tar, low-nicotine cigarettes. *JAMA*, **250**, 2840–2.

Feyerabend, C., Ings, R.M.J. and Russell, M.A.H (1985) Nicotine pharmacokinetics and its application to intake from smoking. *Br. J. Clin. Pharmacol.*, **19**, 239–47.

Gritz, E.R., Baer-Weiss, V., Benowitz, N.L. *et al.* (1981) Plasma nicotine and cotinine concentration in habitual smokeless tobacco users. *Clin. Pharmacol. Ther.*, **30**, 201–9.

Haley, N.J., Sepkovic, D.W. and Hoffmann, D. (1989) Elimination of cotinine from body fluids: Disposition in smokers and nonsmokers. *Am. J. Public Health*, **79**, 1046–8.

Ikushima, S., Muramatsu, I., Sakakibara, Y., *et al.* (1982) The effects of *d*-nicotine and *l*-isomer on nicotinic receptors. *J. Pharmacol. Exp. Ther.*, **222**, 463–70.

Jacob, P., III, Benowitz, N.L., Copeland, J.R. *et al.* (1988a) Disposition kinetics of nicotine and cotinine enantiomers in rabbits and beagle dogs. *J. Pharm. Sci.*, **77**, 396–400.

Jacob, P., III, Benowitz, N.L. and Shulgin, A.T. (1988b) Recent studies of nicotine metabolism in humans. *Pharmacol. Biochem. and Behavior*, **41**, 474–9.

Jacob, P., III, Benowitz, N.L. and Shulgin, A.T. (1988c) Synthesis of optically pure deuterium-labelled nicotine, nornicotine and cotinine. *Journal of Labelled Compounds and Radiopharmaceuticals*, **25**, 1117–28.

Jacob, P., III, Benowitz, N.L., Yu, L. *et al.* (1986) Determination of nicotine-1′-N-oxide by gas chromatography following thermal conversion to 2-methyl-6-(3-pyridyl)tetrahydro-1,2-oxazine. *Anal. Chem.*, **11**, 2218–21.

Jacob, P., III, Shulgin, A., Yu, L. *et al.* (1992) Determination of the nicotine metabolite *trans*-3′-hydroxycotinine in smokers using GC with nitrogen-selective detection or selected ion monitoring. *J. Chromatog.*, **583**, 145–54.

Jacob, P., III, Yu, L., Wilson, M. *et al.* (1991) Selected ion monitoring method for determination of nicotine, cotinine, and deuterium-labelled analogs. Absence of an isotope effect in the clearance of (*S*)-nicotine-3′-3′-d_2 in humans. *Biol. Mass Spectrom.* **20**, 247–52.

Jarvis, M.J., Russell, M.A.H., Benowitz, N.L. *et al.* (1988) Elimination of cotinine from body fluids: Implications for noninvasive measurement of tobacco smoke exposure. *Am. J. Public Health*, **78**, 696–8.

Jarvis, M.J., Tunstall-Pedoc, H., Feyerabend, C., *et al.* (1984) Biochemical markers of smoke absorption and self-reported exposure to passive smoking. *J. Epidem. Community Health*, **38**, 335–9.

Jones, C.J., Schiffman, M.H., Kurman, R. *et al.* (1991) Elevated nicotine levels in cervical lavages from passive smokers. *Am. J. Public Health*, **81**, 378–9.

Kozlowski, L.T., Appel, C.P., Frecker, R.C. *et al.* (1982) Nicotine, a prescribable drug available without prescription. *Lancet*, **1**, 334.

Kozlowski, L.T., Pope, M.A. and Lux, J.E. (1988) Prevalence of the misuse of ultra-low-tar cigarettes by blocking filter vents. *Am. J. Public Health*, **78**, 694–5.

Kyerematen, G.A., Damiano, M.D., Dvorchik, B.H. *et al.* (1982) Smoking-induced changes in nicotine disposition: Application of a new HPLC assay for nicotine and its metabolites. *Clin. Pharmacol. Ther*, **32**, 769–80.

Lee, B.L., Benowitz, N.L. and Jacob, P., III. (1987) Influence of tobacco abstinence on the disposition kinetics and effects of nicotine. *Clin. Pharmacol. Ther.*, **41**, 474–9.

Martin, B.R., Tripathi, H.L., Aceto, M.D. *et al.* (1983) Relationship of the biodisposition of the stereoisomers of nicotine in the central nervous system to their pharmacological actions. *J. Pharmacol. Exp. Ther.*, **226**, 157–63.

Neurath, G.B., Dunger, M., Orth, D. *et al* (1987) *Trans*-3′-hydroxycotinine as a main metabolite in urine of smokers. *Int. Arch. Occup. Environ. Health*, **59**, 199–201.

Pool, W.F., Godin, C.S. and Crooks, P.A. (1985) Nicotine racemization during cigarette smoking. *The Toxicologist*, **5**, 232.

Risner, M.E., Cone, E.J., Benowitz, N.L. *et al.* (1988) Effects of the stereoisomers of nicotine and nornicotine on schedule-controlled responding and physiological parameters of dogs. *J. Pharmacol. Exp. Ther.*, **244**, 807–13.

Rylander, E., Pershagen, G., Curvall, M. *et al.* (1989) Exposure to environmental tobacco smoke and urinary excretion of cotinine and nicotine in children. *Acta Paed. Scand.*, **78**, 449–50.

Shigenaga, M.K., Trevor, A.J. and Castagnoli, N., Jr. (1988) Metabolism dependent covalent binding of (*S*)-[5′-3H]-nicotine to liver and lung microsomal macromolecules. *Drug Metabolism and Disposition*, **16**, 397–402.

Sloan, J.W., Martin, W.R., Hernandez, J. *et al.* (1985) Binding characteristics of (−)- and (+)-nicotine to the rat brain P fraction. *Pharmacol. Biochem. and Behavior*, **23**, 987–93.

Stookey, G.K., Katz, B.P., Olson, B.L. *et al.* (1987) Evaluation of biochemical validation measures in determination of smoking status. *J. Dental Res.*, **66**, 1597–1601.

Wall, M.A., Johnson, J., Jacob, P., III *et al.* (1988) Cotinine in the serum, saliva, and urine of nonsmokers, passive smokers and active smokers. *Am. J. Public Health*, **78**, 699–701.

Sources of inter-individual variability in nicotine pharmacokinetics

11

S. Cholerton, N.W. McCracken and J.R. Idle

11.1 INTRODUCTION

There is a considerable degree of inter-individual variability in the toxicological, physiological and psychological responses to tobacco products. Although these responses are determined by many factors, to a certain extent the differences may be attributed to variation in the pharmacokinetics of nicotine, the major pharmacologically active component of tobacco.

The pharmacokinetics of most drugs and xenobiotics are determined to a greater or lesser extent by the processes of absorption, distribution, metabolism and excretion. In turn these processes are under the influence of a wide range of both endogenous and exogenous factors. Studies in man and many animal species have provided considerable information regarding the role of these factors in the inter-individual variability in the pharmacokinetics of nicotine.

11.2 ABSORPTION OF NICOTINE

The variety of ways in which nicotine-containing products are administered and the pH dependency of nicotine absorption markedly influence the amount of nicotine which enters the systemic circulation. With the exception of the intravenous route, nicotine administration involves absorption through a cell membrane. In the form of the undissociated

Nicotine and Related Alkaloids: Absorption, distribution, metabolism and excretion. Edited by J.W. Gorrod and J. Wahren. Published in 1993 by Chapman & Hall, London. ISBN 0 412 55740 1

base, highly lipid soluble nicotine readily permeates cell membranes, however differences in pH throughout the body determine the actual amount of nicotine absorbed. The many commonly used sites of absorption of nicotine will be considered in turn.

11.2.1 SITES OF NICOTINE ABSORPTION

(a) Lung

The most common route of administration of nicotine is in the form of tobacco smoke which is absorbed through the lungs. Nicotine is rapidly absorbed from inhaled cigarette smoke, reaching the systemic circulation at a rate comparable with intravenous administration (Russell and Feyerabend, 1978). Although there is considerable inter-individual variability in the daily intake of nicotine from cigarettes (Benowitz and Jacob, 1984), smokers maintain relatively constant levels of nicotine in the body (Benowitz *et al.*, 1982b). There is considerable experimental evidence to suggest that smokers have the ability to maintain such steady concentrations of nicotine in the body by modifying the way in which cigarettes are smoked. Processes involved in smoking such as depth of inhalation and frequency of puffs are controlled by the smoker to obtain the desired amount of nicotine (Herning *et al.*, 1983). In a study in which the number of cigarettes available to smokers was reduced from unlimited to five per day, the nicotine yield per cigarette tripled (Benowitz *et al.*, 1986a). Furthermore, smokers who change from high-yield to low-yield nicotine cigarettes derive more nicotine from the low-yield cigarettes than would be predicted from machine determined yield (Benowitz *et al.*, 1986b). Presumably this occurs as a consequence of either increased puff frequency or increased smoking intensity.

(b) Buccal mucosa

The buccal absorption of nicotine from tobacco smoke is highly dependent upon the pH of the smoke. Since smoke from most European cigarettes (flue cured tobacco) is acidic (pH 5.5) the nicotine is in the ionized form and little is absorbed from smoke held in the mouth (Gori *et al.*, 1986) In contrast, smoke from cigars, pipe tobacco and a limited number of European cigarettes (air cured tobacco) is alkaline (pH 8.5), the nicotine is mostly in the un-ionized form and well absorbed through the oral mucosa.

Although the release of nicotine from chewing tobacco and nicotine gum is gradual and continuous over a 30 min chewing period (Benowitz *et al.*, 1987a), the extent of extraction and absorption control the actual intake of nicotine into the systemic circulation. Absorption of nicotine from chewing tobacco and nicotine gum is facilitated by buffering of the preparations to an alkaline pH ($\sim$8.5). The maintenance of an alkaline pH in the mouth should reduce inter-individual variability in absorption. Nevertheless intake of nicotine from gum is highly variable among subjects (McNabb *et al.*, 1982). This can be partly attributed to differences in the extent of extraction of nicotine from gum which is dependent on vigour and duration of chewing. In a study in which chewing time was constant, two-fold variability in the extraction of nicotine from gum was seen in a group of 14 healthy volunteers (Benowitz *et al.*, 1987a).

A greater proportion of nicotine is extracted from 4 mg pieces of gum than from 2 mg pieces. More salivation is associated with the chewing of a 4 mg piece of gum and as a consequence the amount of saliva produced has been proposed as an important factor in determining the extent of extraction of nicotine from gum. Furthermore it has been suggested that individuals manipulate the amount of gum derived nicotine which enters the systemic circulation by swallowing some of the nicotine (Benowitz *et al.*, 1987a).

Because nicotine is released gradually from nicotine gum, the administration of a 4 mg piece of gum per hour results in less variability in peak to trough nicotine concentrations than with one cigarette per hour (Feyerabend and Russell, 1978). The use of chewing tobacco and oral snuff is also associated with prolonged plasma levels of nicotine.

Since oral snuff is not chewed and most preparations are buffered to an alkaline pH, the major factor which influences the amount of nicotine derived from tobacco in this form is the length of time the snuff is left in the mouth.

Although the role of gum disease in the buccal absorption of nicotine has yet to be defined, a recent study suggests that nicotine is rapidly taken up into gingival fibroblasts and then released at a slower rate (Hanes *et al.*, 1991). This may have important implications for the buccal absorption of nicotine in individuals with gingival hyperplasia.

(c) Nasal mucosa

Nicotine, in the form of dry snuff or as a smoking cessation spray, is rapidly absorbed from the nasal mucosa. Peak plasma nicotine concentrations comparable with those of cigarette smoking can be achieved with dry snuff (Russell *et al.*, 1980)

(d) Gastrointestinal tract

Although the recreational use of tobacco does not rely on the oral administration of nicotine, this route of administration is important for certain research purposes (Beckett *et al.*, 1971a, 1972) and also in many cases of poisoning with nicotine (Malizia *et al.*, 1983). As a consequence of pH dependency, absorption of nicotine from the stomach is limited, whereas that from the intestine is extensive. Even so, only 25–30% of the dose absorbed reaches the systemic circulation because of extensive first-pass metabolism of nicotine.

By increasing intragastric pH, more nicotine will be absorbed from the stomach, however this will not increase systemic availability. Only factors which modify first-pass metabolism will alter this pharmacokinetic parameter. At a sufficiently high oral dose of nicotine first-pass metabolism will be saturated and a greater proportion of the dose will reach the systemic circulation. Although relatively rare, cases of fatal oral nicotine ingestion have been reported (Manoguerra and Freeman, 1983; Lavoie and Harris, 1991).

Tobacco enemas used in the treatment of constipation and intestinal parasites have been associated with nicotine poisoning (Oberst and McIntyre, 1953; Garcia-Estrada and Fischman, 1977). Since first-pass metabolism may be less when a drug is given rectally than when administered orally, systemic availability of nicotine from tobacco enemas and the chance of poisoning may be greater than after oral administration. The rectal absorption of nicotine-1′-N-oxide has also been demonstrated and as a consequence of its reduction in the gut, this results in substantial amounts of nicotine and cotinine in the urine within 0.5 h of administration (Beckett *et al.*, 1970).

(e) Skin

In addition to being a common route of administration associated with accidental poisoning (Faulkner, 1933; Gehlbach *et al.*, 1974), the dermal absorption of nicotine from a patch has recently appeared on the market as an adjuvant to smoking cessation. Nicotine appears to be absorbed slowly following transdermal absorption, with peak plasma concentrations achieved after 90 min (Rose *et al.*, 1984). The pH dependency of dermal absorption of nicotine has only been demonstrated in the cat (Travell, 1960), although this is probably the case in other species. In a case of accidental poisoning with nicotine in an individual who had scabies with severe pruritus, nicotine appeared to be particularly well absorbed through the excoriated skin (Benowitz *et al.*, 1987b).

11.3 DISTRIBUTION OF NICOTINE

Route of administration is an important factor in determining the tissue distribution of nicotine. It is often stated that the lag time between cigarette smoking and entry of nicotine into the brain is less than that after intravenous injection. This has never been directly investigated and is based on the assumption that the time to travel from the lungs to the aorta, and subsequently to the brain, is shorter than that from an intravenous injection site. Nevertheless nicotine does enter the brain very quickly. However, brain levels decline rapidly as nicotine is distributed to other parts of the body. Like many basic drugs, nicotine (pKa 7.9) is extensively distributed into virtually all tissues of the body of many mammals, however inter-organ differences in nicotine distribution within the same species and inter-species differences in the distribution of nicotine to certain organs, particularly the brain, do exist (Schmiterlöw *et al.*, 1967; Stålhandske, 1970; Tsujimoto *et al.*, 1975).

There appears to be some debate as to the degree of plasma protein binding of nicotine in humans. Benowitz and co-workers (1982a) determined an average value of 4.9%, whereas other studies have suggested that the plasma protein binding of nicotine may be as high as 20% (Schievelbein, 1984). A value in this range (4.9–20) means that the ratio of bound concentration to the total concentration (*fb*) is $\leqslant 0.2$ and thus nicotine is said to show little or no plasma protein binding (Rowland and Tozer, 1980). As a consequence its distribution will not be significantly affected by disorders which alter plasma protein concentrations. The volume of distribution of nicotine will be very much dependent on body weight and has been estimated at between one and three times body weight in man.

Although disappearance from plasma and elimination of nicotine and its metabolites are not affected by pregnancy in humans, the distribution of nicotine is altered in this state (Klein and Gorrod, 1979). Nicotine has been shown to cross the placenta of several species. In the rabbit there is evidence to suggest that nicotine may penetrate the preimplanted blastocyte and that nicotine levels in the blastocyte may be as much as four times higher than in maternal plasma (Sieber and Fabro, 1971). Similarly, in the rat plasma levels of nicotine are higher in the foetus than in the mother (Mosier and Jansons, 1972), whereas in the mouse, foetal tissue levels of nicotine tend to be lower than in maternal blood or placenta (Tjalve *et al.*, 1968). In a study in which the placental transfer and distribution of nicotine in the pregnant rhesus monkey was investigated, foetal blood levels of nicotine achieved were higher than those in the maternal circulation (Suzuki *et al.*, 1974). Furthermore the disappearance of nicotine from the foetal circulation was slower than that from the mothers. In smoking mothers, nicotine has been detected in both umbilical vein serum and amniotic fluid at concentrations higher than those in maternal serum (Luck and Nau, 1984a). Cotinine has also been detected in the amniotic fluid of both smoking mothers and those exposed to environmental cigarette smoke (Van Vunakis *et al.*, 1974, Andresen *et al.*, 1982). The concentration of nicotine in breast milk has been shown to be concentrated to almost three times that in serum (Luck and Nau, 1984b). In one study ingestion by infants of breast milk from smoking mothers resulted in an average infant serum to maternal serum cotinine concentration ratio of 0.06 (Luck and Nau, 1984a). Not suprisingly there have been cases of infant nicotine poisoning as a result of being nursed by a smoking mother (Abel, 1984).

From a study of 35 women with histologically confirmed cervical intraepithelial neoplasia, there is evidence that nicotine and, to a lesser extent, cotinine are strongly concentrated in cervical mucus when compared with the levels in serum. This mucosal accumulation of nicotine and its metabolite was seen in smokers, nonsmokers and passive smokers (Hellberg *et al.*, 1988). Since the analysis of nicotine in breast fluid from non-lactating women (Petrakis *et al.*, 1978), in hair (Haley and Hoffman, 1985) and in aspirate from ovarian cysts (Hellberg and Nilsson, 1988) has failed to detect any accumulation of nicotine and cotinine when compared to serum levels, concentration in cervical mucus appears to be unique. However, whether this occurs in women with normal histology is not known.

11.4 METABOLISM OF NICOTINE

In most mammalian species, nicotine is rapidly and extensively metabolized primarily in the liver, but there is also evidence to suggest that extrahepatic metabolism of nicotine occurs in the lung and the kidney. The metabolism of nicotine involves both oxidative and conjugating pathways and yields a complex pattern of metabolites (Gorrod and Jenner, 1975; Nakayama, 1988; Kyerematen and Vesell, 1991).

Like that of many drugs and xenobiotics, the metabolism of nicotine is under the influence of a wide range of host factors (Vesell, 1982). These sources of variability in nicotine metabolism can be divided into two types, endogenous and exogenous.

11.4.1 ENDOGENOUS INFLUENCES ON THE METABOLISM OF NICOTINE

The endogenous influences on the metabolism of nicotine occur as a consequence of the pathology and physiology of the individual and are largely genetic and developmental but also include the influence of disease states and the site of metabolism of nicotine.

(a) Genetic regulation of enzymes involved in nicotine metabolism

(i) Cytochromes P450

Cytochromes P450 represent a superfamily of haem-containing monooxygenases which are involved in the metabolism of both endogenous

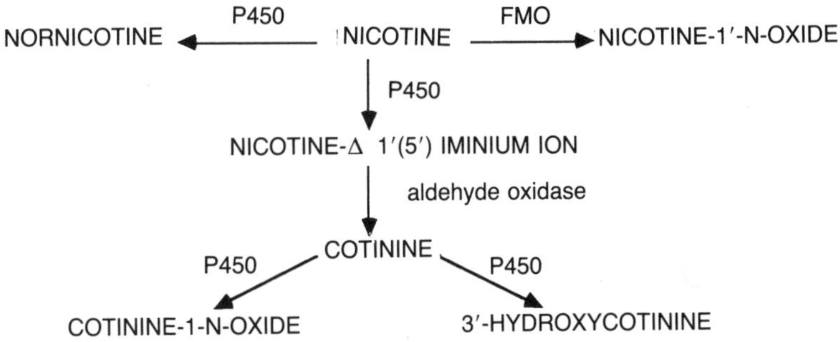

Figure 11.1 The involvement of cytochromes P450, aldehyde oxidases and flavin-containing monooxygenases (FMO) in the production of some major nicotine metabolites.

and xenobiotic compounds. Some P450 isozymes display genetic polymorphism and others show marked variability in hepatic expression (Gonzalez, 1992; Forrester *et al.*, 1992). Such factors will undoubtedly contribute to inter-individual variability in the metabolism of nicotine.

It is now well established that cytochromes P450 are involved in the oxidative metabolism of nicotine to cotinine. This metabolic pathway is a two-step process (Figure 11.1) in which nicotine is converted to nicotine $\Delta^{1'(5')}$ iminium ion (Murphy, 1973; Peterson *et al.*, 1987) followed by the metabolism of the $\Delta^{1'(5')}$ iminium ion to cotinine by cytosolic aldehyde oxidases (Brandänge and Lindblom, 1979; Gorrod and Hibberd, 1982). It is the initial step which is mediated by cytochromes P450 (Williams *et al.*, 1990), however, the role of individual isozymes has yet to be fully elucidated.

Recent studies using cDNA expressed human cytochromes P450 (Table 11.1) suggest that CYP2B6 exhibits high activity with respect to cotinine formation (Flammang *et al.*, 1992; McCracken *et al.*, 1992). CYP2B6 is expressed in human livers at markedly variable levels, with many liver specimens containing low or undetectable levels while others contain moderate levels (Yamano *et al.*, 1989; Forrester *et al.*, 1992). There is evidence from studies in hamsters (McCoy and DeMarco, 1986) and rats (Rüdell *et al.*, 1987; Hammond *et al.*, 1991) that hepatic microsomal C-oxidation of nicotine can be induced by phenobarbitone. Furthermore, in rabbits C-oxidation of nicotine is greatest with CYP2B4, a phenobarbitone inducible isozyme (McCoy *et al.*, 1989). However, in man it is unclear whether inducers such as phenobarbitone or structural mutations in the gene which encodes CYP2B6 are the cause of variability in expression of this P450. Although the inter-individual differences in expression of CYP2B6 have great potential to influence an individual's ability to produce cotinine, it must be remembered that this is not a constitutive enzyme, and as such will be expressed at

Table 11.1 The role of individual human cytochrome P450 isozymes in the C-oxidation of nicotine

P450 isozyme	Nicotine C-oxidation[a]	Cotinine formation[b]
Control	0	0
1A2	0	–
2A6	21	1.5
2B6	166	5.5
2C8	16	0
2C9	46	0
2D6	0	2.5
2E1	50	0
2F1	15	–
3A3	0	–
3A4	0	0
3A5	0	–
4B1	18	–

[a]nmoles product/minute/mg cell lysate protein. (From Flammang *et al.*, 1992.)
[b]pmoles cotinine/minute/mg protein. (From McCracken *et al.*, 1992.)

low or undetectable levels in many individuals, and thus it is unlikely under normal circumstances to contribute to the production of cotinine.

There is evidence to suggest that CYP2A6 has the ability to mediate the C-oxidation of nicotine (Flammang *et al.*, 1992; McCracken *et al.*, 1992). Like CYP2B6, CYP2A6 is also expressed in human liver at markedly variable levels and concordance of expression with CYP2B6 has been demonstrated (Yamano *et al.*, 1990; Forrester *et al.*, 1992). The 7-hydroxylation of coumarin is known to be mediated by CYP2A6 (Yamano *et al.*, 1990) and this reaction shows considerable inter-individual variability within a normal healthy population (Cholerton *et al.*, 1992a). Although an inactive variant has been identified, CYP2A6v, its role in the inter-individual variability of *in vivo* coumarin 7-hydroxylase activity is as yet unclear. Furthermore, there is evidence to suggest that CYP2A6 expression can be induced by agents such as pyrazole.

By far the best studied genetic polymorphism of xenobiotic metabolism is that which affects the metabolism of debrisoquine (Mahgoub *et al.*, 1977). Between 5–10% of a Caucasian population have a defect in the gene which codes for CYP2D6 which renders them poor metabolizers not only of debrisoquine but also of many widely used medicines (Cholerton *et al.*, 1992b). A recent report provides evidence that CYP2D6 has the ability to mediate the conversion of nicotine to cotinine (McCracken *et al.*, 1992). Such an involvement may have far-reaching consequences for the toxicological, physiological and psychological effects of tobacco products.

Two isozymes of the 2C subfamily, CYP2C8 and CYP2C9, are thought to be involved in the C-oxidation of nicotine (Flammang *et al.*, 1992). Mephenytoin and tolbutamide, both substrates of isozymes of the 2C subfamily (Gonzalez, 1992) exhibit polymorphic metabolism (Wilkinson *et al.*, 1989; Scott and Poffenbarger, 1979); however, the individual isozymes responsible for this inter-individual variability in metabolism have yet to be elucidated. The mephenytoin polymorphism was initially thought to reside with CYP2C9, however *S*-mephenytoin hydroxylation has not been observed with this isozyme expressed in several different systems (Relling *et al.*, 1989; Romkes *et al.*, 1991). The tolbutamide polymorphism may reside in CYP2C8 or CYP2C9 since both isozymes mediate the hydroxylation of this hypoglycaemic agent. Variability in hepatic expression of these isozymes may contribute to inter-individual variation in nicotine pharmacokinetics.

CYP2E1 appears to have the ability to mediate the C-oxidation of nicotine (Flammang *et al.*, 1992). This enzyme is constitutively expressed and shows a large degree of inter-individual variation in human livers, which may in part be due to the fact that it is ethanol inducible. The fact that heavy cigarette smoking often accompanies heavy consumption of alcoholic beverages (Maletzky and Klotter, 1974; Griffiths *et al.*, 1976) may provide indirect evidence for the induction of CYP2E1.

Since it appears that several cytochrome P450 isozymes are able to mediate the conversion of nicotine to cotinine in human liver, genetic influences on the expression of these enzymes will play a major role in determining the production of cotinine in a particular individual. Furthermore, since there is evidence to suggest that cotinine can be produced in extra-hepatic tissues, the genetic regulation of the expression of individual P450s in other tissues may also influence the pharmacokinetics of nicotine seen in an individual.

The metabolism of cotinine to 3′-hydroxycotinine and to cotinine-N-oxide are also thought to be cytochrome P450 mediated reactions. Although there is no information regarding the P450 isozymes involved in these metabolic pathways, genetic regulation of these enzymes is likely to influence the formation of these metabolites.

(ii) Flavin-containing monooxygenase (FMO)

Nicotine-1′-N-oxide, the major N-oxidation product of nicotine N-oxidation, has been shown to be produced by liver homogenate (Booth and Boyland, 1970, 1971; Jenner *et al.*, 1973a) and liver microsomes (McCoy and DeMarco, 1986; Nakayama *et al.*, 1987) from a variety of species. Furthermore, the urinary excretion of nicotine-1′-N-oxide has been demonstrated in rats (Kyerematen *et al.*, 1988a), guinea pigs (Nwosu and Crooks, 1988), stumptailed macaques (Seaton *et al.*, 1991), cigarette smokers (Booth and Boyland, 1970) and chewers of nicotine gum (Cholerton *et al.*, 1988). There is considerable evidence from the use of inhibitors (Nakayama *et al.*, 1987; McCoy *et al.*, 1989) and purified enzyme (Damani *et al.*, 1988) to suggest that the N-oxidation of nicotine is mediated by the flavin-containing monooxygenase.

Studies using highly purified preparations of flavin-containing monooxygenase derived from various tissues of mice, rats, guinea pigs and rabbits (Kimura *et al.*, 1983; Tynes *et al.*, 1985; Brodfuehrer and Zannoni, 1986) suggest that tissue and species specific differences in this enzyme system exist. Furthermore, multiple forms of hepatic and pulmonary flavin-containing monooxygenase have been identified in the mouse (Tynes and Philpot, 1987). However, it is the role of the flavin-containing monooxygenase in the clinical condition trimethylaminuria, which provides the best insight into the genetic control of this enzyme in humans.

The flavin-containing monooxygenase is known to mediate the N-oxidation of trimethylamine (Hlavica and Kehl, 1977), a simple aliphatic tertiary amine which is derived from normal dietary components such as choline and carnitine. Trimethylaminuria, known colloquially as 'the fish odour syndrome' occurs as a result of an impaired ability to N-oxidize trimethylamine. It is now well established that an individual's ability to N-oxidize trimethylamine is genetically determined and that this metabolic trait exhibits polymorphism in a white Caucasian population (Al-Waiz *et al.*, 1987a). The average daily urinary excretion of trimethylamine N-oxide and trimethylamine for European white Caucasians consuming a non-fish diet is approximately 50 mg and 1–2 mg respectively (A1-Waiz *et al.*, 1987b). However, affected individuals excrete significantly increased concentrations of free trimethylamine, not only in their urine but also in their breath and sweat. Since trimethylamine has a powerful fish-like smell, these individuals are characterized by an objectionable smell of rotting fish, hence the name 'the fish odour syndrome'.

In a study in which the ability to produce trimethylamine N-oxide from dietary derived trimethylamine was determined in 169 unrelated volunteers and compared to that in two sisters with trimethylaminuria, 95.5 ± 3.3% (mean ± S.D.) of the urinary trimethylamine was in the form of the N-oxide for the healthy population, whereas this value was only 10.5 and 22.2% for the two sisters (Ayesh *et al.*, 1988). The two sisters were then given a dose of nicotine and their urinary output of nicotine-1′-N-oxide and cotinine was compared to that from a group of unaffected volunteers after the same dose of nicotine. The sisters showed a marked deficiency in the ability to produce the N-oxide of nicotine when compared to the controls (Table 11.2). Since individuals

Table 11.2 The urinary excretion of nicotine-1′-N-oxide (NNO) and cotinine (COT) in two trimethylaminuric sisters (S1 and S2) and six unaffected individuals after the administration of nicotine (4 mg) in a chewing gum preparation[a]

	% TMA as TMAO	NNO	COT
S1	10.5	<0.1	2.0
S2	22.2	<0.1	2.4
Unaffected individuals	95.5 ± 3.3[b]	0.88 ± 0.69	0.49 ± 0.29

TMA, trimethylamine; TMAO, trimethylamine N-oxide. Data expressed as mean ± S.D.
[a]From Ayesh *et al.*, 1988.
[b]From Ayesh *et al.*, 1987.

who are deficient in the ability to N-oxidize trimethylamine also have an impairment in the ability to form nicotine-1′-N-oxide, this suggests that N-oxidation of trimethylamine and nicotine is mediated by the same enzyme. Studies of the genetic aspects of the N-oxidation of trimethylamine have lead to the proposal that this reaction is controlled by a single autosomal diallelic locus with one allele for rapid and extensive oxidation and an uncommon variant allele for impaired metabolism. As such, trimethylaminuria is considered to be an inherited metabolic disorder in which affected individuals are homozygous for the allele which confers defective N-oxidation. Although a human flavin-containing monooxygenase (FMO1) has been isolated, this enzyme has been shown to be present in low amounts in human adult liver (Dolphin *et al.*, 1991). The role of FMO1 in the fish odour syndrome is yet to be determined.

Carriers of the fish odour syndrome (heterozygotes) do not display any of the clinical characteristics of trimethylaminuria and can only be differentiated from homozygous unaffected individuals by the administration of a 600 mg oral dose of trimethylamine. Heterozygotes show saturation of N-oxidation capacity at this dose, whereas a 900 mg oral dose of trimethylamine is required to reduce N-oxide production in homozygous unaffected individuals. Using this 'trimethylamine load test', the prevalence of heterozygotes of this metabolic defect in a British population was found to be two out of 169 individuals, although the authors suggest that this may in fact be an underestimate (Al-Waiz *et al.*, 1987a). Although the metabolism of nicotine has yet to be investigated in known heterozygotes for trimethylaminuria, it may be that such individuals have a reduced capacity to metabolize nicotine to its N-oxide when compared to homozygous unaffected subjects. Since nicotine N-oxides can be directly nitrosated to form nitrosamines (Sepkovic *et al.*, 1986; Hecht and Hoffman, 1988), a reduced capacity to N-oxidize nicotine may be distinctly advantageous.

(b) Developmental regulation of nicotine metabolism

(i) Age

Age-related differences in metabolism are often the reason for the young and elderly of several animal species showing increased sensitivity to many drugs. Foetal mouse liver only acquires the ability to metabolize nicotine to cotinine on the day before birth. However, cotinine production starts to rise after birth and increases rapidly until adult levels are reached at age four weeks (Stålhandske *et al.*, 1969). Additional metabolites begin to be produced after three weeks. Ageing female mice (up to 18 months) do not show any reduction in their ability to produce cotinine (Slanina and Stålhandske, 1977). In Sprague–Dawley rats both C- and N-oxidation pathways of nicotine metabolism have been shown to significantly decrease from age 40 d to 100 d (Kyerematen *et al.*, 1988a).

In a study of 85 cigarette smokers, Klein and Gorrod (1978) found that with increasing age the amount of cotinine excreted decreased relative to nicotine N-oxide. This effect was considered to be the result of an age-related alteration in C- rather than N-oxidation and not a consequence of cigarette consumption since it was seen in light, moderate and heavy smokers.

There is evidence to suggest that the hepatic human form of FMO, designated FMO1, is subject to developmental regulation. Northern blot analysis has demonstrated that mRNA for this enzyme is more abundant in foetal than human liver (Dolphin *et al.*, 1991). This may have important implications for unborn children of smoking mothers since, as stated previously, nicotine N-oxides can be directly nitrosated to form nitrosamines.

(ii) Sex

Although there is a genetic component to the sex differences seen in drug metabolism, hormonal influences play a major role in determining these differences.

There is considerable evidence for the presence of sex-specific cytochromes P450 in the rat; such differences have yet to be reported in man. Nevertheless there are several reports of sex differences in the metabolism of nicotine in man. After an intravenous dose of nicotine, female nonsmokers were shown to excrete more unchanged nicotine, less cotinine and show a higher combined recovery of the two than male nonsmokers (Beckett *et al.*, 1971b). Furthermore, in a study of 22 habitual cigarette smokers, the total plasma clearance of nicotine (normalized for body weight) was shown to be lower in females than in males (Benowitz and Jacob, 1984). Since there were no sex differences in daily intake of nicotine, there was an increased tendency for women to be exposed to higher nicotine levels than men. In the largest population study of nicotine metabolism to date, no differences in the ratio of urinary cotinine/nicotine N-oxide were seen between male and female smokers (Klein and Gorrod, 1978). Sex differences in nicotine metabolism seen in stumptailed macaques is similar to those observed in man with significantly lower clearance of nicotine in females when compared with males (Seaton *et al.*, 1991).

In an *in vitro* study several species of male rat were shown to metabolize nicotine faster than their female counterparts (Kyerematen *et al.*, 1988a). However, after a single i.v. dose of [^{14}C]-nicotine, the prolonged plasma half-life observed in female Sprague–Dawley rats was balanced by their larger volume of distribution, no sex difference in plasma clearance was observed. Even so, sex differences were observed *in vivo* such that female rats had lower plasma cotinine, higher urinary recoveries of nicotine and lower total urinary output of metabolites than male rats (Kyerematen *et al.*, 1988a). In addition, sex differences appear to exist in the metabolism of nicotine enantiomers in the rat (Nwosu and Crooks, 1988). After a dose of either *S*-(−)- or *R*-(+)-nicotine, 24 h urine levels of unchanged drug were significantly higher in female than in male rats. After *R*-(+)-nicotine urinary levels of 3′-hydroxycotinine were twice as high in females than males, whereas urinary nicotine N′-oxide was almost twice that in males than it was in females.

Sex differences in the amount of enzyme activity of flavin-containing monooxygenase have been observed in several animal species. In mouse liver flavin-containing monooxygenase activity is repressed by testosterone such that the activity of this enzyme is higher in females than in males (Duffel *et al.*, 1981). In contrast, testosterone appears to positively regulate flavin-containing monooxygenase activity in the male rat liver since hypophysectomy and castration reduce activity (Lemoine *et al.*, 1991). In a study of the sexual dimorphism of nicotine metabolism, the N-oxidation of nicotine by female rat liver was significantly lower than that for males in the four strains studied (Kyerematen *et al.*, 1988a). Although the activity of this pathway is reduced by castration and returned to near normal levels by the administration of testosterone, liver homogenate from castrated males still produces more nicotine N-oxide than that from females (Kyerematen *et al.*, 1988a).

(iii) Species

As with many other drugs, marked species variation exists in the metabolism of nicotine both *in vivo* and *in vitro*. In a study which compared the disposition kinetics of nicotine and cotinine in rabbits and Beagle dogs, plasma cotinine concentrations achieved in the dogs after nicotine administration were so much lower than those in rabbits that the authors suggested that either a metabolite other than cotinine is formed in dogs or that cotinine clearance is much greater in dogs than in rabbits (Jacob *et al.*, 1988). Stumptailed macaques appear to share some similarities with humans in nicotine metabolism in that after a single intravenous dose of nicotine their pattern of metabolism in plasma and urine closely resembles that in man (Seaton *et al.*, 1991). In a study in which the metabolism of nicotine was investigated in human, rat, hamster, guinea pig and mouse hepatocytes, large species differences in both the extent of nicotine metabolism and the pattern of metabolites was observed (Kyerematen *et al.*, 1990a). In human and rat hepatocytes only 30% of the nicotine was metabolized, whereas in those from guinea pigs there was 95% metabolism. Guinea pig liver appears to have a greater capacity to N-oxidize nicotine than that of other species and also shows marked intra-animal variation in extent of formation of both nicotine-1′-N-oxide and cotinine (Jenner and Gorrod, 1973; Jenner *et al.*, 1973a; Kyerematen *et al.*, 1990a). In a study in which relative amounts of flavin-containing monooxygenase in hepatic microsomes were determined by immunochemical staining, those from the pig had considerably higher amounts than the other species considered, although flavin-containing monooxygenase activity was not determined (Dannan and Guengerich, 1982).

Although *S*-(−)-nicotine is the major alkaloid of tobacco, *R*-(+)-nicotine makes up as much as 10% of the total nicotine content of cigarette smoke (Benowitz *et al.*, 1983a). As for many drugs which occur as racemates, there are stereoselective differences in nicotine metabolism between the species. *S*-(−)-nicotine undergoes a considerable degree of metabolism in the liver of the guinea pig, the rabbit and the hamster, with less metabolism in the mouse and a negligible amount in the rat (Jenner and Gorrod, 1973; Jenner *et al.*, 1973a). After intraperitoneal injections of *R*-(+)- and *S*-(−)-nicotine, 3′-hydroxycotinine was the major urinary metabolite in guinea pigs, hamsters and rabbits whereas in the rat cotinine was an important urinary metabolite. Nicotine N-oxide was formed from both enantiomers in the guinea pig and the rat but was not detected in the hamster and the rabbit (Nwosu and Crooks, 1988). In the same study none of the species investigated produced N-methylated metabolites after *S*-nicotine and only the guinea pig produced such metabolites after *R*-nicotine. Both *R*- and *S*-nicotine have been shown to be rapidly metabolized and have similar disposition in the Beagle dog and the rabbit, although *R*-nicotine is metabolized slightly faster than *S*-nicotine in the dog (Jacob *et al.*, 1988). Stereoselectivity in nicotine metabolism has also been demonstrated using hepatocytes. Hamster hepatocytes appear to convert *S*-nicotine faster than *R*-nicotine to cotinine whereas *R*-nicotine is converted faster than *S*-nicotine to nornicotine (Kyerematen *et al.*, 1990a).

There is evidence to suggest that there is stereoselective formation of the diastereoisomers of nicotine N-oxide from the enantiomers of nicotine. *S*-Nicotine is preferentially converted to *R,S-cis*-nicotine-1′-N-oxide in several species whereas *R*-nicotine forms more *S,R-trans*-nicotine-1′-N-oxide in the same species except the rabbit, which gives an excess of the *cis* diastereoisomers from both nicotine enantiomers (Jenner *et al.*, 1973a). Since there is considerable evidence of stereoselectivity in the metabolism of nicotine, the relative proportions of the enantiomers administered will influence the overall pharmacokinetic profile of this alkaloid.

(iv) Strain

Although there is evidence to suggest that there are strain differences with respect to the pharmacodynamics of nicotine (Marks *et al.*, 1989), the literature is somewhat lacking in reports of strain differences in nicotine metabolism. However in a study of the comparative disposition of nicotine and its metabolites in three inbred strains of mice, significant inter-strain differences were seen in the pharmacokinetics of nicotine, cotinine and nicotine-1′-N-oxide (Petersen *et al.*, 1984). Furthermore, when both C- and N-oxidation pathways of nicotine metabolism were determined *in vitro* in four strains of rat, although males metabolized nicotine faster than females, mature (100-day-old) Wistar and Long Evans rats showed the largest sex difference ($>50\%$) for the formation of both cotinine and nicotine N-oxide (Kyerematen *et al.*, 1988a). Young (40-day-old) Fischer rats revealed the least sex difference and it was suggested that this may be due to delayed sexual maturation in the Fischer strain compared with the other three strains.

(v) Ethnic group

In humans, ethnic differences are known to exist in the metabolism of many drugs and other xenobiotics (Kalow *et al.*, 1986; Lou, 1990), thus evidence to suggest that there are inter-ethnic differences in the metabolism of nicotine is not so suprising. When serum cotinine concentrations were determined in a large group of smokers, higher median levels were observed in Blacks (221 ng/ml) than in Whites (170 ng/ml) even though the estimated daily nicotine exposure and serum thiocyanate levels were higher in Whites (Wagenknecht *et al.*, 1990). This ethnic difference in serum cotinine could not be explained by either a bias of reporting of the number of cigarettes consumed or differences in nicotine intake. Similarly, black children exposed to environmental tobacco smoke exhibited significantly higher serum cotinine concentrations than white children after taking into account the number of smokers in the house (Pattishall *et al.*, 1985). In a study of Mexican Americans, more than one in five individuals who reported smoking less than ten cigarettes per day had higher than expected ratios of serum cotinine levels to daily cigarette consumption (Pérez-Stable *et al.*, 1990). Although this led the authors to conclude that these persons were under-reporting their cigarette consumption, the possibility of a genetic cause for this anomaly was discussed.

Ethnic differences in nicotine disposition may have far-reaching consequences, not only because the results may be difficult to rationalize if serum cotinine levels do not match those expected on the basis of their reported tobacco consumption (Idle, 1989) but also since they may be related to racial differences in cancer incidence (Satariono and Swanson, 1988).

(c) Site of metabolism

Although it is generally accepted that for most drugs and other xenobiotics the liver is the primary site of metabolism, several other organs have been shown to have xenobiotic metabolizing ability. The ability of the lung to metabolize nicotine is of particular interest since this is the organ primarily exposed to nicotine in the form of tobacco smoke.

In a study in which the metabolism of nicotine was investigated in the isolated rat liver and lung, the lung exhibited a considerably greater capacity to metabolize nicotine than the liver (Foth *et al.*, 1988). Furthermore, cotinine levels reach steady state in the isolated perfused rat lung which suggests that extensive further metabolism of cotinine is unlikely in this organ (Foth *et al.*, 1991). More than 90% of the total P450 content of the uninduced rabbit lung is composed of two forms, one which is identical to the major phenobarbitone inducible enzyme in the liver, CYP2B4, and the other, CYP4B1, which has been shown to *N*-oxidize a number of aromatic amines (Serabjit-Singh *et al.*, 1979). CYP2B4 is the major form responsible for the C-oxidation of nicotine in rabbit liver (McCoy *et al.*, 1989) and rabbit lung (Williams *et al.*, 1990). Although *S*-nicotine undergoes extensive C-oxidation in the rabbit lung, it is a poor substrate for a pulmonary form of the flavin-containing monooxygenase in this species (Williams *et al.*, 1990). Many other species have been shown to possess lung specific forms of flavin-containing monooxygenase (Tynes and Philpot, 1987) and inter-organ differences in the N-oxidation of nicotine have been reported. Furthermore, there is immunochemical and catalytic evidence to suggest that both rabbit lung and liver possess more than one form of flavin-containing monooxygenase (Lawton *et al.*, 1991).

Liver, lung and kidney from the guinea pig all have the ability to metabolize nicotine to cotinine and to *R*- and *S*-nicotine-1′-N-oxides, however hepatic tissue is the most active (Booth and Boyland, 1971). More *R*- than *S*-nicotine-1′-N-oxide is produced by all three tissues and this trend is further exaggerated as nicotine concentrations decrease.

In rats, mice and rabbits flavin-containing monooxygenase concentration is highest in hepatic tissue, although in male mice (but not female) enzyme concentration was nearly as high in lung and kidney as in liver (Dannan and Guengerich, 1982). A renal form of flavin-containing monooxygenase has been purified which appears to be immunologically related to the hepatic enzyme but exhibits some minor differences with respect to substrate specificity (nicotine data not available), amino acid composition, peptide analysis and NH_2-terminal sequence data (Venkatesh *et al.*, 1991).

Although rat liver flavin-containing monooxygenase appears to be positively regulated by testosterone, the hormonal regulation of flavin-containing monooxygenase activity in rats appears to be tissue specific. Hypophysectomy reduces both hepatic and pulmonary flavin-containing monooxygenase activity in male rats but elevates this activity in the kidney (Lemoine *et al.*, 1991).

In the rat there is evidence to suggest that tissue specific sex differences in nicotine metabolism exist. In mature (100-day-old) Long Evans and Wistar rats hepatic nicotine metabolizing activity is greater in males than in females whereas renal activity is greater in females than in males (Kyerematen *et al.*, 1988a).

(d) Disease

As noted previously (Svensson, 1987; Kyerematen and Vesell, 1991), there is very little information regarding the effect of disease on the pharmacokinetics of nicotine. By following basic pharmacokinetic principles, it is reasonable to suggest that for a drug such as nicotine which exhibits hepatic blood flow dependent elimination, diseases which alter blood flow to the liver, such as cirrhosis, acute viral hepatitis and congestive heart failure, may cause substantial alterations in nicotine pharmacokinetics.

In a unique study with respect to the effect of disease on the metabolism of nicotine, the ratio of cotinine to nicotine 1′-N-oxide in 24 h urine samples from smoking patients with cancer of the urinary bladder were compared to those in healthy smoking controls (Gorrod *et al.*, 1974) The ratio was seen to be significantly higher in cancer patients than in the control group (Figure 11.2); however, whether this was a consequence of the disease and/or its treatment or indicates a metabolic predisposition to smoking-induced bladder cancer has yet to be established.

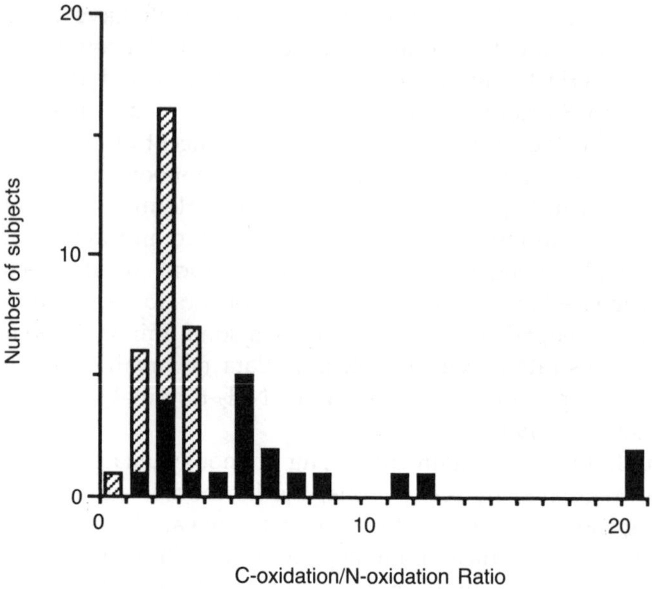

Figure 11.2 Distribution of cotinine/nicotine 1′-N-oxide ratios between a control group (n = 21; hatched bars) and patients with bladder cancer (n = 23). (Drawn from the data of Gorrod *et al.*, 1974.)

11.4.2 EXOGENOUS INFLUENCES ON THE METABOLISM OF NICOTINE

The exogenous influences on the metabolism of nicotine occur as a consequence of exposure of the individual to drugs and other xenobiotics including alcohol and tobacco, and also as a result of the dietary habits of the individual. The exogenous influences can thus be considered environmental in nature.

(a) Exposure to drugs and other xenobiotics

(i) Alcohol

Heavy cigarette smoking is often accompanied by heavy consumption of alcoholic beverages (Maletzky and Klotter, 1974; Griffiths *et al.*, 1976) and several studies have established that cancers of the head and neck are significantly increased in men who are heavy users of both alcohol and tobacco (Rothman and Keller, 1972; Schmidt and Popham, 1981). Since smokers have the ability to regulate fairly constant plasma nicotine levels, it has been suggested that the increased tobacco usage associated with heavy drinking may be due to compensation for increased nicotine metabolism in the presence of alcohol. Evidence to support the assumption that ethanol is an inducer of nicotine metabolism comes from a study by Adir and co-workers (1980) in which ethanol was orally administered chronically to rats for a period of 12.5 d. After a single intravenous dose of [methyl-^{14}C]-nicotine, this pretreatment resulted in significantly lower plasma levels of total radioactivity and significantly lower plasma levels of nicotine and cotinine than in control animals. Furthermore the apparent volume of distribution and total plasma clearance of nicotine and the apparent volume of distribution and rate of production of cotinine were all significantly increased by alcohol pretreatment. After chronic administration of ethanol to hamsters for 28 d, although hepatic microsomes demonstrated significantly higher activity with respect to the N-oxidation of nicotine when compared to controls, there was no significant difference in C-oxidation (McCoy and DeMarco, 1986).

In contrast, in studies where acute doses of alcohol were administered, alcohol appears to have an inhibitory effect on the metabolism of nicotine (Schüppel and Domdey-Bette, 1987; Domdey-Bette and Schüppel, 1987). Total plasma clearance of nicotine was significantly decreased but the plasma concentrations of cotinine were unaffected by alcohol in rabbits. Furthermore with increasing doses of alcohol, the rate of CO_2 exhalation by rats after the administration of [methyl-^{14}C]-nicotine declined. In the isolated perfused rat liver, alcohol has been shown to substantially decrease the clearance of nicotine, and in rat liver homogenate alcohol was shown to inhibit the conversion of nicotine to cotinine.

Recent studies using purified rabbit cytochromes P450 and cDNA-expressed human cytochromes P450 have demonstrated the ability of CYP2E1 to mediate the C-oxidation of nicotine (McCoy *et al.*, 1989; Flammang *et al.*, 1992). This ethanol inducible form of P450 in humans was seen to be one-third as active as the most active form CYP2B6, whereas in the rabbit the ethanol inducible form was only one-fifth as active as CYP2B4 or CYP2C3, the most active forms in this species. The ability of the ethanol inducible form of P450 in humans to metabolize nicotine to the iminium ion may have important toxicological implications for those individuals who are both heavy smokers and heavy users of alcohol, since this electrophilic metabolite can bind covalently to cellular macromolecules.

(ii) Drugs and chemicals

The effects of cigarette smoking, and thus exposure to nicotine, on the pharmacokinetics of many commonly used therapeutic agents have been reviewed in detail (Jusko, 1978; Miller, 1989, 1990). In contrast there is only a limited number of drugs whose influence on the pharmacokinetics of nicotine have been investigated. Phenobarbitone is probably the most commonly studied drug known to perturb the metabolism of nicotine and the effects of this inducing agent on the metabolism of nicotine have been studied in man and many other species.

Studies *in vivo* in the phenobarbitone pretreated rat demonstrate markedly increased (nine-fold) hepatic first-pass extraction of nicotine after oral administration but only a two-fold increase in the clearance of intravenously administered nicotine when compared to controls (Foth *et al.*, 1990). The authors suggested that this difference was probably due to extrahepatic metabolism of nicotine. In male stumptailed macaques, phenobarbitone pretreament accelerated the rates of production of eight distinct nicotine metabolites but not those of nornicotine and nicotine-1′-N-oxide (Seaton *et al.*, 1991).

Phenobarbitone pretreatment of mice resulted in enhanced metabolism of nicotine to cotinine as determined by the relative amounts of urinary nicotine and cotinine (Stålhandske, 1970). Although studies in isolated perfused rat liver and lung have demonstrated that phenobarbitone pretreatment causes a marked increase in both the hepatic and the pulmonary elimination of nicotine, tissue dependent degrees of induction are seen (Rüdall *et al.*, 1987; Foth *et al.*, 1991). When compared to control values, nicotine clearance by the lung was increased two-fold and that by the liver eight-fold in pretreated animals (Foth *et al.*, 1988). The biliary excretion of nicotine and its metabolites from the isolated perfused rat liver are increased almost three-fold by phenobarbitone-pretreatment (Foth *et al.*, 1988).

Phenobarbitone pretreated human hepatocytes obtained from non-smoking obese individuals demonstrated a two-fold increase in the ability to convert nicotine to cotinine when compared to untreated controls (Kyerematen *et al.*, 1990a). Hepatic microsomes from rabbits (McCoy *et al.*, 1989), mice (Stålhandske, 1970), rats (Nakayama *et al.*, 1982) and hamsters (McCoy and DeMarco, 1986) pretreated with phenobarbitone show an increase in the rate of C-oxidation of nicotine when compared to controls. Similarly phenobarbitone significantly induces nicotine metabolism in hepatocytes isolated from rats, guinea pigs and mice, however an increased production of cotinine was not observed in pretreated hamster hepatocytes (Kyerematen *et al.*, 1990a). Kyerematen and co-workers (1990a) suggest that this apparent lack of induction may be due to further metabolism of cotinine in this species. There is evidence to suggest that phenobarbitone produces a ten-fold increase in the elimination half-lives and the clearance values of cotinine in the isolated perfused rat liver. In contrast, in the isolated perfused rat lung, phenobarbitone does not induce further metabolism of cotinine (Foth *et al.*, 1991).

The use of phenobarbitone as an inducing agent has shed some light on the individual cytochromes P450 which are involved in the conversion of nicotine to cotinine. Studies in rabbits (McCoy *et al.*, 1989), rats (Nakayama *et al.*, 1982) and guinea pigs (Nakayama *et al.*, 1985) have shown that phenobarbitone inducible hepatic forms of cytochrome P450 have considerable nicotine C-oxidation ability. Induction of sex specific forms of cytochromes P450 may account for the greater nicotine metabolic activity in pretreated male rat hepatocytes when compared with the same from females (Kyerematen *et al.*, 1990a). This sex difference in nicotine metabolism was not observed in untreated animals.

The N-oxidation of nicotine appears to be either unaffected by phenobarbitone pretreatment as seen in stumptailed macaques (Seaton *et al.*, 1991), in hamster liver microsomes (McCoy and DeMarco, 1986) and rat hepatocytes (Kyerematen *et al.*, 1990a) or inhibited by administration of this barbiturate as seen in conscious rats (Foth *et al.*, 1990), rabbit microsomes (McCoy *et al.*, 1989) and guinea pig hepatocytes (Kyerematen *et al.*, 1990a).

Pretreatment of rats with 3-methylcholanthrene or β-naphthoflavone did not produce any change in the specific activity of hepatic nicotine oxidase (Nakayama *et al.*, 1982). Similar pretreatment had no significant effect on the rate of nicotine elimination or the rate of cotinine formation in either the isolated perfused rat liver or lung (Foth *et al.*, 1991). Although inducers of cytochromes P450 of the 1A subfamily (previously known as cytochrome P448) appear not to have any pronounced effects on the disposition of nicotine in hepatic microsomes or isolated organs from the rat, hepatocytes from β-naphthoflavone pretreated male rats show a significantly greater ability to produce both nornicotine and nicotine-1′-N-oxide and a significantly greater total amount of nicotine metabolized than control hepatocytes (Kyerematen *et al.*, 1990a). Although there were no significant differences in nicotine metabolism in hepatocytes from pretreated hamsters and guinea pigs when compared with controls, the use of hepatocytes from mice pretreated with β-naphthoflavone increased the formation of both nicotine-1′-N-oxide and 3′-hydroxycotinine and reduced the amount of unchanged nicotine remaining at the end of the incubation period. Hepatocytes from male rats pretreated with Arochlor 1254 were shown to have an increased ability to produce both cotinine and nornicotine when compared to control hepatocytes and also to increase the total amount of nicotine metabolized.

Using guinea pig microsomes fortified with partially purified aldehyde oxidase, metyrapone or 1-octylamine inhibited *S*-cotinine formation by 90% but had no effect on the production of the N′-oxide. In contrast, the flavin-containing monooxygenase inhibitor methimazole inhibited all N′-oxide production but had no effect on the production of *S*-cotinine (Nakayama *et al.*, 1987). In a study in which the effect of various chemical inhibitors on the metabolism of nicotine was determined in rabbit lung microsomes, N-hydroxyamphetamine totally inhibited the formation of nornicotine and the iminium ion and inhibited N-oxidation by 57%, whereas α-methylbenzylaminobenzotriazole (a suicide substrate which is specific for CYP2B4 and CYP1A1 in rabbit lung) totally inhibited production of the iminium ion and inhibited N-oxidation by 41% but had no effect on the formation of nornicotine (Williams *et al.*, 1990). A specific inhibitor of CYP2B4, norbenzphetamine, reduced the formation of the iminium ion by 84% and the N-oxide by 59% but reduced the production of nornicotine only slightly. 1-Octylamine was seen to inhibit 100% of the metabolism to nornicotine and the iminium ion but only slightly inhibited the N-oxidation of nicotine.

In healthy male volunteers, the H_2-antagonist cimetidine has been shown to decrease the total and the metabolic clearances of nicotine by 30 and 27%, whereas after ranitidine the clearances were decreased by 10 and 7%, respectively (Bendayan *et al.*, 1990). Since cimetidine, and to a lesser extent ranitidine, are known to inhibit the cytochrome P450 mediated metabolism of many drugs (Somogyi and Muirhead, 1987), inhibition of nicotine metabolism was suggested as the likely cause of the altered nicotine pharmacokinetics. Further evidence to support the notion that cimetidine inhibits the metabolism of nicotine was obtained from a study in stumptailed macaques in which this H_2-antagonist increased the amounts of nicotine excreted unchanged, and prolonged the half-lives of nicotine and four of its metabolites (Seaton *et al.*, 1991). Since smokers are known to regulate their nicotine intake to obtain fairly constant plasma nicotine levels, the reduction in the total clearance of nicotine in man by cimetidine prompted Bendayan and co-workers (1990) to propose the investigation of the use of cimetidine as a smoking reduction or cessation aid.

The antithyroid agent carbimazole, a known inhibitor of the flavin-containing monooxygenase, has been shown to reduce the N-oxidation of nicotine administered in the form of a chewing gum preparation to healthy volunteers (Cholerton *et al.*, 1988). These observations in man with cimetidine and carbimazole raise questions as to the effects of other commonly used therapeutic agents on the pharmacokinetics of nicotine.

(iii) Tobacco

The ability of tobacco smoking to alter the pharmacokinetics of many therapeutically important compounds is well established and in most cases appears to be due to induction of drug metabolizing enzymes (Jusko, 1978; Miller, 1989, 1990). Similarly, there is evidence to suggest that the metabolism of nicotine is accelerated in smokers when compared to nonsmokers. In a study in which habituated smokers and naive smokers smoked a cigarette containing a standardized amount of nicotine, the mean plasma nicotine $t_{1/2}\beta$ was significantly shorter in the smokers than the nonsmokers (Kyerematen *et al.*, 1983). When [^{14}C]-nicotine was administered intravenously, the elimination half-life of nicotine was shorter in smokers than in nonsmokers (Kyerematen *et al.*, 1982, 1983). Furthermore the elimination rate constant of nicotine was elevated and the volume of distribution was diminished in smokers (Kyerematen *et al.*, 1982). Since clearance and area under the plasma concentration–time curve were unaffected by smoking, Kyerematen and co-workers (1982) concluded that smoking exerts an inductive effect on nicotine metabolism which is independent of the decrease in volume of distribution. Since the elimination half-life of cotinine has also been shown to be shorter in smokers than in nonsmokers, this suggests that the further metabolism of cotinine is also induced in cigarette smokers (Kyerematen *et al.*, 1982). A recent study has demonstrated faster urinary and plasma elimination of both nicotine and cotinine in smokers when compared with nonsmokers, with the opposite trend observed for the urinary excretion of a metabolite postulated to be cotinine $\Delta^{2',3'}$-enamine (Kyerematen *et al.*, 1990b). The half-life of cotinine in smokers has also been shown to be less than half that in individuals exposed to environmental tobacco smoke (Sepkovic *et al.*, 1986; Haley *et al.*, 1989). This may have important toxicological consequences since components of tobacco smoke will remain in the bodies of nonsmokers longer than for smokers. The observations on the effect of cigarette smoking on the metabolism of nicotine *in vivo* have also been seen in human hepatocytes. Production of cotinine was significantly elevated and nicotine remaining unchanged was significantly reduced when nicotine metabolism by hepatocytes from smokers was compared to those from nonsmokers (Kyerematen *et al.*, 1990a). Hepatocytes from ex-smokers also exhibited a trend towards induction of nicotine metabolism.

When a single intravenous dose of nicotine was administered to male and female smokers and nonsmokers, the urinary recoveries of nicotine and cotinine from male smokers fell into two groups (Beckett *et al.*, 1971b). In one group there was lower recovery of both alkaloids than was observed in male nonsmokers and in the other there was a similar recovery of nicotine but a higher recovery of cotinine than in male nonsmokers. Although female smokers were seen to excrete less nicotine than female nonsmokers, they excreted more cotinine. The occurrence of two distinct groups within the male smokers and the sex differences which were seen may indicate both genetic and sex differences in the expression of cytochromes P450 in these groups.

Since cigarette smoking appears to have an inductive effect on nicotine metabolism, Lee and co-workers (1987) hypothesized that tobacco abstinence would slow nicotine metabolism. However, in a study in which smokers abstained from smoking for one week, both total and non-renal clearance of nicotine were increased (Lee *et al.*, 1987). This suggests that, in addition to the inductive effect of cigarette smoking on nicotine metabolism, there is also an inhibitory effect. There is evidence to suggest that pretreatment with nicotine can inhibit the metabolism of nicotine in mouse liver homogenate (Stålhandske and Slanina, 1970).

Unlike cigarette smoking, chronic tobacco intake in the form of pipe smoking and snuff appears not to induce the metabolism of nicotine (Kyerematen *et al.*, 1983). This may be due to the fact that the pipe smoker and the snuff dipper are either not exposed or exposed to a lesser extent to polycyclic aromatic hydrocarbons.

Species differences appear to exist with respect to the effects of nicotine pretreatment on nicotine metabolism *in vivo*. In a study in which mice, rats, guinea pigs, rabbits and dogs were pretreated with nicotine for 10 d, variable degrees of induction of nicotine oxidase activity was observed in all species except the guinea pig, in which there was no effect (Nakashima *et al.*, 1980).

(b) Dietary habits

Nicotine has been shown to be present in *Solanaceae* plants such as potatoes and also in black tea leaves and instant tea preparations (Sheen, 1988; Davis *et al.*, 1991). It appears that this is not simply residual nicotine from its use as an insecticide but that this alkaloid is actually synthesized in these plants since it is found in the leaf, stem and root as well as the fruit (Sheen, 1988). Irrespective of the source, the ingestion of nicotine associated with the consumption of such foodstuffs has major implications since it may compromise the use of urinary cotinine as a biomarker of exposure to environmental tobacco smoke particularly in vegans and vegetarians (Idle, 1989).

In a study in which the effect of a high protein meal on nicotine disposition was determined in healthy smoking subjects, 50 min after the meal hepatic blood flow was seen to increase by 31% (Lee *et al.*, 1989). When the same meal was consumed during a steady state infusion of nicotine, an 18% decrease in blood levels of nicotine were detected when compared to control values. Although the study demonstrates that food can accelerate the metabolism of nicotine, the authors suggest that the small decrease in blood nicotine levels is unlikely to be the primary reason for the increased desire to smoke a cigarette after a meal.

In a study in which the metabolic interaction between caffeine and cigarette smoking was investigated, a slight tendency towards greater cigarette consumption and higher plasma nicotine levels were observed during administration of low-dose caffeine when compared to the no caffeine situation (Brown and Benowitz, 1989). It has been suggested that the undefined association between nicotine disposition and caffeine consumption may be elucidated if total plasma clearance measurements were made (Kyerematen and Vesell, 1991).

11.5 EXCRETION OF NICOTINE

11.5.1 URINARY pH AND FLOW RATE

As for the absorption of nicotine, the excretion of this alkaloid is also pH dependent. At a pH of 5.5 or less, nicotine is almost completely ionized and thus cannot be reabsorbed through the renal tubule. In a group of subjects with acidic urinary pH (4.8 ± 0.2), the percentage of an oral dose of nicotine recovered unchanged ranged from 9.2 to 27.9 (Beckett *et al.*, 1972). Under similar conditions, an average of 23% was excreted in the urine unchanged when nicotine was administered intravenously (Rosenberg *et al.*, 1980). When urinary pH was made alkaline (>7) in these individuals, unchanged nicotine recovery was reduced to 2%, since un-ionized nicotine is readily reabsorbed through the renal tubule. In the presence of acidic wine (pH 4.9) and less acidic urine (pH 6.7), renal clearance may vary from 43–598 ml/min (Benowitz *et al.*, 1983b). Such differences in renal clearance result in greater total clearances for nicotine under conditions of acidic when compared with alkaline urine. Furthermore since the volume of distribution of nicotine appears to be unaffected by urinary pH, terminal half-lives for nicotine are shorter under acidic conditions compared with alkaline conditions (Rosenberg *et al.*, 1980). Urinary acidification has been shown to increase cigarette consumption in man and to increase nicotine self-administration in rats (Benowitz and Jacob, 1985; Lang, 1980).

Since the renal clearance of nicotine is so highly dependent on urinary pH, many volunteers are subjected to rigorous acidification or alkalinization procedures prior to and during trials of nicotine metabolism. There is some dispute as to whether the urinary excretion of nicotine is affected by urinary flow rate. Beckett and co-workers (1971a) concluded that urinary excretion was unaffected by urine flow, however Feyerabend and Russell (1978) found it to be proportional to the rate of urine flow under acidic conditions.

There is evidence to suggest that the urinary excretion of cotinine may be dependent on both urinary pH and flow rate. Triggs (1967) found that when urine was adjusted to an acid pH the excretion of cotinine was affected by changes in urinary output. Furthermore, Beckett and co-workers (1972) demonstrated that urinary cotinine excretion was enhanced under conditions of acidic urinary pH and was futher increased by a high urine flow rate. In contrast, Matsukura *et al.* (1979) found that there was no correlation between urinary cotinine excretion, urinary pH or urine flow after smoking cigarettes.

After its oral administration, the rate of excretion and urinary recovery of the highly water soluble weakly basic, nicotine-1′-N-oxide, appear to be unaffected by changes in either urinary pH or urine flow rate (Jenner *et al.*, 1973b).

11.5.2 DRUGS

Since the renal elimination of nicotine involves glomerular filtration, renal tubular secretion and urinary pH dependent tubular reabsorption, all three mechanisms represent potential sites at which drugs can act to affect the urinary excretion of nicotine.

Since the urinary excretion of nicotine is pH dependent, urinary pH is often adjusted for *in vivo* studies of the metabolism of nicotine. Urinary alkalinization is achieved by the administration of sodium bicarbonate and acidification by the administration of ammonium chloride using dosage regimens suitable for the particular demands of the study.

In a study in which the effect of cimetidine and ranitidine on the hepatic and renal elimination of nicotine in humans was investigated, mean renal clearances of nicotine were decreased by 53 and 47% respectively (Bendayan *et al.*, 1990). Since both H_2-antagonists are basic drugs eliminated by active tubular secretion, Bendayan *et al.* (1990) suggest that an interaction with nicotine might occur at this site which results in reduced renal clearance of nicotine. Although at normal urinary pH the effect of this interaction would be negligible, in individuals with acid urine it would affect the pharmacokinetics of nicotine since under such conditions renal clearance significantly contributes to the total clearance of nicotine.

11.6 IMPLICATIONS OF INTER-INDIVIDUAL VARIABILITY IN NICOTINE PHARMACOKINETICS

From the above, it is clear that the processes of absorption, distribution, metabolism and excretion can be affected by a myriad of host and environmental factors which will, in turn, result in individual variability in the pharmacokinetics of a dose of nicotine. Although differences within a group of individuals in all four processes affect the pharmacokinetics of nicotine, it seems likely that it is those factors which determine the way in which an individual metabolizes nicotine that will have the greatest implications for certain aspects of the pharmacology and the toxicology of nicotine.

Exposure to nicotine containing products is commonly assessed by measurement of cotinine in biological fluids such as blood, urine and saliva (Sepkovik and Haley, 1985; Barlow and Wald, 1988; Van Vunakis *et al.*, 1989). Inter-individual variability in the metabolism of nicotine to cotinine and cotinine to further metabolites, as determined by genetic and environmental influences on the expression of xenobiotic metabolizing enzymes, casts considerable doubt on the use of cotinine as either a qualitative or quantitative biomonitor of exposure to nicotine. A chemically reactive intermediate formed during the conversion of nicotine to cotinine has been shown to bind covalently to hepatic and pulmonary macromolecules (Shigenaga *et al.*, 1988). It has been suggested that metabolism dependent alkylation of biomacromolecules of this type may contribute to some of the reported irreversible lesions caused by long-term exposure to tobacco products. If this is the case, then an individual's ability to produce the reactive metabolite and also to further metabolize this reactive species, will determine the extent of covalent binding and the toxicity of nicotine in that individual (Gorrod, 1981).

Further evidence to suggest that inter-individual variability in the metabolism of nicotine may contribute to its toxicity comes from a study of individuals with smoking-induced bladder cancer (Gorrod *et al.*, 1974). As previously mentioned, the ability to produce cotinine appeared to predispose individuals to this disease.

An area of nicotine pharmacology which has so far been beyond the scope of this chapter is pharmacodynamics. It is now recognized that nicotine is the tobacco derived compound which is responsible for the maintenance of the smoking habit. Although some individuals experience little success when attempting to kick this habit, others find the process relatively easy. Furthermore, certain individuals experience such unpleasant effects during their first exposure to cigarette smoke that they do not take up the habit, others never even attempt to smoke. Although the role of inter-individual variability in the pharmacokinetics of nicotine has yet to be determined in such individuals, this surely presents a fascinating area for future studies.

ACKNOWLEDGEMENTS

This work was supported by the Smokeless Tobacco Research Council, New York. Dr Cholerton is Bayer Lecturer in Pharmacogenetics.

REFERENCES

Abel, E.L. (1984) Smoking and pregnancy. *J. Psychoactive Drugs*, **16**, 327–8.

Adir, J., Wildfeurer, W. and Miller, R.P. (1980) Effect of ethanol pretreatment on the pharmacokinetics of nicotine in rats. *J. Pharmacol. Exp. Ther.*, **212**, 274–9.

Al-Waiz, M., Ayesh, R., Mitchell, S.C., *et al.* (1987a) A genetic polymorphism of the N-oxidation of trimethylamine in humans. *Clin. Pharmacol. Ther.*, **42**, 588–94.

Al-Waiz, M., Ayesh, R., Mitchell, S.C., *et al.* (1987b) The relative importance of N-oxidation and N-demethylation in the metabolism of trimethylamine in man. *Toxicology*, **43**, 117–21.

Andersen, B.D., Ng, K.J., Iams, J.D. *et al.* (1982) Cotinine in amniotic fluids from passive smokers. *Lancet*, **i**, 791–2.

Ayesh, R., Al-Waiz, M., Crothers, M.J. *et al.* (1988) Deficient nicotine N-oxidation in two sisters with trimethylaminuria. *Br. J. clin. Pharmacol.*, **25**, 664–5P.

Barlow, R.D. and Wald, N.J. (1988) Use of urinary cotinine to estimate exposure to tobacco smoke. *J. Am. Med. Assoc.*, **259**, 1808.

Beckett, A.H., Gorrod, J.W. and Jenner, P. (1970) Absorption of (−)-nicotine-1′-N-oxide in man and its reduction in the gastrointestinal tract. *J. Pharm. Pharmacol.*, **22**, 722–3.

Beckett, A.H., Gorrod, J.W. and Jenner, P. (1971a) The analysis of nicotine-1′-N-oxide in urine in the presence of nicotine and cotinine and its application to the study of *in vivo* nicotine metabolism in man. *J. Pharm. Phamacol.*, **23**, 55S–61S.

Beckett, A.H. Gorrod, J.W. and Jenner, P. (1971b) The effect of smoking on nicotine metabolism *in vivo* in man. *J. Pharm. Pharmacol.*, **23**, 62S–7S.

Beckett, A.H., Gorrod, J.W. and Jenner, P. (1972) A possible relation between pKa_1 and lipid solubility and the amounts excreted in urine of some tobacco alkaloids given to man. *J. Pharm. Pharmacol.*, **24**, 115–20.

Bendayan, K., Sullivan, J.T., Shaw, C. *et al.* (1990) Effect of cimetidine and ranitidine on the hepatic and renal elimination of nicotine in humans. *Eur. J. Clin. Pharmacol.*, **38**, 165–9.

Benowitz, N.L. and Jacob, P. III (1984) Daily intake of nicotine during cigarette smoking. *Clin. Pharmacol. Ther.*, **35**, 499–504.

Benowitz, N.L. and Jacob, P. III (1985) Nicotine renal excretion rate influences nicotine intake during cigarette smoking. *J. Pharmacol. Exp. Ther.*, **234**, 153–5.

Benowitz, N.L., Jacob, P. III, Jones, R.T. *et al.* (1982a) Interindividual variability in the metabolism and cardiovascular effects of nicotine in man. *J. Pharmacol. Exp. Ther.*, **221**, 368–72.

Benowitz, N.L., Kuyt, F. and Jacob, P. III (1982b) Circadian blood nicotine concentrations during cigarette smoking. *Clin. Pharmacol. Ther.*, **32**, 758–64.

Benowitz, N.L., Hall, S.M., Herning, R.I. *et al.* (1983a) Smokers of low-yield cigarettes do not consume less nicotine. *N. Engl. J. Med.*, **309**, 139–142.

Benowitz, N.L., Kuyt, F. and Jacob, P. III (1983b) Cotinine disposition and effects. *Clin. Pharmacol. Ther.*, **309**, 139–142.

Benowitz, N.L., Jacob, P. III, Kozlowski, L.T. *et al.* (1986a) Influence of smoking fewer cigarettes on exposure to tar, nicotine and carbon monoxide. *N. Engl. J. Med.*, **315**, 1310–3.

Benowitz, N.L., Jacob, P. III, Yu, L. *et al.* (1986b) Reduced tar, nicotine and carbon monoxide exposure while smoking ultra low- but not low yield cigarettes. *J. Am. Med. Assoc.*, **256**, 241–6.

Benowitz, N.L., Jacob, P. III, and Savanapridi, C. (1987a) Determinants of nicotine intake while chewing nicotine polacrilex gum. *Clin. Pharmacol. Ther.*, **41**, 467–73.

Benowitz, N.L., Lake, T., Keller, K.H., *et al.* (1987b) Prolonged absorption with development of tolerance to toxic effects after cutaneous exposure to nicotine. *Clin. Pharm. Ther.*, **42**, 119–20.

Booth, J. and Boyland, E. (1970) Enzymic oxidation of (−)-nicotine by guinea pig tissues *in vitro*. *Biochem. Pharmacol.*, **19**, 733–42.

Booth, J. and Boyland, E. (1971) The metabolism of nicotine into two optically-active stereoisomers of nicotine-1′-oxide by animal tissues *in vitro* and by cigarette smokers. *Biochem. Pharmacol.*, **20**, 407–15.

Brandänge, S. and Lindblom, L. (1979) The enzyme 'aldehyde oxidase' is an iminium oxidase. Reaction with nicotine $\Delta^{1',(5')}$ iminium ion. *Biochem. Biophys. Res. Commun.*, **91**, 991–6.

Brodfuehrer, J.I. and Zannoni, V.G. (1986) Flavin-containing monooxygenase and ascorbic acid deficiency. *Biochem. Pharmacol.*, **36**, 3161–7.

Brown, C.P. and Benowitz, N.L. (1989) Caffeine and cigarette smoking. Behavioural, cardiovascular and metabolic interactions. *Pharmacol. Biochem. Behav.*, **43**, 565–70.

Cholerton, S., Ayesh, R., Idle, J.R., *et al.* (1988) The pre-eminence of nicotine N-oxidation and its diminution after carbimazole administration. *Br. J. clin. Pharmacol.*, **26**, 652–3P.

Cholerton, S., Idle, M.E., Vas, A., *et al.*, (1992a) Comparison of a novel thin-layer chromatographic-fluoresence detection method with a spectrofluorometric method for the determination of 7-hydroxycoumarin in human urine. *J. Chromatogr.*, **575**, 325–30.

Cholerton, S., Daly, A.K. and Idle, J.R. (1992b) The role of individual human cytochromes P450 in drug metabolism and clinical response. *Trends Pharmacol. Sci.* **13**, 434–439.

Damani, L.A., Pool, W.P., Crooks, P.A. *et al.* (1988) Stereoselectivity in the N-oxidation of nicotine isomers by flavin-containing monooxygenase. *Mol. Pharmacol.*, **33**, 702–5.

Dannan, G.P. and Guengerich, F.P. (1982) Immunochemical comparison and quantitation of microsomal flavin-containing monooxygenase in various hog, mouse, rat, rabbit, dog and human tissues. *Mol. Pharmacol.*, **22**, 787–94.

Davis, R.A., Stiles, M.F., DeBethizy, J.D., *et al.* (1991) Dietary nicotine: A source of urinary cotinine. *Food Chem. Toxicol.*, **29**, 821–7.

Dolphin, C., Shephard, E.A., Povey, S. *et al.* (1991) Cloning, and chromosomal mapping of a human flavin-containing monooxygenase (FMOI) primary sequence. *J. Biol. Chem.*, **266**, 12 379–85.

Domdey-Bette, A. and Schüppel, R. (1987) Ethanol inhibits the cotinine formation in the isolated perfused rat liver and *in vitro*, in *International Symposium on Nicotine*, (Eds M.F. Rand and K. Thurau) ICSU, Gold Coast, pp38–9.

Duffel, M.W., Graham, J.M. and Ziegler, D.M. *et al.* (1981) Changes in dimethylaniline, N-oxidase activity of mouse liver and kidney induced by steroid sex hormones. *Mol. Pharmacol.*, **19**, 134–9.

Faulkner, J.M. (1933) Nicotine poisoning by absorption through the skin. *J. Am. Med. Assoc.*, **100**, 1664–5.

Feyerabend, C. and Russell, M.A.H. (1978) Effect of urinary pH and nicotine excretion rate on plasma nicotine during cigarette smoking and chewing nicotine gum. *Br. J. Clin. Pharmacol.*, **5**, 293–7.

Flammang, A.M., Gelboin, H.V., Aoyama, T. *et al.* (1992) Nicotine metabolism by cDNA-expressed cytochrome P450s. *Biochem. Archives*, **8**, 1–8.

Forrester, L.M., Henderson, C.J., Glancey, M.J. *et al.* (1992) Relative expression of cytochromes P450 in human liver and association with the metabolism of drugs and xenobiotics. *Biochem. J.*, **281**, 359–68.

Foth, H., Rüdell, U., Ritter, G. *et al.* (1988) Inhibitory effect of nicotine on benzo(a)pyrene elimination and marked pulmonary metabolism of nicotine in isolated perfused rat lung. *Klin. Wochenschr.*, **665**, 98–104.

Foth, H., Walther, U.I. and Kahl, G.F. (1990) Increased hepatic nicotine elimination after phenobarbitone induction in the conscious rat. *Toxicol. Appl. Pharmacol.*, **105**, 382–92.

Foth, H., Looschen, H., Neurath, H., *et al.* (1991) Nicotine metabolism in the isolated perfused lung and liver of phenobarbital- and benzoflavone-treated rats. *Arch. Toxicol.*, **65**, 68–72.

Garcia-Estrada, H. and Fischman, C.M. (1977) An unusual case of nicotine poisoning. *Clin. Toxicol.*, **10**, 391–3.

Gehlbach, S.H., Williams, W.A., Perry, L.D. *et al.* (1974) Green tobacco sickness: an illness of tobacco harvesters. *J. Am. Med. Assoc.*, **229**, 1880–3.

Gonzalez, F.J. (1992) Human cytochromes P450: Problems and prospects. *Trends Pharmacol. Sci.*, **13**, 346–352.

Gori, G.B., Benowitz, N.L. and Lynch, C.J. (1986) Mouth versus deep airway absorption of nicotine in cigarette smokers. *Pharmacol. Biochem. Behav.*, **25**, 1181–4.

Gorrod, J.W. (1981) Covalent binding as an indicator of drug toxicity, in *Testing for Toxicity* (Ed J.W. Gorrod) Taylor & Francis Ltd., London, pp. 77–93.

Gorrod, J.W. and Hibberd, A.R. (1982) The metabolism of nicotine-$\Delta^{1'(5')}$-iminimum ion, *in vivo* and *in vitro*. *Eur. J. Drug Metab. Pharmacokin.*, **7**, 293–8.

Gorrod, J.W. and Jenner, P. (1975) The metabolism of tobacco alkaloids, in *Essays in Toxicology vol 6*, Academic Press Inc., San Francisco, pp. 35–78.

Gorrod, J.W., Jenner, P., Keysell, G.R. *et al.* (1974) Oxidative metabolism of nicotine by cigarette smokers with cancer of the urinary bladder. *J. Natl. Cancer Inst.*, **52**, 1421–4.

Griffiths, R.R., Bigelaw, G.E. and Liebson, I. (1976) Facilitation of human tobacco self-administration by ethanol: a behavioural analysis. *J. Exp. Anal. Behav.*, **25**, 279–92.

Haley, N.J. and Hoffman, D. (1985) Analysis for nicotine and cotinine in hair to determine cigarette smoker status. *Clin. Chem.*, **31**, 1598–600.

Haley, N.J., Sepkovic, D.W. and Hoffman, D. (1989) Elimination of cotinine from body fluids: disposition in smokers and nonsmokers. *Am. J. Pub. Health*, **79**, 1046–8.

Hammond, D.K., Bjercke, R.J., Langone, J.J. *et al.* (1991) Metabolism of nicotine by rat liver cytochromes P450. Assessment utilizing monoclonal antibodies to nicotine and cotinine. *Drug Metab. Disp.*, **19**, 804–8.

Hanes, P.J., Schuster, G.S. and Lubas, S. (1991) Binding, uptake and release of nicotine by human gingival fibroblasts. *J. Periodontol.*, **62**, 147–52.

Hecht, S.S. and Hoffman, D. (1988) Tobacco-specific nitrosamines an important group of carcinogens in tobacco and tobacco smoke. *Carcinogenesis*, **9**, 875–84.

Hellberg, D. and Nilsson, S. (1988) Smoking and cancer of the ovary (letter) *N. Engl. J. Med.*, **318**, 782–3.

Hellberg, D., Nilsson, S., Haley, N.J. *et al.* (1988) Smoking and cervical intraepithelial neoplasia: Nicotine and cotinine in serum and cervical mucus in smokers and nonsmokers. *Am. J. Obstet. Gynecol.*, **158**, 190–3.

Herning R.I., Jones, R.T., Benowitz, N.L. *et al.* (1983) How a cigarette is smoked determines blood nicotine levels. *Clin. Pharmacol. Ther.*, **33**, 84–90.

Hlavica, P. and Kehl, M. (1977) Studies on the mechanism of hepatic microsomal N-oxide formation. The role of cytochrome P450 and mixed-function amine oxidase in the N-oxidation of N,N-dimethylaniline. *Biochem. J.*, **164**, 487–96.

Idle, J.R. (1989) Titrating exposure to tobacco smoke using cotinine – a minefield of misunderstandings. *J. Clin. Epidemiol.*, **43**, 313–71.

Jacob, P. III, Benowitz, N.L., Copeland, J.R. *et al.* (1988) Disposition kinetics of nicotine and cotinine enantiomers in rabbits and Beagle dogs. *J. Pharm. Sci.*, **77**, 396–400.

Jenner, P. and Gorrod, J.W. (1973) Comparative *in vitro* metabolism of some tertiary N-methyl tobacco alkaloids in various species. *Res. Commun. Chem.*, **6**, 829–43.

Jenner, P., Gorrod, J.W. and Beckett, A.H. (1973a). Species variation in the metabolism of *R*-(+)- and *S*-(−)- nicotine by αC- and N-oxidation *in vitro*. *Xenobiotica*, **3**, 573–80.

Jenner, P., Gorrod, J.W. and Beckett, A.H. (1973b) The absorption of nicotine-1′-N-oxide and its reduction in the gastrointestinal tract in man. *Xenobiotica*, **3**, 341–9.

Jusko, W.J. (1978). Role of tobacco smoking in pharmacokinetics. *J. Pharmacokinet. Biopharm.*, **6**, 7–39.

Kalow, W., Goedde H.V. and Agarwal, D.P. (Eds) (1986) *Ethnic Differences in Reactions to Drugs and Xenobiotics*, Alan R Liss, New York.

Kimura, T., Kodama, M and Nagata, C. (1983) Purification of mixed function amine oxidase from rat liver microsomes. *Biochem. Biophys. Res. Commun.*, **110**, 640–5.

Klein, A.E. and Gorrod, J.W. (1978) Age as a factor in the metabolism of nicotine. *Eur. J. Drug Metab. Pharmacokinet.*, **3**, 51–8.

Klein, A.E. and Gorrod, J.W. (1979) Metabolism of nicotine in cigarette smokers during pregnancy. *Eur. J. Drug Metab. Pharmacokinet.*, **2**, 87–93.

Kyerematen, G.A. and Vesell, E.S. (1991) Metabolism of nicotine. *Drug Metab. Rev.*, **23**, 3–41.

Kyerematen, G.A., Damiano, M.D., Dvorchik, B.H. *et al.* (1982) Smoking-induced changes in nicotine disposition: Application of a new HPLC assay for nicotine and its metabolites. *Clin. Pharmacol. Ther.*, **32**, 769–80.

Kyerematen, G.A., Dvorchik, B.H. and Vesell, E.S. (1983) Influence of different forms of tobacco intake on nicotine elimination in man. *Pharmacology*, **26**, 205–9.

Kyerematen, G.A., Owens, G.F., Chattopadhyay, B. *et al.* (1988a) Sexual dimorphism of nicotine metabolism and distribution in the rat. Studies *in vivo* and *in vitro*. *Drug Metab. Dispos.*, **16**, 823–8.

Kyerematen, G.A., Taylor, L.H., deBethizy, J.D. *et al.* (1988b) Pharmacokinetics of nicotine and twelve metabolites in the rat. Application of a new radiometric HPLC assay. *Drug Metab. Dispos.*, **16**, 125–9.

Kyerematen, G.A., Morgan, M., Warner, G. *et al.* (1990a) Metabolism of nicotine by hepatocytes. *Biochem. Pharmacol.*, **40**, 1747–56.

Kyerematen, G.A., Morgan, M.L., Chattopadhyay, G. *et al.* (1990b) Disposition of nicotine and eight metabolites in smokers and nonsmokers. Identifica-tion in smokers of two metabolites that are longer lived than cotinine. *Clin. Pharmacol. Ther.*, **48**, 641–51.

Lang, W.J. (1980) Factors influencing the self-administration of nicotine and other drugs by rats. *Proceedings of the Australian Physiological and Pharmacological Society*, **11**, 33–6.

Lavoie, F.W. and Harris, T.M. (1991) Fatal nicotine ingestion. *J. Emerg. Med.*, **9**, 133–6.

Lawton, M.P., Kronbach, T., Johnson, E.F. *et al.* (1991) Properties of expressed and native flavin-containing monooxygenase: Evidence of multiple forms in rabbit liver and lung. *Mol. Pharmacol.*, **40**, 692–8.

Lee, B.L., Benowitz, N.L. and Jacob, P. III (1987) Influence of tobacco abstinence on the disposition kinetics and effects of nicotine. *Clin. Pharmacol. Ther.*, **41**, 474–9.

Lee, B.L., Jacob, P. III, Jarvik, M.E. *et al.* (1989) Food and nicotine metabolism. *Pharmacol. Biochem. Behav.*, **9**, 621–5.

Lemoine, A., Williams, D.E., Cresteil, T. *et al.* (1991) Hormonal regulation of microsomal flavin-containing monooxygenase: Tissue-dependent expression and substrate specificity. *Mol. Pharmacol.*, **40**, 211–7.

Lou, Y.C. (1990) Differences in drug metabolism polymorphism between Orientals and Caucasians. *Drug Metab. Rev.*, **22**, 451–475.

Luck, W. and Nau, H. (1984a). Exposure of the foetus, neonate and nursed infant to nicotine and cotinine from maternal smoking. *N. Engl. J. Med.* **311**, 672–3.

Luck, W. and Nau, H. (1984b) Nicotine and cotinine concentrations in serum and milk of nursing smokers. *Br. J. clin. Pharmacol.*, **18**, 9–15.

Mahgoub, A., Idle, J.R., Dring, L.G. *et al.* (1977) Polymorphic hydroxylation of debrisoquine in man. *Lancet*, **ii**, 584–6.

Maletzky, B.M. and Klotter, J. (1974) Smoking and alcoholism. *Am. J. Psychiat.*, **131**, 445–52.

Malizia, E., Andreucci, G., Alfani, F. *et al.* (1983) Acute intoxication with nicotine alkaloids and cannabinoids in children from ingestion of cigarettes. *Human Toxicol.*, **2**, 315–6.

Manoguerra, A.S. and Freeman, D. (1983) Acute poisoning from ingestion of *Nicotiana glauca*. *J. Toxicol. Clin. Toxicol.*, **19**, 861–4.

Marks, M.J., Stitzel, J.A. and Collins, A.C. (1989) Genetic influences on nicotine responses. *Pharmacol. Biochem. Behav.*, **33**, 667–78.

Matsukura, S., Sakamoto, N., Seino, Y., *et al.* (1979) Cotinine excretion and daily cigarette smoking in habituated smokers. *Clin. Pharmacol. Ther.*, **25**, 555–61.

McCoy, G.D., and DeMarco, G.J. (1986) Characterization of hamster liver nicotine metabolism II. Differential effects of ethanol or phenobarbital-pretreatment on microsomal N and C oxidation. *Biochem. Pharmacol.*, **35**, 4590–2.

McCoy, G.D., DeMarco, G.J. and Koop, D.R. (1989) Microsomal nicotine metabolism: A comparison of relative activities of six purified rabbit cytochrome P450 isozymes. *Biochem. Pharmacol.*, **38**, 1185–8.

McCracken, N.W., Cholerton, S. and Idle, J.R. (1992) Cotinine formation by cDNA-expressed human cytochromes P450. *Med. Sci. Res.*, **20**, 877–8.

McNabb, M.E., Ebert, R.V. and McCusker, K. (1982) Plasma nicotine levels produced by chewing nicotine gum. *J. Am. Med. Assoc.*, **248**, 865–8.

Miller, L.G. (1989) Recent developments in the study of the effects of cigarette smoking on clinical pharmacokinetics and clinical pharmacodynamics. *Clin. Pharmacokinet.*, **17**, 90–108.

Miller, L.G. (1990) Cigarettes and drug therapy: pharmacokinetic and pharmacodynamic considerations. *Clin. Pharm.*, **9**, 125–35.

Mosier, H.D. and Jansons, R.A. (1972) Distribution and fate of nicotine in the rat fetus. *Teratology*, **6**, 303–11.

Murphy, P.J. (1973) Enzymatic oxidation of nicotine to nicotine $\Delta^{1'(5')}$ iminium ion. *J. Biol. Chem.*, **248**, 2796–800.

Nakashima, T., Nakanishi, Y., Tanaka, Y. *et al.* (1980) Studies on nicotine metabolism in various animals (3). *J. Nara Med. Ass.*, **31**, 212–7.

Nakayama, H. (1988) Nicotine metabolism in mammals, *Drug Metab. Drug Interact.*, **6**, 95–121.

Nakayama, H., Nakashima, T. and Kurogochi, Y. (1982). Heterogeneity of hepatic nicotine oxidase. *Biochim. Biophys. Acta*, **715**, 254–7.

Nakayama, H., Nakashima, T. and Kurogochi, Y. (1985) Cytochrome P450 dependent nicotine oxidation by liver microsomes of guinea pigs. *Biochem. Pharmacol.*, **34**, 2281–6.

Nakayama, H., Fujihara, S., Nakashima, T. *et al.* (1987) Formation of two major nicotine metabolites in livers of guinea-pigs. *Biochem. Pharmacol.*, **36**, 4313–7.

Nwosu, C.G. and Crooks, P.A. (1988) Species variation and stereoselectivity in the metabolism of nicotine enantiomers. *Xenobiotica*, **18**, 1361–72.

Oberst, B.B. and McIntyre, R.A. (1953) Acute nicotine poisoning. *Pediatrics*, **11**, 338–40.

Pattishall, E.N., Strope, G.L., Etzel, R.A. *et al.* (1985) Serum cotinine as a measure of tobacco smoke exposure in children. *Am. J. Dis. Child.*, **139**, 1101–4.

Pérez-Stable, E.J., Marin, B.V.O., Marin, G. *et al.* (1990) Apparent underreporting of cigarette consumption among Mexican American smokers. *Am. J. Pub. Health*, **80**, 1057–61.

Petersen, D.R., Norris, K.J. and Thompson, J.A. (1984). A comparative study of the disposition of nicotine and its metabolites in three inbred strains of mice. *Drug Metab. Dispos.*, **12**, 725–31.

Peterson, L.A., Trevor, A.J. and Castagnoli, N. (1987) Stereochemical studies on the cytochrome P450 catalysed oxidation of (S)-nicotine to the (S)-nicotine $\Delta^{1'(5')}$-iminium species. *J. Med. Chem.*, **30**, 249–54.

Petrakis, N.L., Gruenke, L.D., Beelen, T.C. *et al.* (1978) Nicotine in breast fluid of non-lactating women. *Science*, **199**, 303–5.

Relling, M.V., Aoyama, T., Gonzalez, F.J. *et al.* (1990) Tolbutamide and mephenytoin hydroxylation by human cytochrome P450s in the 2C subfamily. J. Pharmacol. Exp. Ther., **252**, 442–7.

Romkes, M., Faletto, M.B., Blaisdell, J.A. *et al.*, (1991) Cloning and expression of complementary DNAs for multiple members of the human cytochrome P450IIC subfamily. *Biochemistry*, **30**, 3247–55.

Rose, J.E., Jarvik, M.E. and Rose, K.D. (1984) Transdermal administration of nicotine. *Drug and Alcohol Dependence*, **13**, 209–13.

Rosenberg, J., Benowitz, N.L., Jacob, P. III *et al.* (1980) Disposition kinetics and effects of intravenous nicotine. *Clin. Pharmacol. Ther.*, **28**, 517–22.

Rothman, K. and Keller, A. (1972) The effect of joint exposure to alcohol and tobacco on risk of cancer of the mouth and pharynx. *J. Chron. Dis.*, **25**, 711–16.

Rowland, M. and Tozer, T.N. (1980). *Clinical Pharmacokinetics. Concepts and Applications*, Lea & Febiger, Philadelphia.

Rüdell, U., Foth, H. and Kahl, G.F. (1987). Eightfold induction of nicotine elimination in perfused rat liver by pretreatment with phenobarbital. *Biochem. Biophys. Res. Commun.*, **148**, 192–8.

Russell, M.A.H. and Feyerabend, C. (1978) Cigarette smoking: dependence on high nicotine boli. *Drug Metab. Rev.*, **8**, 29–57.

Russell, M.A.H., Jarvis, M.D. and Feyerabend, C. (1980) A new age for snuff. *Lancet*, **i**, 474–5.

Satariono, W.A. and Swanson, G.M. (1988) Racial differences in cancer incidence: the significance of age-specific patterns. *Cancer*, **62**, 2640–53.

Schievelbein, H. (1984) Nicotine resorption and fate, in *Nicotine and the tobacco smoking habit* (Ed D.J. Balfour), Pergamon Press, Oxford, pp. 1–15.

Schmidt, W. and Popham, R.E. (1981) The role of drinking and smoking in mortality from cancer and other causes in male alcoholics. *Cancer*, **47**, 1031–41.

Schmiterlöw, C.G., Hansson, E., Andersson, G., *et al.* (1967) Distribution of nicotine in the central nervous system. *Ann. NY Acad. Sci.*, **142**, 2–14.

Schüppel, R. and Domdey-Bette, A. (1987) The pharmacokinetics of nicotine and the $^{14}CO_2$-exhalation rate from radiolabelled nicotine as affected by ethanol, in *International Symposium on Nicotine*, (Eds M.F. Rand and K. Thurau), ICSU, Gold Coast, pp 34–5.

Scott, J. and Poffenbarger, P.L. (1979) Pharmacogenetics of tolbutamide metabolism in humans. *Diabetes*, **28**, 41–51.

Seaton, M., Kyerematen, M., Morgan, M. *et al.* (1991) Nicotine metabolism in stumptailed macaques, *Macaca arctoides. Drug Metab. Disp.*, **19**, 946–54.

Sepkovic, D.W. and Haley, N.J. (1985) Biomedical applications of cotinine quantitation in smoking related research. *Am. J. Public Health*, **75**, 663–5.

Sepkovic, D.W., Haley, N.J. and Hoffmann, D. (1986) Elimination from the body of tobacco products by smokers and passive smokers. *J. Am. Med. Assoc.*, **256**, 863.

Sepkovic, D.W., Haley, N.J., Axelrad, C.M. *et al.* (1986) Short term studies on the *in vivo* metabolism of N-oxides of nicotine in rats. *J. Toxicol. Environ. Health*, **18**, 205–14.

Serabjit-Singh, C.J., Wolf, C.R. and Philpot, R.M. (1979) The rabbit pulmonary monooxygenase system. Immunochemical and biochemical characterization of enzyme components. *J. Biol. Chem.*, **254**, 9901–7.

Sheen, S.J. (1988) Detection of nicotine in foods and plant materials. J. *Food Sci.*, **53**, 1572–3.

Shigenaga, M.K., Trevor, A.J. and Castagnoli, N. (1988) Metabolism-dependent covalent binding of (*S*)-[5-^{3}H] nicotine to liver and lung microsomal macromolecules. *Drug Metab. Dispos.*, **16**, 397–402.

Sieber, S.M. and Fabro, S. (1971) Identification of drugs in the preimplanted blastocyte and in the plasma, uterine secretions and urine of the pregnant rabbit. *J. Pharmacol. Exp. Ther.*, **178**, 65–75.

Slanina, P. and Stålhandske, T. (1977) *In vitro* metabolism of nicotine in livers of ageing mice. *Arch. Int. Pharmacodyn. Ther.*, **226**, 258–62.

Somogyi, A. and Muirhead, M. (1987) Pharmacokinetic interactions of cimetidine. *Clin. Phamacokinet.*, **12**, 321–56.

Stålhandske, T. (1970) Effects of increased liver metabolism of nicotine on its uptake, elimination and toxicity in mice. *Acta Physiol. Scand.*, **80**, 222–34.

Stålhandske, T. and Slanina, P. (1970) Effect of nicotine treatment on the metabolism of nicotine in the mouse liver *in vitro*. *Acta Pharmacol. et Toxicol.*, **28**, 75–80.

Stålhandske, T., Slanina, P., Tjälve, H. *et al.* (1969) Metabolism *in vitro* of ^{14}C-nicotine in livers of foetal, newborn and young mice. *Acta Pharmacol. Toxicol.*, **27**, 363–80.

Suzuki, K., Horiguchi, T., Comas-Urrutia, A.C. *et al.* (1974) Placental transfer and distribution of nicotine in the pregnant Rhesus monkey. *Am. J. Obstet. Gynecol.*, **199**, 253–4.

Svensson, C.K. (1987) Clinical pharmacokinetics of nicotine. *Clin. Pharmacokin.*, **12**, 30–40.

Tjälve, H., Hansson, E. and Schmiterlöw, C.G. (1968) Passage of ^{14}C nicotine and its metabolites into mice foetuses and placentae. *Acta Pharmacol. Toxicol.*, **26**, 539–55.

Travell, J. (1960) Absorption of nicotine from various sites. *Ann. NY Acad. Sci.*, **90**, 13–30.

Triggs, E.J. (1967) Ph.D Thesis, University of London

Tsujimoto, A., Nakashima, T., Tanino, S. *et al.* (1975) Tissue distribution of [^{3}H] nicotine in dogs and rhesus monkeys. *Toxicol. Appl. Pharmacol.*, **32**, 21–31.

Tynes, R.E. and Philpot, R.M. (1987) Tissue and species dependent expression of multiple forms of mammalian microsomal flavin-containing monooxygenase. *Mol. Pharmacol.*, **31**, 569–74.

Tynes, R.E., Sabourin, P.J. and Hodgson, E. (1985) Identification of distinct hepatic and pulmonary forms of microsomal flavin-containing monooxygenase in the mouse and rabbit. *Biochem. Biophys. Res. Commun.*, **126**, 1069–75.

Van Vunakis, H., Langone, J.J. and Milunsky, A. (1974) Nicotine and cotinine in the amniotic fluid of smokers in the second trimester of pregnancy. *Am. J. Obs. Gynecol.*, **120**, 64–6.

Van Vunakis, H., Tashkin, D.P., Rigas, B., *et al.* (1989) Relative sensitivity and specificity of salivary and serum cotinine in identifying tobacco-smoking status of self-reported nonsmokers and smokers of tobacco and/or marijuana. *Arch. Envir. Health*, **44**, 53–8.

Venkatesh, K., Levi, P.E. and Hodgson, E. (1991) The flavin-containing monooxygenase of mouse kidney. A comparison with the liver enzyme. *Biochem. Pharmacol.*, **42**, 1411–20.

Vesell, E.S. (1982) On the significance of host factors that affect drug disposition. *Clin. Pharmacol. Ther.*, **31**, 1–7.

Wagenknecht, L.E., Cutter, G.R., Haley, N.J. *et al.* (1990) Racial difference in serum cotinine levels among smokers in the coronary artery risk development in (young) adults study. *Am. J. Public Health*, **80**, 1053–6.

Wilkinson, G.R., Guengerich, F.P. and Branch, R.A. (1989) Genetic polymorphism of S-mephenytoin hydroxylation. *Pharmacol. Ther.*, **43**, 53–76.

Williams, D.E., Shigenaga, M.K. and Castagnoli, N. (1990) The role of cytochromes P450 and flavin-containing monooxygenase in the metabolism of (*S*)-nicotine by rabbit lung. *Drug Metab. Dispos.*, **18**, 418–28.

Yamano, S., Nhamburo, P.T., Aoyama, T. *et al.* (1989) cDNA cloning and sequence and cDNA-directed expression of human P450 IIB1: identification of a normal and two variant cDNAs derived from the *CYP2B* locus of chromosome 19 and differential expression of the IIB mRNAs in human liver. *Biochemistry*, **28**, 7342–8.

Yamano, S., Tatsuno, J. and Gonzalez, F.J. (1990) The *CYP2A3* gene product catalyses coumarin 7-hydroxylation in human liver microsomes. *Biochemistry*, **29**, 1322–9.

Physiologically based pharmacokinetic modelling of nicotine 12

S.L. Schwartz, M.R. Gastonguay, D.E. Robinson and N.J. Balter

12.1 PHYSIOLOGICALLY BASED PHARMACOKINETIC MODELLING

Pharmacokinetic models are used to describe or predict the time-dependent changes in drug concentration in blood and tissues, i.e., concentration × time (C × t) curves. There are two basic approaches to pharmacokinetic modelling. In the first, empirical or compartmental models are derived using curve-fitting programs to provide a mathematical description of C × t data. In the second, physiologically based pharmacokinetic (PBPK) models are constructed based on anatomical and physiological data from animals and humans, incorporating physicochemical data on a specific drug or chemical. Unlike empirical or compartmental pharmacokinetic models, PBPK models are physiologically realistic, bringing a biological basis to the mathematical description of a compound's pharmacokinetics. Until relatively recently, the utility of PBPK modelling had been limited by the computational intensity of the PBPK approach and the availability of adequate computer hardware and software. In recent years, however, PBPK modelling has become a practical tool for the study of a variety of pharmacological and toxicological problems (Bischoff *et al.*, 1971; Gerlowski and Jain, 1983).

PBPK models provide a basis for computer simulation of the dynamic behaviour of molecules as they are absorbed, distributed and eliminated by the body. The theory and applications of PBPK modelling are well

Nicotine and Related Alkaloids: Absorption, distribution, metabolism and excretion. Edited by J.W. Gorrod and J. Wahren. Published in 1993 by Chapman & Hall, London. ISBN 0 412 55740 1

documented in the literature (Andersen, 1989; D'Souza and Boxenbaum, 1988; Gerlowski and Jain, 1983), so only a brief description is given here.

A PBPK model is comprised of physiologically relevant tissue compartments with real volumes, connected by real blood flows. For reasonably lipophilic compounds, distribution within each compartment is limited by blood flow, although membrane-limited processes can also be accounted for within the framework of a PBPK model. The other parameters influencing the instantaneous compartmental concentration of the compound, including the tissue/plasma partition coefficient, tissue clearance, tissue binding, or saturable metabolism can be incorporated into the model. Linear or non-linear mass-balanced differential equations, which take all of these parameters into account, describe the time-dependent change in the amount of compound in each compartment; the time-dependent changes in blood reflect the balance of the changes occurring in each of the tissue compartments. Model simulations require the simultaneous numerical solution of a series of these equations, which together describe the entire system. For the models described in this chapter, simulations were accomplished using the PBPK specific software CMATRIX*, which was developed in this laboratory for the design of multicompartmental physiological pharmacokinetic models.

12.2 POTENTIAL UTILITY OF A PBPK MODEL OF NICOTINE

Nicotine is a particularly good candidate for PBPK modelling. There is an extensive literature on nicotine chemistry, pharmacology and pharmacokinetics in both animals and man, which can be applied to the development of PBPK models. Several models have been published (deBethizy and Andersen, 1990; Gabrielsson and Bondesson, 1987; Schwartz *et al.*, 1987). A PBPK model of nicotine in the rat has been developed and recently expanded to include a pharmacodynamic model of nicotinic receptor dynamics (deBethizy and Andersen, 1990; Plowchalk and deBethizy, 1991; Plowchalk *et al.*, 1992). Gabrielsson and Bondesson (1987) used PBPK models of nicotine and cotinine to study the clearance and distribution of these compounds in the rat.

The nicotine PBPK models established in our laboratory have been applied to several areas of interest related to nicotine pharmacokinetics and pharmacodynamics, including prediction of target tissue concentrations, coupling of pharmacokinetic and pharmacodynamic models, and development of biomonitoring protocols for the retrospective estimation

* CMATRIX software was developed by R.T. Ball, O. Skliar and S.L. Schwartz, Georgetown University School of Medicine, Washington, DC 20007, USA.

of nicotine exposure. These applications are discussed in greater detail in this chapter. They illustrate the major value of PBPK modelling, i.e., its ability to enable researchers to make *a priori* predictions of drug pharmacokinetics (and response) and to provide a means for predicting answers to experimental 'what if?' scenarios.

12.3 DEVELOPMENT OF THE PBPK MODEL OF NICOTINE AND COTININE IN HUMANS

A description of the creation of the nicotine, cotinine and nicotine–cotinine models provides an illustration of the PBPK model development process (Schwartz *et al.*, 1987; Robinson *et al.*, 1992). These models were created by first establishing and validating PBPK models for nicotine and cotinine separately. The PBPK models for nicotine (NIC model) and cotinine (COT model) were then linked to form a PBPK model which simultaneously described the disposition of nicotine and its biotransformation product, cotinine (NIC-COT model, Figure 12.1).

The specific requirements of the models include the designation of

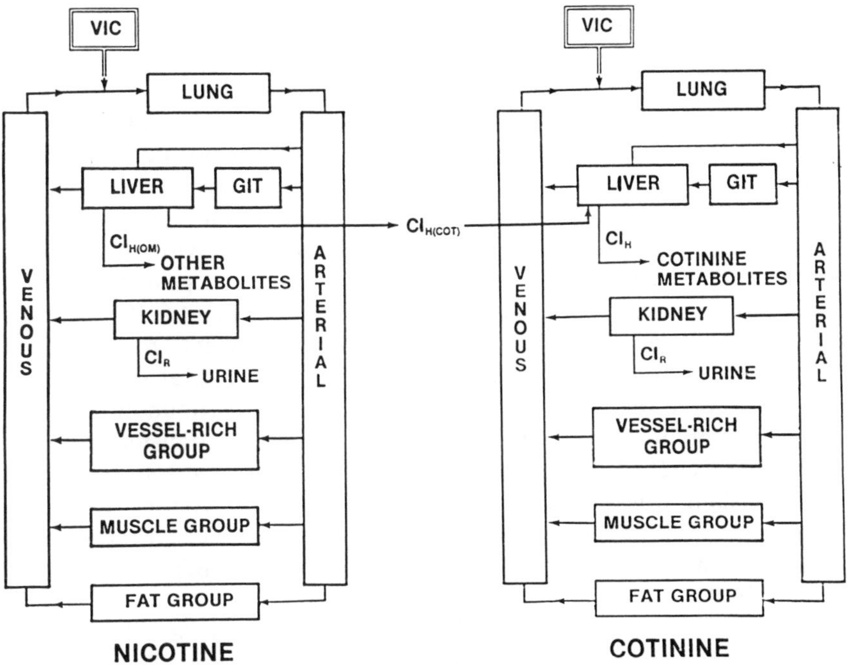

Figure 12.1 Conceptual physiologically based pharmacokinetic model for nicotine and cotinine. (From Robinson *et al.*, 1992.)

compartments, compartmental volumes and blood flow rates, tissue/blood partition coefficients and elimination rates. Based on the physicochemical properties of nicotine and cotinine, flow-limited distribution was assumed. Decisions regarding compartment designation were based on tissue perfusion (because of the flow-limited distribution), ease of biological sampling of the tissue compartment, and tissue function. Specific compartments for liver and kidney were included to provide compartments for the metabolism and excretion of nicotine and cotinine. The gastrointestinal tract tissue (GIT) was included as a compartment because flow/tissue volume relationships are more accurately depicted by recognizing the sequential blood flow between the gastrointestinal tract and liver, and to allow the model to accommodate oral dosing, if desired. Other high perfusion tissues, including the brain, heart, spleen, sex organs and adrenal glands were grouped into one compartment, the vessel-rich group (VRG). The muscle compartment included all medium perfusion tissues, while low perfusion tissues were represented by the fat compartment. The lung served as the circulatory link between arterial and venous blood and can be modified to simulate inhalation exposures to nicotine. All compartments are linked through arterial supply and venous outflow, except for the sequential blood flow from the gastrointestinal tract to the liver. Elimination occurs from both the liver and kidney compartments and is represented by hepatic and renal clearance, respectively.

Previous experience with PBPK models led to the observation that introduction of the dose directly into the venous blood compartment to simulate the intravenous administration of drugs is not physiologically accurate (Gastonguay *et al.*, 1990). This is because the model treats the dose as if it were instantaneously distributed throughout the venous blood compartment. As a result model simulations with this type of drug input result in an overestimation of the actual peak blood concentration (Figure 12.2). A venous infusion compartment (VIC) has been incorporated into the PBPK models to circumvent this problem (Robinson *et al.*, 1992). The VIC provides a physiological site for drug administration and allows for some systemic distribution prior to venous sampling. Drug administered via the VIC is transferred to the pulmonary circulation via a first-order process ($k = 4.1\ \text{min}^{-1}$). This rate constant was derived from regional flow and volume values. A volume of 1.5 l for the VIC was chosen based on published estimates of arm, heart and lung capacities in man (Abramson, 1967; Quiring, 1949; Rothe, 1984).

Generic NIC and COT models were developed to describe nicotine and cotinine pharmacokinetics in the general population. Therefore, the physiological parameters for each model compartment (volume and blood flow) were based on population values for a 70 kg man (Williams

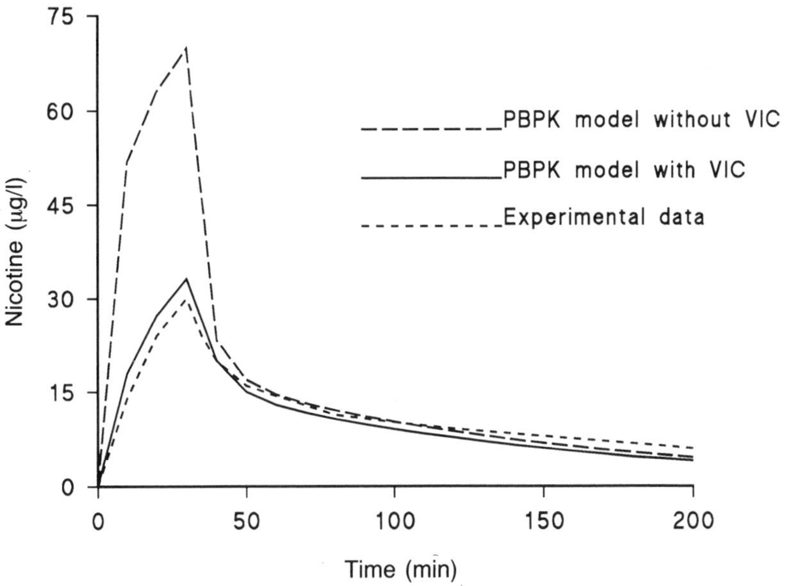

Figure 12.2 Effect of venous infusion compartment on NIC PBPK model simulation of experimental infusion of nicotine. Experimental data are the result of an intravenous nicotine infusion of 2 μg/kg/min for 30 min in smokers (Benowitz *et al.*, 1982). NIC PBPK model simulations are shown with (solid line) and without (long-dashed line) the incorporation of a venous infusion compartment.

and Leggett, 1989; Leggett and Williams, 1991; Davis and Mapleson, 1981). The compound-specific model parameters for nicotine and cotinine, including tissue/blood partition coefficients (R), and rates of drug metabolism and excretion were estimated as follows.

Initial values of R for nicotine and cotinine in each compartment were estimated based on steady-state distribution data for each of these compounds in animals (Benowitz, 1988; Gabrielsson and Bondesson, 1987). The initial estimates were adjusted empirically to achieve optimal simulations of experimental data. Model values for hepatic and renal clearances of nicotine and cotinine were based on an average of the reported literature values in man. NIC and COT model parameters are summarized in Table 12.1. Simulated concentrations of nicotine and cotinine in the venous and arterial compartments of the NIC and COT models are reported as μg/l whole blood. Whole blood concentrations are generally comparable to plasma and serum concentrations since plasma protein binding of both nicotine and cotinine is minimal and blood/plasma ratios for these compounds are near unity (Benowitz *et al.*, 1983, Schievelbein, 1982).

The NIC and COT PBPK models were validated by simulating pub-

Table 12.1 Standard parameters for nicotine–cotinine model

Organ/tissue compartment	Tissue volume (litres)	Blood flow (litres/min)	Partition coeffient[a]
Nicotine model[b]			
Arterial blood	1.4	6.1	1.0
Venous blood	4.0	6.1	1.0
Lung	0.6	6.1	2.0
GI tract	2.4	1.25	2.0
Liver	1.5	0.3	9.0
Kidney	0.3	1.25	15.0
Muscle	34.4	1.65	2.5
Fat	10.0	0.3	1.0
Vessel-rich tissues	1.55	1.35	3.0
Cotinine model[c]			
Arterial blood	1.4	6.1	1.0
Venous blood	4.0	6.1	1.0
Lung	0.6	6.1	1.0
GI tract	2.4	1.25	1.0
Liver	1.5	0.3	2.0
Kidney	0.3	1.25	2.0
Muscle	34.4	1.65	1.5
Fat	10.0	0.3	0.5
Vessel-rich tissues	1.55	1.35	1.5

[a]Ratio of tissue concentration/effluent venous concentration.
[b]Cl_H = 1.09 l/min (80% to cotinine; 20% to other metabolites); Cl_R = 0.17 l/min.
[c]Cl_H = 0.065 l/min; Cl_R = 0.010 l/min.

lished clinical pharmacokinetic experiments (Benowitz *et al.*, 1983, Curvall *et al.*, 1990a; DeSchepper *et al.*, 1987; Jacob *et al.*, 1988). Values for nicotine and cotinine clearance, volume of distribution and half-life were calculated from PBPK-simulated blood decay curves using both compartmental (Sebalt and Kreeft, 1987) and non-compartmental (Notari, 1987) methods. Compartmental analysis of NIC model simulations yielded a total systemic clearance of 1.56 l/min, an apparent volume of distribution of 207 l and an elimination half-life of 92 min. Non-compartmental analysis of the same simulated data resulted in a clearance of 1.26 l/min, a volume of 288 l and a half-life of 104 min. These parameters are well within the range of the published pharmacokinetic parameters for nicotine in man. The COT model simulated the published data equally well, based on comparison with average published pharmacokinetic parameters for cotinine. A clearance of 0.067 l/min, an apparent volume of distribution of 73 l and an elimination half-life of 12.6 h were determined by compartmental analysis of the COT model simulations. Non-compartmental analysis yielded nearly identical results.

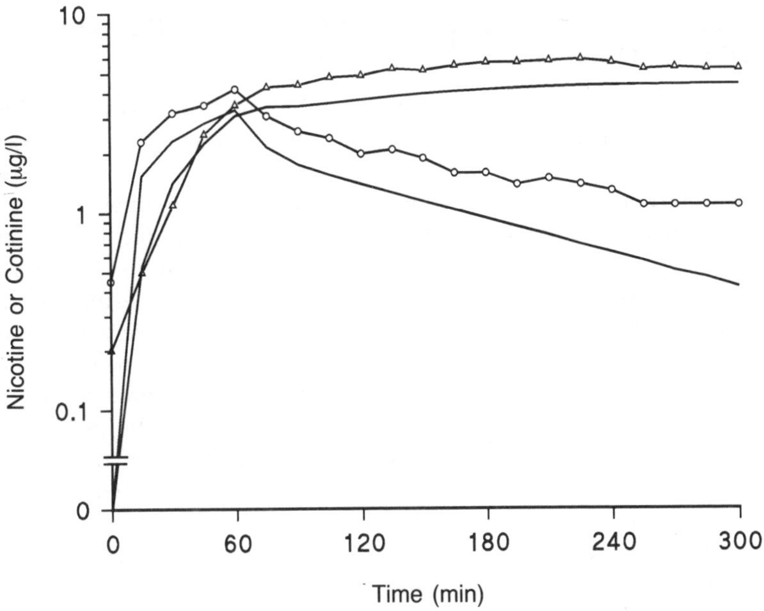

Figure 12.3 Experimental and simulated nicotine and cotinine levels following nicotine infusion. The NIC–COT model was used to simulate an intravenous infusion of nicotine (10 µg/min for 60 min) in smokers (Curvall *et al*, 1990b). Open circles and open triangles represent mean plasma levels of nicotine and cotinine, respectively; simulated levels are represented by solid lines (modified from Robinson *et al*, 1992).

A combined PBPK model of nicotine and cotinine (NIC–COT model) was created by joining the NIC and COT models through the liver compartment (Figure 12.1). Original NIC and COT model parameter values were left unchanged, however the total hepatic clearance of nicotine was divided between cotinine and other metabolites; the cotinine mass was transferred to the cotinine liver, and the mass of other metabolites was transferred to a metabolite holding compartment so that mass balance was maintained within the system. An estimated 70 to 85% of administered nicotine is converted to cotinine (Benowitz, 1988). Accordingly, the hepatic clearance of nicotine in the NIC–COT model was divided between clearance to cotinine and other metabolites in a ratio of 80:20.

The NIC–COT model was validated by simulating the dosing protocols of published studies in which nicotine was infused and plasma levels of nicotine and metabolically produced cotinine were determined (Curvall *et al.*, 1982, 1990b; Kyerematen *et al.*, 1990; Scherer *et al.*, 1988). The result of one of these simulations is shown in Figure 12.3.

12.4 APPLICATION OF A MULTI-METABOLITE PBPK MODEL OF NICOTINE TO EXPOSURE ASSESSMENT

PBPK models that incorporate both parent compound and metabolite(s) provide a tool for examining new approaches to retrospective exposure assessment. In the case of nicotine, for example, cotinine has been used as a biomarker of nicotine exposure from active smoking as well as exposure to environmental tobacco smoke. However, nicotine or cotinine biomonitoring can provide, at best, only semi-quantitative information about exposure, and has only limited ability to estimate dose, duration and time since termination of exposure. However, if biomonitoring includes not only nicotine and cotinine, but other nicotine metabolites as well, a more accurate estimate of the amount of nicotine exposure and time since last exposure to nicotine may be possible. To examine the feasibility of this approach, a PBPK model of nicotine and three metabolites has been developed (Gastonguay *et al.*, 1990).

The first step in the development of a multi-metabolite PBPK model of nicotine was to determine which nicotine metabolites would be appropriate biomarkers. Three metabolites of nicotine were selected based on favourable pharmacokinetics and the current availability of sensitive analytical assays for these compounds. Cotinine, a long-lived metabolite of nicotine with a half-life of approximately 10 to 20 h, has been the metabolite of choice for biomarker studies of nicotine exposure (Curvall and Enzell, 1986; Curvall *et al.*, 1990b; Galeazzi *et al.* 1985; Curvall *et al.*, 1990a). The relatively terminal metabolites nicotine-1′-N-oxide (NNO) and 3′-hydroxycotinine (3OHC) have elimination half-lives ranging from 2 to 4 and 5 to 8 h, respectively (Adlkofer *et al.*, 1988; Scherer *et al.*, 1988; Kyerematen, *et al.* 1990), making them good choices for a multi-metabolite study of nicotine exposure.

The multi-metabolite PBPK model of nicotine, cotinine, NNO and 3OHC was constructed by adding separate physiological flow models for NNO and 3OHC to the already validated NIC–COT PBPK model. This was accomplished by mathematically transferring the NNO produced in the liver of the nicotine flow system to the liver of the NNO system, and by similarly transferring the 3OHC produced in the liver of the cotinine system to the liver of the 3OHC system. The same tissue compartments used in the NIC–COT PBPK model were incorporated into the NNO and 3OHC PBPK models. Clearance values for 3OHC and NNO were obtained from published pharmacokinetic studies (Kyerematen *et al.*, 1990; Scherer *et al*, 1988). Tissue/blood partition coefficients were obtained from animal tissue distribution experiments or estimated based on the physicochemical properties of the metabolites. Parameters were empirically adjusted and the model was validated through simulation of clinical studies of nicotine exposure in

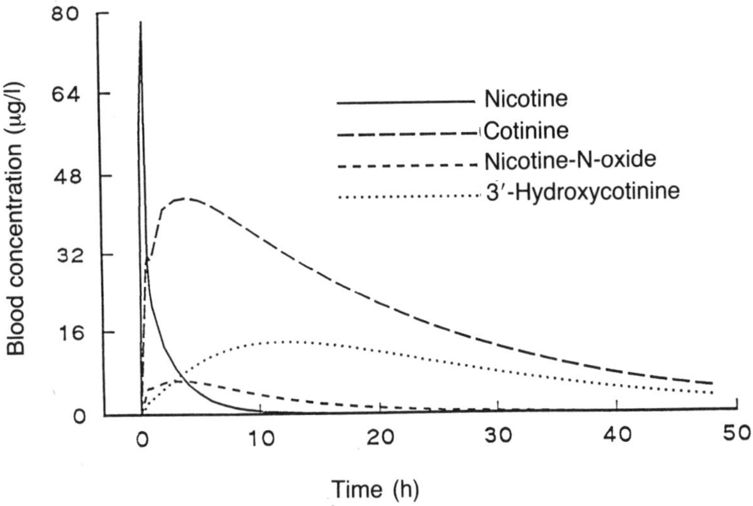

Figure 12.4 NIC multi-metabolite PBPK model simulation of blood levels of nicotine and three metabolites after nicotine exposure. Simulation of an intravenous infusion of nicotine with the multi-metabolite model resulted in the levels shown for nicotine, cotinine, nicotine-1′-N-oxide and 3′-hydroxycotinine.

which plasma levels of NIC and its metabolites were determined (Kyerematen *et al.*, 1990; Scherer *et al.*, 1988).

Preliminary simulations of nicotine exposure with the multi-metabolite PBPK model reveal a time-dependent pattern of parent and metabolite concentration in both blood and urine. It is clear from the results shown in Figure 12.4 that the relative concentrations of nicotine and the three metabolites in blood are unique at any point in time. The same can be said about the relative amounts of these four compounds found in the urine at any given time (Figure 12.5). Therefore, it is possible that one blood sample and/or a single urine collection would be sufficient to determine the time since last exposure to nicotine. The multi-metabolite model could also be used to simulate nicotine exposures of varying magnitudes and durations. If the resulting simulated concentration profiles predict unique markers of the amount, duration and/or time since nicotine exposure, this would provide a framework for the design of clinical studies to test protocols for improved exposure biomarker studies. Such studies would be far too complicated and expensive to conduct without the guidance derived from the PBPK modelling studies.

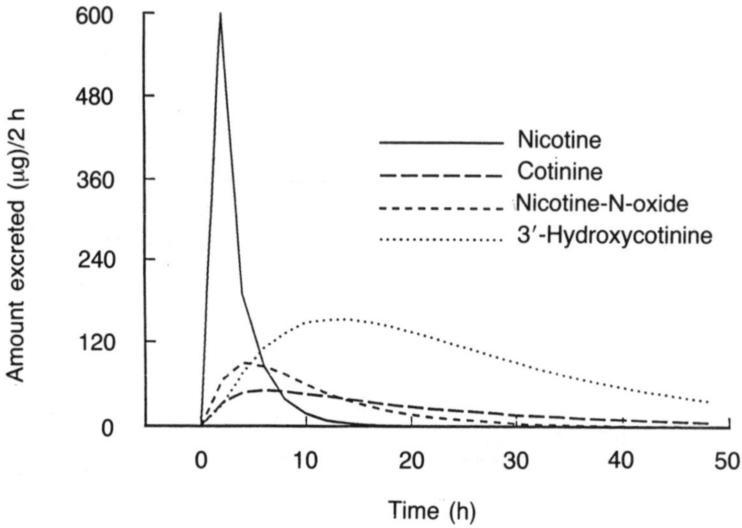

Figure 12.5 NIC multi-metabolite PBPK model simulation of urinary excretion of nicotine and three metabolites after nicotine exposure. Simulation of an intravenous infusion of nicotine with the multi-metabolite model resulted in urinary excretion of nicotine, cotinine, nicotine-1′-N-oxide and 3′-hydroxycotinine as shown. Results are expressed as amount of nicotine or metabolite excreted over a 2 h interval.

12.5 A LOW LEVEL NICOTINE EXPOSURE MODEL

Much of the data used to develop pharmacokinetic models of nicotine is derived from studies in smokers or studies designed to simulate smoking (Benowitz, 1982, 1988; Benowitz and Jacob, 1985; Feyerabend *et al.*, 1985), representing a pharmacokinetic database rich with information on exposure to relatively high levels of nicotine. Recently, studies designed to simulate the low level nicotine exposure similar to that associated with exposure to environmental tobacco smoke have suggested that the pharmacokinetic properties of nicotine may exhibit concentration dependence. At nicotine exposure levels simulating smoking, the elimination half-life of nicotine ranges from 1 to 4 h (Benowitz, 1988), but at low level exposure (peak nicotine plasma concentrations of approximately 1 to 2 μg/l) the observed elimination half-life of nicotine is significantly prolonged (Lewis *et al.*, 1990). One possible explanation for this observation is the existence of a high affinity, low capacity tissue binding compartment. Because the binding is low capacity, it influences the pharmacokinetics of nicotine only at low concentrations; at higher (active smoking) nicotine concentrations, the maximal binding

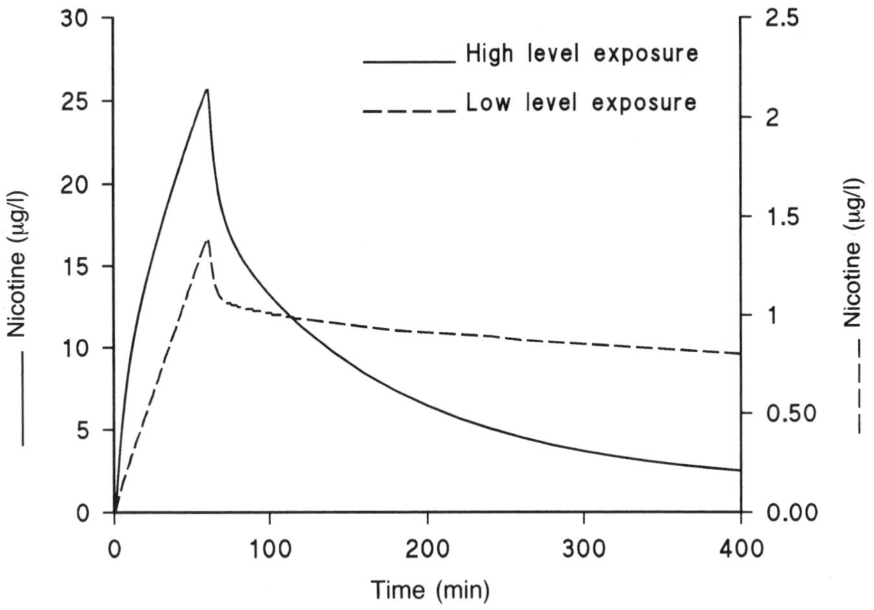

Figure 12.6 Simulation with a low level exposure model of nicotine. A low level exposure model was used to simulate blood levels of nicotine after both low (600 μg over 60 min, dashed line) and high (4200 μg over 60 min, solid line) level exposures.

capacity is overwhelmed and, therefore, there is no apparent effect on the pharmacokinetics.

To examine this possibility, a tissue binding compartment for nicotine was incorporated into the standard NIC PBPK model. The binding compartment was assumed to be anatomically associated with the lung in this model, but incorporation of binding into any high perfusion tissue would yield similar results. All other compartments and parameters were left unchanged. The accumulation of nicotine in the putative binding compartment was described by a saturable binding isotherm with a maximal binding capacity of 10 μg and a dissociation constant of 0.05 μg/l. Simulations with the nicotine binding compartment model illustrated a concentration dependent change in nicotine pharmacokinetics. Adjustments of model parameters and subsequent simulations of low level nicotine exposure resulted in elimination half-lives of up to 11.5 h, while high level exposure simulations with the same model resulted in an elimination half-life of 2.5 h (Figure 12.6). This is not conclusive evidence that a binding compartment for nicotine exists as described, but the model simulations provide a plausible hypothesis that could be tested experimentally.

12.6 PBPK AND NICOTINE PHARMACODYNAMICS

The majority of the published pharmacokinetic analyses of plasma C × t data following intravenous administration of nicotine to volunteers utilize classic compartmental (Kyerematen *et al.*, 1982, 1990; Porchet *et al.*, 1987, 1988) or model independent (Benowitz *et al.*, 1982; Feyerabend *et al.*, 1985) methods. Such methods do not provide information about tissue concentrations of nicotine and are thus limited in their ability to provide insight into the relationship between nicotine dose and target tissue concentration. PBPK models, on the other hand, simulate tissue as well as blood concentrations, providing information on the relationship between nicotine dose and target tissue concentration. Coupled with data from pharmacodynamic experiments, PBPK modelling could also describe the relationship between nicotine dose and pharmacological response.

Until recently, the majority of pharmacokinetic–pharmacodynamic (PK–PD) models have been developed by connecting a classic compartmental pharmacokinetic model to a pharmacodynamic model that relates response to the concentration of agonist in the central pharmacokinetic compartment (Sheiner, 1985). In the classic PK–PD approach, parameters are generated *a posteriori* by simultaneously or sequentially fitting empirical pharmacokinetic and pharmacodynamic data to equations by least-squares non-linear regression. Porchet *et al.* (1988) have taken this design one step further by combining a classic compartmental PK model with a parameter dependent PD model that describes the development of acute tolerance to the cardioaccelerating effect of nicotine.

Although a compartmental pharmacokinetic model of nicotine may provide useful information for PK–PD modelling of nicotine, the combination of a PD model with a PBPK model should provide even more information because of the predictions about target tissue drug concentration that the PBPK model will provide. Simulations with the NIC PBPK model and the multi-metabolite model could describe the concentration *vs.* time relationship of nicotine and its metabolites in the target tissue or, ideally, at the receptor site or biophase. Combining this concentration data with an appropriate pharmacodynamic model would result in the description of the dose-related pharmacological response to nicotine.

The cardioaccelerating response to nicotine is a prime candidate for PK–PD modeling because of the non-linear, non-monotonic relationship between plasma concentration and an effect that is easily measurable (Benowitz, 1988; Porchet *et al.*, 1987, 1988). For example, the plot of heart rate during an infusion of nicotine *vs.* plasma nicotine concentration results in a clockwise loop over time (proteresis). Possible explanations for this observation are: (1) the development of acute tolerance; (2)

the existence of distributional disequilibrium between nicotine concentrations in the blood and at the site of action; or (3) the formation of an antagonistic metabolite. A physiologically based pharmacokinetic–pharmacodynamic (PBPK–PD) model of nicotine has potential use in distinguishing between these possibilities.

To examine the feasibility of the PBPK–PD approach for nicotine, the standard NIC PBPK model was linked to a pharmacodynamic model that was originally described by Ariens and Simonis (1964), and later adapted by Porchet *et al.* (1988). The NIC PBPK–PD model generates a response (increased heart rate) which is proportional to the concentration of nicotine in a putative site of action compartment (Figure 12.7). Tolerance is manifested in attenuation of the response by a hypothetical non-competitive antagonist which is generated by a first-order process from the concentration of nicotine at the site of action in the PBPK model. Rate constants describing the formation and elimination of the hypothetical antagonist define the time course of acute tolerance. In developing the pharmacodynamic component of the PBPK–PD model, the Porchet *et al.* (1988) model parameters, S (maximal effect/C_{50}) and C_{50} (concentration of agonist producing half-maximal effect), had to be adjusted by the tissue–blood partition coefficient (R) for the appropriate target tissue since, as presented by Porchet *et al.*, these values represented the concentration–effect relationship of nicotine in the central pharmacokinetic compartment. The parameter describing the offset of tolerance in the Porchet *et al.* (1988) model, $k_{ant,0}$, was used without adjustment in the NIC PBPK–PD model.

The PBPK model and the pharmacodynamic model were linked by the concentration of nicotine in the site-of-action compartment to form a functional PBPK–PD model of nicotine's effect on heart rate. This is in contrast to previous PK–PD models which linked pharmacokinetics to pharmacodynamics via the concentration of drug in the central compartment of a classic pharmacokinetic model. In order to compare the PBPK–PD model with the classic PK–PD model, the clinical PK–PD experiments (Porchet *et al.*, 1988) were simulated. In these experiments, two nicotine infusions separated by 1, 2 or 3.5 were administered to male smokers. Blood nicotine levels and heart rate were monitored for 5 h after beginning the initial infusion. The NIC PBPK–PD model was able to adequately simulate these experiments (Figure 12.8).

To explore the importance of assigning a particular tissue as the target tissue in the NIC PBPK–PD model, pharmacodynamic response was simulated in models that generated effect from different target tissue compartments. When brain was assigned as the target tissue, the resulting PBPK–PD model was able to reproduce published clinical and model data using the pharmacodynamic model parameters described by Porchet *et al.* (1988), modified as discussed. Blood concentration *vs.* time

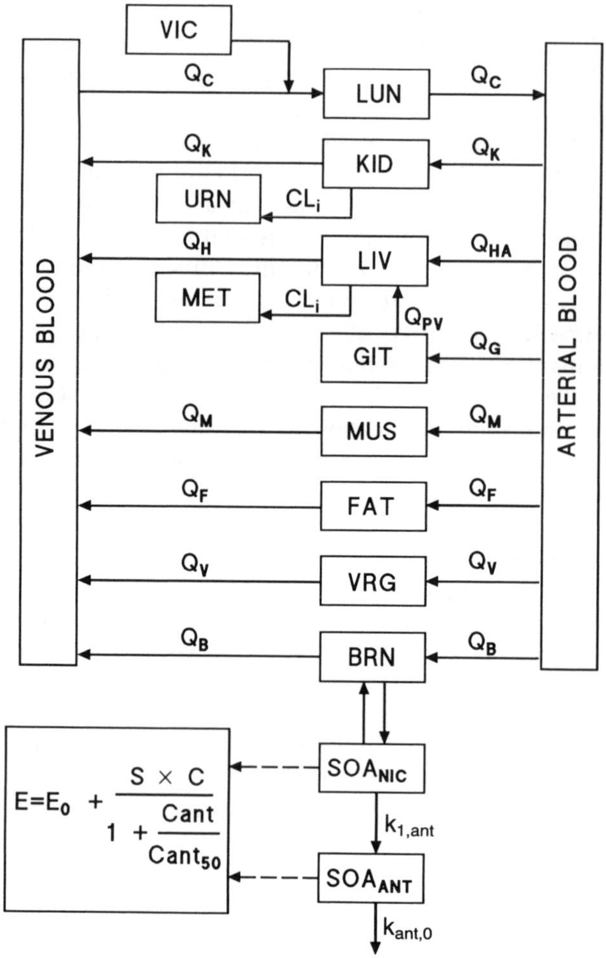

Figure 12.7 Conceptual physiologically based pharmacokinetic and pharmacodynamic model for nicotine.

data generated from nicotine PBPK–PD model simulations were nearly superimposable with the pharmacokinetic data from the published experiments (Porchet *et al.*, 1988). However, when target tissues other than the brain were employed, further adjustment of the PBPK–PD model was necessary in order to simulate the results of the published experiments. For example, when the compartment representing the vessel-rich group of tissues was designated the target tissue, a small empirical adjustment of the bidirectional first-order rate constant linking the target tissue and site of action compartments was required for the model simulations to fit the experimental data. When other tissues, such as muscle or fat, were designated the target tissue, model simulations

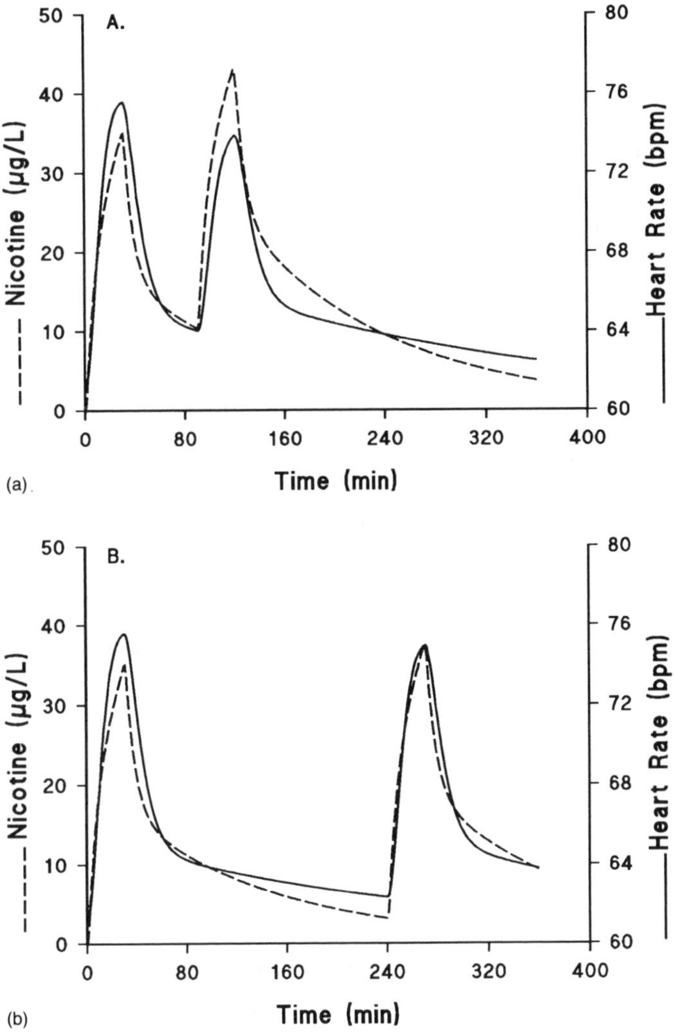

Figure 12.8 NIC PBPK–PD model simulation of a published pharmacokinetic–pharmacodynamic experiment in man. The NIC PBPK model was used to simulate a published experimental protocol (Porchet *et al.*, 1988). Resulting heart rate (solid lines) and blood concentration (dashed line) are shown. (a) Simulation of two intravenous infusions of nicotine (5250 μg over 30 min, each) separated by 1 h. (b) Simulation of two intravenous infusions of nicotine (5259 μg over 30 min, each) separated by 3.5 h.

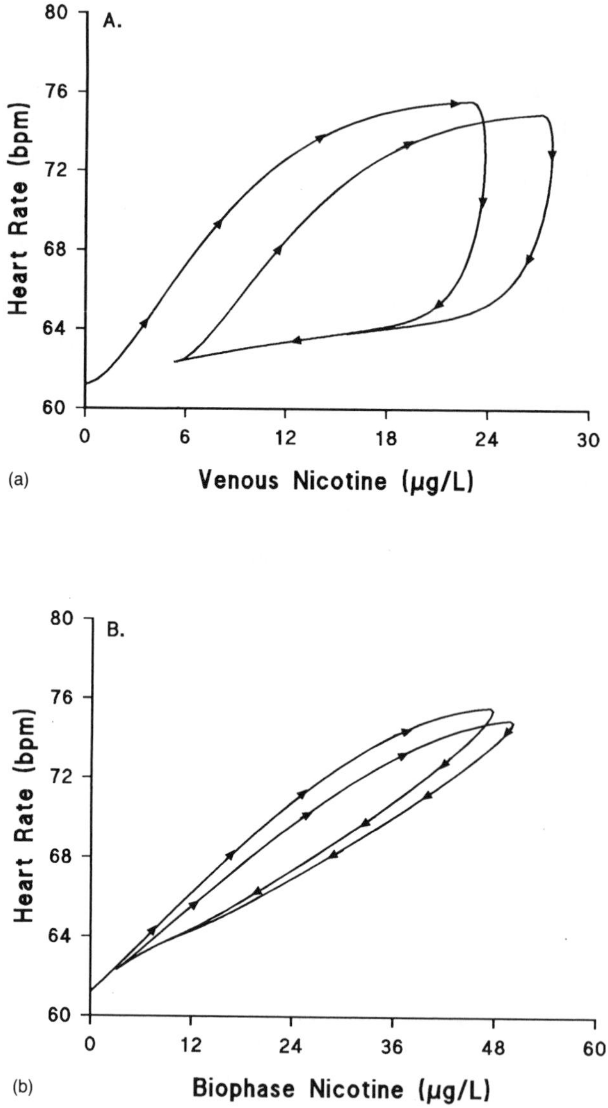

Figure 12.9 NIC PBPK–PD model simulation of nicotine concentration-effect proteresis. NIC PBPK–PD model simulation of successive nicotine infusions separated by 1 h (Porchet *et al.*, 1988) resulted in an apparent proteresis. Concentrations in two of the PBPK–PD model compartments were plotted *vs.* effect to compare apparent proteresis. Arrows indicate progression of time. (a) Proteresis observed when simulated venous nicotine concentration is plotted *vs.* heart rate; (b) proteresis observed when simulated biophase concentration is plotted *vs.* heart rate.

would not fit the published experimental data unless the pharmacodynamic model parameters reported by Porchet *et al.* were significantly altered. Although the target tissue(s) for nicotine's cardioaccelerating effect cannot yet be definitively determined by this approach, these preliminary experiments suggest that the site of action for this effect is probably in a highly perfused, rapidly equilibrating tissue compartment.

The nicotine concentration-effect relationship resulting from the brain-based PBPK–PD model simulations was also examined. With this model, the proteresis resulting from the plot of heart rate *vs.* forearm vein nicotine concentration is partially collapsed when effect is plotted *vs.* biophase concentration of nicotine (Figure 12.9). The results suggest that a distributional disequilibrium between venous blood and biophase nicotine concentration is responsible for a large portion of the observed proteresis. A time dependent concentration-effect relationship, such as acute pharmacodynamic tolerance, could account for the remaining proteresis that is observed.

The development of the NIC PBPK-PD model demonstrates that a physiological approach can be applied to PK–PD modelling. Unlike other biologically-based PK-PD models of nicotine that are designed to study nicotinic receptor dynamics (Plowchalk *et al.*, 1992), the NIC PBPK–PD model is capable of modelling a functional response to nicotine. Thus, the NIC PBPK–PD model has the potential to predict response to various nicotine exposure scenarios. Additionally, the results of the NIC PBPK–PD model simulations demonstrate that PBPK–PD models can provide information that is otherwise unobtainable with the classic PK–PD approach.

12.7 SUMMARY

Examples provided in this chapter illustrate the multi-faceted utility of PBPK modelling techniques for studying nicotine pharmacology, and suggest investigational strategies for studies of nicotine pharmacokinetics, pharmacodynamics and exposure assessment. Like any theoretically based modelling approach, PBPK modelling can serve as a mechanism for prediction in both current applications and future experimentation (Sheiner, 1989). Hopefully, the continued development of models, such as the ones described in this chapter, will facilitate the fruitful collaboration between theoretical and empirical scientists.

ACKNOWLEDGEMENTS

M.R.G. is supported by a pre-doctoral I.R.T.A. fellowship from the National Institute on Drug Abuse, Addiction Research Center, Baltimore, MD, USA.

REFERENCES

Abramson, D.I. (1967) *Circulation in the extremities*, Academic Press, New York.

Adlkofer, F., Scherer, G., Jarczyk, L. *et al.* (1988) Pharmacokinetics of 3′-hydroxycotinine, in *Pharmacology of Nicotine*, ICSU Symposium Series, **9**, 25–8.

Andersen, M.E. (1989) Tissue dosimetry, physiologically based pharmacokinetic modeling, and cancer risk assessment. *Cell Biol. and Toxicol.*, **5**, 405–16.

Ariens, E.J., and Simonis, A.M. (1964) A molecular basis for drug action. The interaction of one or more drugs with different receptors. *J. Pharm. Pharmacol.*, **16**, 289–312.

Benowitz, N.L. (1988) Pharmacokinetics and pharmacodynamics of nicotine, in *Pharmacology of Nicotine*, ICSU Symposium Series, **9**.

Benowitz, N.L. and Jacob, P. III. (1985) Metabolism, pharmacokinetics and pharmacodynamics of nicotine in man, in *Tobacco Smoking and Nicotine: A Neurobiological Approach*, (Eds E.T. Iwamoto and L. David), Plenum Press, New York.

Benowitz, N.L., Jacob, P. III, Jones, R.T. *et al.* (1982) Interindividual variability in the metabolism and cardiovascular effects of nicotine in man. *J. Pharmacol. Exp. Ther.*, **221**, 368–72.

Benowitz, N.L., Kuyt, F., Jacob, P. III *et al.* (1983) Cotinine disposition and effects. *Clin. Pharmacol. Ther.*, **34**, 604–11.

Bischoff, K.B., Dedrick, R.L., Zaharko, D.S. *et al.* (1971) Methotrexate pharmacokinetics. *J. Pharm. Sci.*, **60**, 1128–33.

Curvall, M., Elwin, C.E., Kazemi-Vala, E. *et al.* (1990a) The pharmacokinetics of cotinine in plasma and saliva from non-smoking healthy volunteers. *Eur. J. Clin. Pharmacol.*, **38**, 281–7.

Curvall, M., and Enzell, C.R. (1986) Monitoring absorption by means of determination of nicotine and cotinine. *Arch. Toxicol. Suppl.*, **9**, 88–102.

Curvall, M., Kazemi-Vala, E., and Enzell, C.R. (1982) Simultaneous determination of nicotine and cotinine in plasma using capillary column gas chromatography with nitrogen sensitive detection. *J. Chromatogr. Biomed. Appl.*, **232**, 283–93.

Curvall, M., Kazemi-Vala, E., Enzell, C.R. *et al.* (1990b) Simulation and evaluation of nicotine intake during passive smoking: cotinine measurements in body fluids of nonsmokers given intravenous infusions of nicotine. *Clin. Pharmacol. Ther.*, **47**, 42–9.

Davis, N.R. and Mapleson, W.W. (1981) Structure and quantification of a physiological model of the distribution of injected agents and inhaled anaesthetics. *Br. J. Anaesth.*, **53**, 399–404.

deBethizy, J.D., and Andersen, M.E. (1990) A physiologically based pharmacokinetic (PB-PK) model for nicotine in the rat (Abstract). *Toxicologist*, **10**, 862.

DeSchepper, P.J., Hecken, A.V., Daenens, P. *et al.* (1987) Kinetics of cotinine after oral and intravenous administration to man. *Eur. J. Clin. Pharmacol.*, **31**, 583–58.

D'Souza, R.W. and Boxenbaum, H. (1988) Physiological pharmacokinetic models: some aspects of theory, practice and potential. *Toxicol. Ind. Health*, **4**, 151–71.

Feyerabend, C., Ings, R.M.J., and Russell, M.A.H. (1985) Nicotine pharmacokinetics and its application to intake from smoking. *Br. J. Clin. Pharmacol.*, **19**, 239–47.

Gabrielsson, J. and Bondesson, U. (1987) Constant-rate infusion of nicotine and cotinine. I. A physiological pharmacokinetic analysis of the cotinine disposition, and effects on clearance and distribution in the rat. *J. Pharmacokin. Biopharm.*, **15**, 583–99.

Galeazzi, R.L. Daenens, P. and Gugger, M. (1985) Steady-state concentration of cotinine as a measure of nicotine-intake by smokers. *Eur. J. Clin. Pharmacol.* **28**, 301–4.

Gastonguay, M.R., Balter, N.J. and Schwartz, S.L. (1990) A physiologically-based pharmacokinetic (PBPK) model of nicotine and three metabolites (Abstract). *Pharmacologist*, **32**, 141.

Gerlowski, L.E. and Jain, R.K. (1983) Physiologically based pharmacokinetic modeling: principles and applications. *J. Pharm. Sci.*, **72**, 1103–25.

Jacob, P., III, Benowitz, N.L. and Shulgin, A.T. (1988) Recent studies of nicotine metabolism in humans. *Pharmacol. Biochem. Behav.*, **30**, 249–53.

Kyerematen, G.A., Damiano, M.D., Dvorchik, B.H. *et al.* (1982) Smoking-induced changes in nicotine disposition: application of a new HPLC assay for nicotine and its metabolites. *Clin. Pharmacol. Ther.*, **32**, 769–80.

Kyerematen, G.A., Morgan, M.L., Chattopadhyay, B. *et al.* (1990) Disposition of nicotine and eight metabolites in smokers and nonsmokers: identification in smokers of two metabolites that are longer lived than cotinine. *Clin. Pharmacol. Ther.*, **48**, 641–51.

Leggett, R.W. and Williams, L.R. (1991) Suggested reference values for regional blood volumes in humans. *Health Physics*, **60**, 139–54.

Lewis, E.A., Tang, H., Gunther, K. *et al.* (1990) Use of urine nicotine and cotinine measurements to determine exposure of nonsmokers to sidestream tobacco smoke. Indoor Air '90. International Conference on Indoor Air Quality and Climate. Ottawa, 151–6.

Notari, R.E. (1987) *Biopharmaceutics and clinical pharmacokinetics*. 4th ed. Marcel Dekker Inc., New York, 49.

Plowchalk, D.R. and deBethizy, J.D. (1991) A PB–PK model for nicotine tissue and plasma kinetics in the Sprague–Dawley rat (Abstract). *Toxicologist*, **11**, 280.

Plowchalk, D.R., Fluhler, E.N. Lipiello, P.M. *et al.* (1992) A biologically-based model for nicotinic receptor dynamics in the rat brain (Abstract). *Toxicologist*, 293.

Porchet, H.C., Benowitz, N.L. and Sheiner, L.B. (1988) Pharmacodynamic model of tolerance: application to nicotine. *J. Pharmacol. Exp. Ther.*, **244**, 231–6.

Porchet, H.C., Benowitz, N.L., Sheiner, L.B. *et al.* (1987) Apparent tolerance to the acute effect of nicotine results in part from distribution kinetics. *J. Clin. Invest.*, 80, 1466–71.

Quiring, D.P. (1949) *Collateral circulation: anatomical aspects*, Lea and Febiger, Philadelphia.

Robinson, D.E., Balter, N.J. and Schwartz, S.L. (1992) A physiologically-based pharmacokinetic model for nicotine and cotinine in man. *J. Pharmacokin. Biopharm.*, **20**, 591–609.

Rothe, C.F. (1984) Venous system: physiology of the capacitance vessels, in *Handbook of Physiology*, Section 2 Volume III (Ed. J.T. Shepherd), Waverly Press, Baltimore, 397.

Scheiner, L.B. (1985) Modeling pharmacodynamics: Parametric and nonparametric approaches, in *Variability in Drug Therapy: Description Estimantion and Control*, (Ed. M. Rowland, *et al.*), Raven Press, New York.

Scheiner, L.B. (1989) Clinical pharmacology and the choice between theory and empiricism. *Clin. Pharmacol. Ther.*, **6**, 605–15.

Scherer, G. Jaczy, L. Heller, W.D. *et al.* (1988) Pharmacokinetics of nicotine, cotinine, and 3′-hydroxycotinine in cigarette smokers. *Klin. Wochenschr.*, **66** (suppl XI), 5–11.

Schievelbein, H. (1982) Nicotine, resorption and fate. *Pharmacol. Ther.*, **18**, 233–48.

Schwartz. S.L., Ball, R.T. and Witorsch, P. (1987) Mathematical modelling of nicotine and cotinine as biological markers of environmental tobacco smoke exposure. *Tox. Let.*, **35**, 53–8.

Sebalt, R.J. and Kreeft, J.H. (1987) Efficient pharmacokinetic modeling of complex clinical dosing regimens: the universal elementary dosing regimen and computer algorithm EDFAST. *J. Pharm. Sci.*, **76**, 93–100.

Williams, L.R. and Leggett, R.W. (1989) Resting values for resting blood flow to organs of man. *Clin. Phys. Physiol. Meas.*, **10**, 187–217.

New methods for probing the disposition of nicotine in humans

13

J. Wahren, J. Gabrielsson, M. Curvall and E. Kazemi Vala

13.1 INTRODUCTION

For many centuries nicotine has been consumed by humans in different forms of tobacco. More recently nicotine itself has become available as a pharmaceutical agent in the form of nicotine-containing chewing gum or nicotine preparations for transdermal absorption. The importance of nicotine in human physiology is related to its pleasurable effects as well as to its possible role in tobacco-related disorders. It may be assumed that, as with most pharmacological agents, there is a relationship between plasma concentration of nicotine and its pharmacological effects. It follows that the factors which determine the disposal of nicotine may influence the effects of nicotine during tobacco use as well as its effects when nicotine is used as an adjuvant to smoking cessation. A detailed understanding of whole body and regional disposal of nicotine is of interest with regard to the basic physiology and pharmacology of nicotine. It is noteworthy that literature is almost totally devoid of information concerning the influence of disease on the pharmacokinetics of nicotine. Thus, data with regard to the influence of, e.g. liver disease or renal disease on nicotine disposal are not available but may well be of clinical interest.

Studies of nicotine's disposal in humans have traditionally been based primarily on plasma concentration–time curves after nicotine adminis-

Nicotine and Related Alkaloids: Absorption, distribution, metabolism and excretion. Edited by J.W. Gorrod and J. Wahren. Published in 1993 by Chapman & Hall, London. ISBN 0 412 55740 1

tration into the systemic circulation (Rosenberg *et al.*, 1980; Benowitz *et al.*, 1990; Curvall *et al.*, 1990). Relatively few studies, particularly in humans, report on regional uptake and utilization of nicotine in different tissues and organs despite the fact that techniques for such studies are available. The purpose of this presentation is to review and exemplify how regional intravascular catheterization techniques, positron emission tomography (PET) and nuclear magnetic resonance (NMR) techniques may be used in the study of regional distribution of nicotine in humans.

13.2 REGIONAL CATHETERIZATION

Nicotine is generally considered to be eliminated from the circulation primarily by the liver and the kidneys (Svensson, 1987). Yet few, if any, studies have dealt with the regional disposal of nicotine, particularly in humans. Intravascular catheterization techniques afford the possibility of carrying out direct measurements of net exchange across individual organs or tissues. Such techniques were first employed by Forssman in 1929 and subsequently developed by Cournand and Ranges into clinically applicable procedures for cardiac diagnostic work in 1941. In the case of the liver a hepatic vein can be catheterized from an arm vein or a femoral vein using fluoroscopic control. Likewise, a renal vein catheter may be inserted from the femoral vein using percutaneous technique and fluoroscopic control. A catheter will also have to be introduced into either the brachial or femoral artery for arterial blood sampling and determination of the arterio–venous concentration difference. The blood flow to the liver may be measured using constant rate intravenous infusion of indocyanine green dye, an indicator which is eliminated from the blood solely by the liver. Likewise, renal blood flow may be measured after infusion of paraminohippurate (PAH) which is excreted by the kidneys. By measuring simultaneously the blood flow and the arterio–venous concentration difference of nicotine or its metabolites, it becomes possible to determine by the direct Fick technique the net exchange across the splanchnic or renal vascular beds. These procedures require aseptic techniques and access to a fluoroscopic unit, but carry little risk of complications with an experienced investigational team.

In view of the relative paucity of data concerning liver and kidney disposal of nicotine in humans we have undertaken a study involving intravenous infusion of nicotine to healthy subjects and simultaneous measurements of nicotine uptake to the splanchnic area and the kidneys. The participants were examined on two different occasions receiving infusions of nicotine at two different rates in randomized order. The subjects (age 25 ± 3 years, $n = 6$) were all habitual smokers and they discontinued smoking 24 h prior to the studies. They came to the laboratory in the morning and were given a light, standardized breakfast and subsequently a light lunch. Two intravenous lines were established, one via an antecubital vein for nicotine infusion and one in the contralateral arm for blood sampling. Baseline blood samples were drawn and at zero time nicotine infusion either 0.25 or 0.75 μg/min/kg was begun and continued for 8 h. The hepatic and renal veins and an artery were catheterized after 6 h and intravenous infusions of PAH and indocyanine green dye were begun. Four sets of hepatic venous, renal venous and arterial blood samples were drawn at 10 min intervals during the last 30 min of the study period. Regional blood flows were calculated from PAH and indocyanine green dye concentrations, and splanchnic and renal uptake of nicotine was determined as the product of nicotine arterio-venous concentration difference and blood flow. Water was provided *ad libitum* throughout the study period. Nicotine concentration was analysed using a capillary column gas chromatography technique (Curvall *et al.*, 1982).

Measurements of nicotine levels were carried out in both plasma and whole blood. The results demonstrated that the plasma concentration was greater than that for whole blood and that quantitatively significant amounts of nicotine were carried by the blood cells. The ratio of plasma to blood nicotine concentrations was 1.14–1.22. Thus, approximately 65% of nicotine was transported in plasma and 35% via blood cells within the range of concentrations encountered in the study, in agreement with previous reports (Schievelbein, 1981; Kyerematen and Vesell, 1991). Consequently, whole blood rather than plasma concentrations of nicotine will be employed in the calculations of regional nicotine exchange presented below.

The nicotine concentrations during the course of nicotine infusion are shown in Figure 13.1. The basal, pre-infusion nicotine concentration was 3.5 ± 0.8 ng/ml and it rose to an average level of 13.1 ± 0.9 ng/ml after 8 h of nicotine infusion at the lower rate and to 34.9 ± 1.9 ng/ml during infusion at the higher rate. The data indicate that the nicotine concentrations rose in direct proportion to the rate of infusion as previously observed (Curvall *et al.*, 1990). Moreover, the regional measurements indicated that a relative steady state was achieved with regard to arterial nicotine concentrations and regional exchange during the last 30 min of the study period both at the low and the high infusion rate.

The arterio-venous concentration differences for nicotine across the splanchnic tissues and the kidneys may be expressed as fractions of the simultaneous arterial concentration, thereby reflecting the fractional extraction of nicotine by the different tissues. The results showed that 68–70% of the arterial nicotine levels was extracted by the splanchnic tissues while no more than 10–20% was extracted by the kidney at the low and high infusion rates, respectively. The regional blood flow

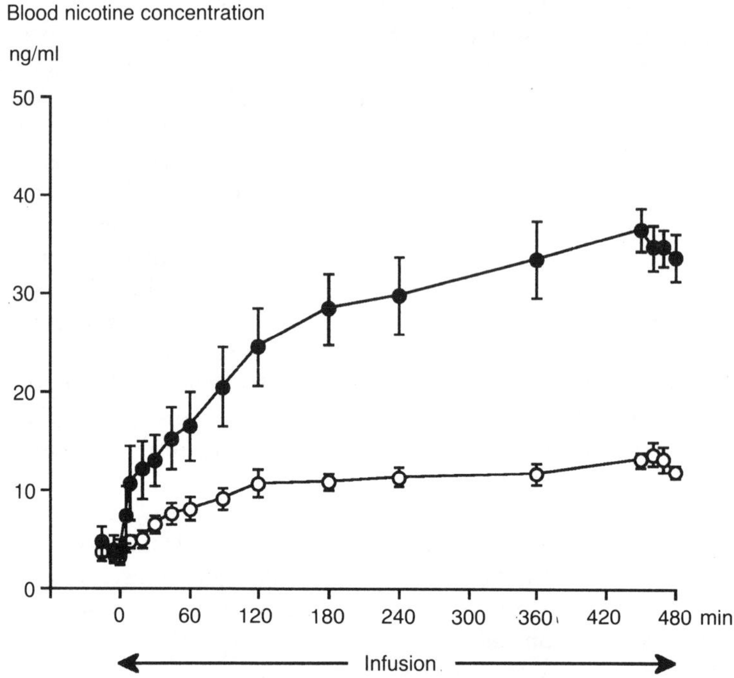

Figure 13.1 Whole blood nicotine concentrations in the basal state and during nicotine infusion for 8 h at 0.25 μg/min/kg (open symbols) or 0.75 μg/min/kg (solid symbols). Mean values ±S.E. are given.

values were no different at the two nicotine infusion rates; it was 1.6 ± 0.1 1/min for the liver and 1.9 ± 0.1 1/min for the kidneys. While the former value is within the normal range the latter is approximately 30% higher than previously published data using the same technique (Björkman *et al.*, 1989). The possibility may thus be considered that nicotine exerts a vasodilating effect in the kidney.

Net splanchnic and renal uptakes for nicotine are presented in Figures 13.2 and 13.3. At the lower nicotine infusion rate splanchnic uptake amounted to 13.0 ± 0.6 μg/min and rose to 39.3 ± 2.8 μg/min at the three-fold greater rate of nicotine infusion. Renal uptake was 4.8 ± 0.8 μg/min at the lower rate of infusion and rose by 35% to 6.5 ± 1.9 μg/min during infusion at the higher rate. No directly comparable previous data appear to be available but the present results are in good agreement with indirect estimates of hepatic and renal uptake based on clearance data (Rosenberg *et al.*, 1980; Benowitz *et al.*, 1990).

As indicated in Figures 13.2 and 13.3 the contributions made by plasma and blood cells, respectively, to nicotine transport were different for the splanchnic tissues and the kidneys. In the case of the splanchnic

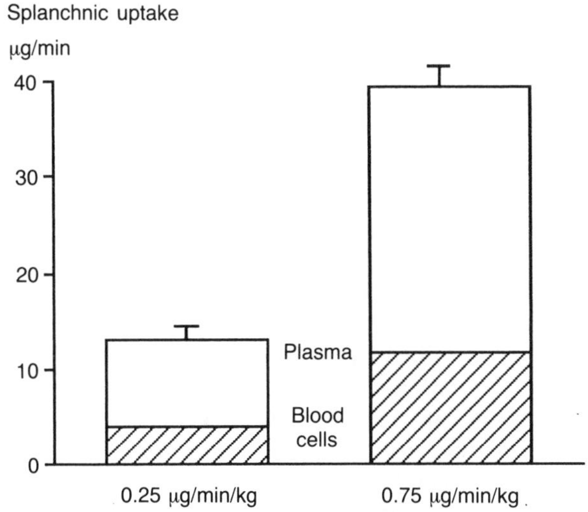

Figure 13.2 Splanchnic uptake of nicotine during steady-state conditions after 8 h of nicotine infusion at 0.25 or 0.75 μg/min/kg. The amount of nicotine taken up from plasma is indicated by the open part of the columns and that from blood cells is denoted by the hatched part. Mean values ±S.E. are presented.

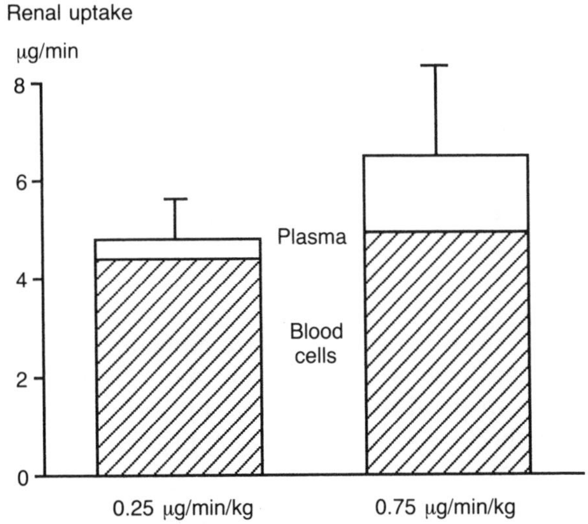

Figure 13.3 Renal uptake of nicotine during steady state conditions after 8 h of nicotine infusion at 0.25 or 0.75 μg/min/kg. The amount of nicotine taken up from plasma is indicated by the open part of the columns and that from blood cells is denoted by the hatched part. Mean values ±S.E. are given.

region nicotine uptake occurred from plasma (70%) and blood cells (30%) largely in direct proportion to the amount of nicotine transported by plasma and cells at both the lower and the higher nicotine concentrations. In contrast, the data indicate that renal nicotine uptake involved blood cell transport (76–92%) to a greater extent than transfer from plasma (8–24%) at both infusion rates. Moreover, renal nicotine uptake rose by no more than 35% when the arterial nicotine concentration increased three-fold during infusion at the higher infusion rate. The reason that nicotine is taken up more avidly from blood cells than from plasma in the kidneys is unclear. Nor is it apparent why renal nicotine uptake fails to increase in proportion to the rise in arterial concentration during infusion at the higher rate. A limitation of urine flow appears unlikely in view of the liberal supply of water during the study. The possibility of a pH limitation may be considered (Rosenberg *et al.*, 1980) but there is no immediate reason to believe that the present subjects developed alkalosis during the study period. Finally, further studies will be required to elucidate the transport kinetics for renal nicotine handling and to determine at which concentration level nicotine transport becomes saturated.

The relative contributions by liver and kidney in the disposal of nicotine may be calculated from the present data. The nicotine infusates were analysed with regard to nicotine concentration and it could be calculated that the true infusion rates were in fact 0.25 and 0.74 μg/min/kg. Under steady-state conditions the rate of infusion equals the rate of elimination. At the lower infusion rate the splanchnic uptake accounted for 64% and renal uptake for 24% of the whole body nicotine disposal. The combined liver and kidney disposal was thus 88% of the total and the remaining unaccounted 12% may possibly reflect elimination via the lungs. The corresponding data for the higher infusion rate were 64% hepatic uptake, 12% renal uptake and 24% unaccounted for. These findings thus emphasize the quantitative importance of the liver in human nicotine disposal.

In summary the present study of regional nicotine disposal confirms that significant nicotine transport occurs not only in plasma but also by blood cells. Direct measurements in intact humans for the first time demonstrate that the liver is quantitatively more important than the kidneys in nicotine disposal and that renal but not hepatic uptake of nicotine appears to become saturated at rising nicotine concentrations.

13.3 POSITRON EMISSION TOMOGRAPHY (PET)

Positron emission tomography (PET) techniques have proved useful in the study of blood flow and oxygen consumption in regions of the brain. The positron-emitting nucleides ^{11}C, ^{15}O and ^{18}F have been used and a high degree of spatial resolution can be achieved. More recently, PET techniques have also been employed to study central neuronal activity of the brain (Frackowiak *et al.*, 1981). Thus, the muscarinic receptors of the brain have been visualized using ^{11}C-benztropine (Dewey *et al.*, 1990). PET technique also lends itself to examination of tissue binding of nicotine. *S*(−)^{11}C-nicotine may be synthesized using nornicotine (Långström *et al.*, 1982) and an uptake of ^{11}C-nicotine to different regions of the brain has been demonstrated (Nordberg *et al.*, 1991). The high affinity binding sites for nicotine show a characteristic regional distribution in the brain with a high content in the thalamus, putamen and caudate nucleus, intermediate in cortical areas and hypothalamus, and low in the hippocampus and medulla oblongata (Nordberg *et al.*, 1989). The distribution of ^{11}C-nicotine in the brain as examined with PET technique agrees well with the distribution of nicotine receptors measured by *in vitro* binding techniques (Nybäck *et al.*, 1989). Characteristic reductions in the number of brain nicotine receptors have been demonstrated in patients with Alzheimer's disease (Nordberg *et al.*, 1989).

Whole body PET scanning systems are now available and the technique thus offers the possibility of examining regional disposal of nicotine not only in the brain but also in liver and kidney. Such studies do not appear to have been undertaken as yet but should help improve our understanding of renal and hepatic nicotine physiology.

13.4 NUCLEAR MAGNETIC RESONANCE (NMR) TECHNIQUES

Nuclear magnetic resonance (NMR) spectroscopy has been employed to determine nicotine metabolites in plasma and urine of smokers (Caldwell *et al.*, 1992; Parviainen *et al.*, 1990). In these studies small bore, high field strength equipment has been used for proton and carbon spectroscopy of small volume samples. Current technology has made available whole body scanning systems with magnetic field strengths of 1.5–2.0 T for medical diagnostic imageing and spectroscopy procedures. While it appears unlikely that the sensitivity of NMR spectroscopy will be sufficient to determine blood or tissue nicotine concentrations in the range of 5–20 ng/ml, other possibilities may be considered. Transient alterations in the concentration of paramagnetic deoxyhaemoglobin may be detected by NMR imageing of the human brain and possibly also of liver and kidney. With this technique functional activation maps of the human visual cortex have been obtained (Frahm *et al.*, 1992) with a spatial resolution comparable to or even better than can be achieved by positron emission tomography. This technique of so-called functional imageing provides a means of detecting differences in tissue oxygenation as reflected by varying amounts of paramagnetic deoxyhaemoglobin in the regional blood pool, primarily the venous capacitance vessels. This technique thus allows a non-invasive examination of tissue anatomy and physiological response. In view of the fact that nicotine exposure by cigarette smoking or nicotine infusion is known to stimulate oxygen consumption (Hofstetter *et al.*, 1986) it is likely that high resolution functional imageing by NMR may become a useful technique for examining tissue disposal and nicotine effects in different regions of the body in the future.

REFERENCES

Benowitz, N.L., Porchet, H., Sheiner, L. *et al.* (1990) Pharmacokinetics, metabolism, and pharmacodynamics of nicotine, in *Nicotine Psychopharmacology: Molecular, Cellular and Behavioral Aspects*, (eds S. Wonnacott, M.A.H Russell and I.P. Stolerman), Oxford University Press, Oxford, pp. 112–57.

Björkman, O., Gunnarsson, R., Felig, P. *et al.* (1989) Splanchnic and renal exchange of infused fructose in insulin-deficient type I diabetic patients and healthy controls. *J. Clin. Invest.*, **83**, 52–9.

Caldwell, W.S., Greene, J.M., Byrd, G.D. *et al.* (1992) Characterization of the glucuronide conjugate of cotinine: a previously unidentified major metabolite of nicotine in smokers' urine. *Chem. Res. Toxicol.*, **5**, 280–5.

Curvall, M., Kazemi Vala, E. and Enzell, C.R. (1982). Simultaneous determination of nicotine and cotinine in plasma using capillary column gas chromatography with nitrogen-sensitive detection. *J. Chromatog. Biomed. Appl.*, **232**, 283–93.

Curvall, M., Kazemi Vala, E., Enzell, C.R. *et al.* (1990) Simulation and evaluation of nicotine intake during passive smoking: Cotinine measurements in body fluids of nonsmokers given intravenous infusions of nicotine. *Clin. Pharmacol. Ther.*, **47**, 42–9.

Dewey, S.L., Volkow, N.D., Logan, J. *et al.* (1990) Age-related decreases in muscarinic cholinergic receptor binding in the human brain measured with positron emission tomography (PET). *J. Neurosci. Res.*, **27**, 569–75.

Frackowiak, R.S.J., Pozzilli, C., Legg, N.J. *et al.* (1981) Regional cerebral oxygen supply and reutilization in dementia. A clinical and physiological study in oxygen-15 and positron emission tomography. *Brain*, **104**, 753–78.

Frahm, J., Merboldt, K.-D. and Hänicke, W. (1993) Functional MRI of human brain activation at high spatial resolution. *Magn. Res. Med.*, **29**, 139–44.

Hofstetter, A., Schutz, Y., Jequier, E. *et al.* (1986) Increased 24-hour energy expenditure in cigarette smokers. *New Engl. J. Med.*, **314**, 79–82.

Kyerematen, G.A. and Vesell, E.S. (1991). Metabolism of nicotine. *Drug Metab. Rev.*, **23**, 3–41.

Långström, B., Antoni, G., Halldin, C. *et al.* (1982) The synthesis of some ^{11}C-labelled alkaloids. *Chemica Scripta*, **20**, 46–8.

Nordberg, A., Nilsson-Håkansson, L., Adem, A. *et al.* (1989) The role of the nicotinic receptors in the pathophysiology of Alzheimer's disease. *Prog. Brain Res.*, **79**, 353–62.

Nordberg, A., Hartvig, P., Lilja, A. *et al.* (1991) Nicotine receptors in the CNS as visualized by positron emission tomography, in *Cholinergic Basis for Alzheimer Therapy*, (eds R. Becker and E. Giacobini), Birkhäuser, Basel, pp. 107–15.

Nybäck, H., Nordberg, A., Långström, B. *et al.* (1989) Attempts to visualize nicotinic receptors in the brain of monkey and man by positrion emission tomography. *Prog. Brain Res.*, **79**, 313–9.

Parviainen, M.T., Puhakainen, E.V., Laatikainen, R. *et al.* (1990) Nicotine metabolites in the urine of smokers. *J. Chromatog.*, **525**, 193–202.

Rosenberg, J., Benowitz, N.L., Peyton, J. *et al.* (1980) Disposition kinetics and effects of intravenous nicotine. *Clin. Pharmacol. Therap.*, **28**, 517–22.

Schievelbein, H. (1982). Nicotine, resorption and fate. *Pharmac. Ther.*, **18**, 233–48.

Svensson, C.K. (1987). Clinical pharmacokinetics of nicotine. *Clin. Pharmacokin.*, **12**, 30–40.

Index

Note: in an attempt to avoid an overlong string of entries under *Nicotine*, entries relating to pharmacokinetic mechanisms of nicotine are entered under the corresponding mechanism. Entries beginning with a letter followed by a hyphen (e.g. C-Oxidation and N-Oxidation) appear at the beginning of the block of entries for the first letter.